ION EXCHANGE
AND SOLVENT EXTRACTION

ION EXCHANGE AND SOLVENT EXTRACTION

A SERIES OF ADVANCES

Volume 11

EDITED BY

Jacob A. Marinsky
Department of Chemistry
State University of
New York at Buffalo
Buffalo, New York

Yizhak Marcus
Department of Inorganic Chemistry
The Hebrew University
Jerusalem, Israel

CRC Press
Taylor & Francis Group
Boca Raton London New York

CRC Press is an imprint of the
Taylor & Francis Group, an **informa** business

CRC Press
Taylor & Francis Group
6000 Broken Sound Parkway NW, Suite 300
Boca Raton, FL 33487-2742

First issued in paperback 2019

ISBN-13: 978-0-8247-8472-0 (hbk)
ISBN-13: 978-0-367-40246-4 (pbk)

Visit the Taylor & Francis Web site at
http://www.taylorandfrancis.com

and the CRC Press Web site at
http://www.crcpress.com

Preface

This volume of *Ion Exchange and Solvent Extraction* presents four
important paths developed for the consideration and interpretation
of the ion-exchange phenomenon. It was felt that their educated
and sophisticated presentation in one volume would provide the
interested reader an excellent opportunity for examination, evaluation, and comparison of state-of-the-art ion-exchange theory.
For this reason, meaningful demonstration of the advantages of an
approach, wherever possible, and extension of the boundaries of
its application for this purpose were sought from the authors.

Chapter 1, by Steven A. Grant and Philip Fletcher, presents an
overview of the chemical thermodynamics of cation-exchange reactions,
highlighting recent developments in theory and practice. Particular
importance is placed on liquid-phase- and solid-phase-activity coefficient models. The major liquid-phase-activity coefficient models are
reviewed emphasizing their application to mixed electrolyte aqueous
solutions. Because the development of excess Gibbs-energy-based
solid-phase-activity coefficient models is in a much more rudimentary

stage, their review is restricted to appropriate mathematical forms,
a survey of the mathematical models by which they may be evaluated,
and the statistical techniques employed to optimize the design of
cation-exchange experiments.

The simple, three-parameter model, used with so much success
for the correlation of ion-exchange phenomena, is reviewed in Chapter 2 by Erik Högfeldt, who introduced it. The model, based on the
Guggenheim zeroth approximation, provides an acceptable fit to all
kinds of ion-exchange data through the use of a few parameters with
simple physical meaning. Its only drawback is the absence of any
predictive properties.

In Chapter 3, Wolfgang H. Höll, Matthias Franzreb, Jürgen Horst,
and Siegfried H. Eberle provide an excellent description of the development and application of surface complexation theory to the ion-
exchange phenomenon. The equilibria of representative organic ion-
exchange resins described with the surface complexation model are
shown to compare favorably with the experimental results.

Further insight with respect to the surface complexation model is
provided by Garrison Sposito in Chapter 4. His description of metal-
natural colloid surface reactions and their consideration by surface
complexation modeling complements Chapter 3.

The Gibbs-Donnan-based analysis of ion exchange and related
phenomena is presented by Jacob A. Marinsky in Chapter 5. It is
claimed that with the Gibbs-Donnan approach a much more realistic
picture of the physical aspects of ion-exchange phenomena is forth-
coming. Insights inaccessible to the other models are inherent in the
Gibbs-Donnan model.

Chapter 6, which considers the influence of humic substances on
the uptake of metal ions by naturally occurring materials, is authored
by James H. Ephraim and Bert Allard. Their development of this
topic compares the applicability of both the Gibbs-Donnan and surface
complexation model in their attempts to analyze the various observations made.

Jacob A. Marinsky
Yizhak Marcus

Contributors to Volume 11

Bert Allard Department of Water and Environmental Studies, Linköping University, Linköping, Sweden

Siegfried H. Eberle Institute for Radiochemistry, Kernforschungs-zentrum Karlsruhe, Karlsruhe, Germany

James H. Ephraim Department of Water and Environmental Studies, Linköping University, Linköping, Sweden

Philip Fletcher Schlumberger Cambridge Research Limited, Cambridge, England

Matthias Franzreb Division of Water Technology, Institute for Radiochemistry, Kernforschungszentrum Karlsruhe, Karlsruhe, Germany

Steven A. Grant Geochemical Sciences Branch, Cold Regions Research and Engineering Laboratory, Hanover, New Hampshire

Erik Högfeldt Department of Inorganic Chemistry, The Royal Institute of Technology, Stockholm, Sweden

Wolfgang H. Höll Division of Water Technology, Department of Demineralization and Ion Exchange, Institute for Radiochemistry, Kernforschungszentrum Karlsruhe, Karlsruhe, Germany

Jürgen Horst Institute for Radiochemistry, Kernforschungszentrum Karlsruhe, Karlsruhe, Germany

Jacob A. Marinsky Department of Chemistry, State University of New York at Buffalo, Buffalo, New York

Garrison Sposito Department of Soil Science, University of California at Berkeley, Berkeley, California

Contents

Contents of Other Volumes

1

Chemical Thermodynamics of Cation Exchange
Reactions: Theoretical and Practical Considerations

STEVEN A. GRANT   Cold Regions Research and Engineering
Laboratory, Hanover, New Hampshire

PHILIP FLETCHER   Schlumberger Cambridge Research Limited,
Cambridge, England

I. INTRODUCTION

The ion-exchange reaction consists of the replacement of an electro-
statically sorbed ion by a like-charged ion from solution. The
exchangeable ions may be either positively or negatively charged;
the ion exchanger may be solid or liquid; and the equilibrating
solution may be aqueous or a mixture of solvents. Given the many
chemical environments in which the ion-exchange reaction occurs, it
is difficult to write clearly about "general" ion-exchange-reaction
thermodynamics. With some loss of generality, this chapter will be
restricted to cation-exchange reactions on solid cation exchangers
bathed by aqueous electrolyte solutions. This chapter will not neglect consideration of the commonly observed sorption by the cation
exchanger of water and neutral salts. It is hoped that the reader
will be able to adopt directly any useful ideas found here to his or
her particular research problem.

The measurement and chemical-thermodynamic characterization
of cation-exchange reactions is commonly employed to understand
the process or the environment in which the process occurs. The
objective of any formulation of the chemical thermodynamics of

cation-exchange reactions is to calculate from experimentally measured cation-exchange data values of thermodynamic functions
(typically exchange equilibrium constants and solid-phase activity
coefficients) which characterize the distribution of [exchanging ions]
between the solid and liquid phases. It is essential to distinguish
between ion-exchange models, such as those for predicting behavior
using molecular or empirical assumptions, and chemical-thermodynamic formulations designed to *quantify* experimental measurements.

This chapter does not attempt to repeat what has been presented in the several excellent reviews of chemical thermodynamics
of cation-exchange reactions [1,2]. Most of this chapter concen-
trates on liquid-phase and solid-phase activity coefficients and the
models which have been developed to estimate their values. The
two sections following this introduction describe the chemical-thermodynamic treatment only, considering the alternative conventions and
techniques that have been used for dealing with both solid and
liquid phases. The chemical thermodynamics of concentrated electrolyte solutions is described in order to facilitate the use of chemical-thermodynamic data to predict or characterize equilibria of multispecies cation-exchange reactions.

Later sections consider the compact description of measured
chemical-thermodynamic properties of solid cation exchangers with
so-called excess Gibbs energy models. Emphasis is placed on experimental design and the statistical techniques necessary to choose
suitable models and estimate model parameters. This chapter is concerned only with the properties of solid cation exchange materials.
However, ligand exchange and the properties of liquid cation exchangers are equally accessible to chemical-thermodynamic characterization.

II. CHEMICAL THERMODYNAMICS OF CATION-EXCHANGE
REACTIONS

A chemical-thermodynamic treatment of cation-exchange reactions
requires the unambiguous definition of *reaction stoichiometries* as
well as *standard states*, *reference states*, and *reference functions*

for the reactants and products. These definitions are needed to
derive the set of equations with which thermodynamically sound
values can be calculated from experimental cation-exchange data [3].

A. Exchange Components and Reaction Stoichiometries

The cation-exchange reaction can be described by the following
general equation:

$$\nu_j M_j^{z_j+} (ads) + \nu_i M_i^{z_i+} (aq) \rightleftharpoons \nu_j M_j^{z_j+} (aq) + \nu_i M_i^{z_i+} (ads) \qquad (1)$$

where ν_i and ν_j represent stoichiometric numbers of the ions, $M_i^{z_i+}$
and $M_j^{z_j+}$ (z_i and z_j being the charge numbers of the subscripted
cations) and (aq) and (ads) define cations to be in either the liquid
phase or the solid (exchanger) phase, respectively. The equation
is correct and valid in multicomponent as well as simple binary systems. However, such representation of a reaction between cations
disguises the fact that any valid chemical-thermodynamic treatment
requires reactions to be between components which exist as independent variables and to which chemical potentials can be assigned.
On this basis components must be neutral species rather than single
ions. For the solid phase, components must contain anionic-ex-
changer material in neutral combinations with a quantity of cationic
charge. In the cation-exchange reaction represented by Eq. (1)
the reaction stoichiometry can be represented two ways:

1. An ion $M_i^{z_i+}$ and a hypothetical exchanger species X^- carrying
   one mole of anionic charge, one possible component has the composition $M_i X_{z_i}$, which represents one mole of ion i in association
   with a solid cation exchange carrying z_i moles of negative charge.
2. A similar possible cation-exchanger combination with a composition
   $M_{i(1/z_i)} X$ representing one mole of positive charge in association
   with a cation exchanger carrying a like amount of negative
   charge.

Both are neutral and acceptable as thermodynamic components. However, their reactions with ions in solution have different stoichiometries.

The first reacts in accordance with the equation

$$z_i M_j X_{z_j} + z_j M_i^{z_i^+} (aq) \rightleftharpoons z_i M_j^{z_j^+} (aq) + z_j M_i X_{z_i} \tag{2}$$

whereas the second reacts with the stoichiometry:

$$M_{j(1/z_j)} X + \frac{1}{z_i} M_i^{z_i^+} (aq) \rightleftharpoons \frac{1}{z_j} M_j^{z_j^+} (aq) + M_{i(1/z_i)} X \tag{3}$$

Understandably these different reaction stoichiometries result in different exchange equilibrium constants and thermodynamic function values. These differences should be borne in mind when interpreting published cation-exchange-reaction thermodynamic data.

B. Component Activities and Exchange Equilibrium Constants

To calculate an exchange equilibrium constant for a given reaction stoichiometry it is necessary to define the activities of the reactants and products. The activity is a measure of the difference in Gibbs energy of a component in a given system at a particular temperature and pressure and of the same component in an arbitrarily chosen standard state. By following the development presented by Guggenheim [4] the *absolute activity* λ_i of a component "i" is

$$RT \ln \lambda_i \stackrel{def}{=} \mu_i \tag{4}$$

where μ_i is the partial molar Gibbs energy or chemical potential ($J\ mol^{-1}$); R, the molar gas constant ($J\ K^{-1} mol^{-1}$); and ln, the base-e, or natural logarithm and T, thermodynamic temperature (K). If Eq. (4) is applied to a component in its actual state and a standard state (denoted by $\ominus$) and the two resulting equations subtracted, the following expression is obtained:

$$\mu_i - \mu_i^{\ominus} = RT \ln \left(\frac{\lambda_i}{\lambda_i^{\ominus}} \right) \tag{5}$$

By defining the ratio of absolute activities as the activity a_i, the common expression for activity is obtained:

$$\mu_i = \mu_i^{\ominus} + RT \ln a_i \tag{6}$$

The activity is seen to be a dimensionless parameter which has the value of unity for any component in its standard state.

From the second law of thermodynamics a reactive mixture will be at equilibrium when the sum of the Gibbs energies of r reactants equals the sum of the Gibbs energies of p products; i.e.,

$$\sum_p \nu_p \mu_p - \sum_r \nu_r \mu_r = 0 \tag{7}$$

It is convenient to define the component $M_i X_{z_i}$ as a type V (for Vanselow) component and represent it by the symbol V_i. Similarly, $M_{i(1/z_i)}X$ is a type G (for Gapon) component represented by G_i. For reactions between type V components, applying (6) and (7) to (2) gives

$$0 = z_j \mu^{\ominus}_{V_i} + z_j RT \ln a_{V_i} + z_i \mu^{\ominus}_{j(aq)} + z_i RT \ln a_{j(aq)}$$

$$- z_i \mu^{\ominus}_{V_j} - z_i RT \ln a_{V_j} - z_j \mu^{\ominus}_{i(aq)} - z_j RT \ln a_{i(aq)} \tag{8}$$

where $\mu^{\ominus}_{V_i}$ and a_{V_i} correspond to the standard chemical potential and activity of component V_i in the exchanger and $\mu^{\ominus}_{i(aq)}$ and $a_{i(aq)}$ represent the standard chemical potential and single ion activity for ion M_i in solution.*

Collecting the standard chemical potentials and logarithmic terms yields

$$z_j \mu^{\ominus}_{V_i} + z_i \mu^{\ominus}_{j(aq)} - z_i \mu^{\ominus}_{V_j} - z_j \mu^{\ominus}_{i(aq)} = -RT \ln \left(\frac{a_{j(aq)}^{z_i} a_{V_i}^{z_j}}{a_{i(aq)}^{z_j} a_{V_j}^{z_i}} \right) \tag{9}$$

---

*Single-ion activities cannot be defined rigorously within the framework of chemical thermodynamics. It would be proper here to present (8) with mean-ionic activities rather than single-ion activities. For solutions with several anionic species, retention of thermodynamically rigorous mean-ionic activities would require representation of (8) for each anion-cation combination. While this is the proper method in which to calculate the various thermodynamic quantities presented here, the benefit derived from this chemical fundamentalism is marginal. In anticipation of later discussions of cation-exchange reactions in multicomponent solutions, single-ion activities are introduced here.

The logarithmic function in the right-hand side of (9) introduces
the type V exchange equilibrium constant defined as

$$\,_j^i K_V \stackrel{\text{def}}{=} \left(\frac{a_{j(aq)}^{z_i} a_{V_i}^{z_j}}{a_{i(aq)}^{z_j} a_{V_j}^{z_i}} \right) \tag{10}$$

The sum of the standard chemical potentials is the standard reaction
Gibbs energy for the type V cation-exchange reaction, and

$$\,_j^i \Delta \bar{G}_V^{\ominus} = -RT \ln \,_j^i K_V \tag{11}$$

The convention employed here represents the thermodynamic functions
($\,_j^i \Delta \bar{G}_V^{\ominus}$ and $\,_j^i K_V$) with a superscript to denote the solid-phase "product"
ion and a subscript to denote the solid-phase "reactant" ion.* This is
a convenient notation since the reaction stoichiometry can be deduced
from the symbols for the constants. Similar expressions can be derived
using (3) for reactions between type G components; i.e.,

$$\,_j^i K_G = \left(\frac{a_{j(aq)}^{1/z_j} a_{G_i}}{a_{i(aq)}^{1/z_i} a_{G_j}} \right) \tag{12}$$

so that

$$\,_j^i \Delta \bar{G}_G^{\ominus} = -RT \ln \,_j^i K_G \tag{13}$$

These variables will have values different from those for the type V
components since their values are clearly dependent on the stoichiometry of reaction in addition to the choice of standard states for the
components.

With (2), (3), (9), (10), and (12), it is easy to show that the
two exchange equilibrium constants are related by

$$\,_j^i K_V = z_i z_j \,_j^i K_G \tag{14}$$

*The current IUPAC recommendations for ion-exchange-reaction nomen-
clature are presented in H. M. N. H. Irving, *Pure Appl. Chem.*, 29:
619 (1972).

C. Standard States, Reference States, and Reference Functions

With the cation-exchange-reaction stoichiometry and the resultant exchange equilibrium constants explicitly defined, the standard states, reference states and reference functions for the reactants and products must be defined. Their definition for cations in aqueous solutions will be discussed in Sec. III.A.

For components in the solid phase it is convenient to define the reference function as *any measurable quantity which approximates a thermodynamic activity* [5].

Any measurable quantity may serve as a reference function so long as the thermodynamic component approaches a defined reference state, the reference function approaches the component's activity uniformly. This requirement can be written as

$$\lim_{\xi \to \xi^\rho} \frac{r(\xi)}{a_i} = 1 \tag{15}$$

where $r(\xi)$ is the reference function and ξ^ρ is the reference state.

Mole fraction, molality and concentration are examples of reference functions. The most widely used reference function for solid or liquid mixtures is the mole fraction x_i. For a mixture comprised of N components the mole fraction is defined by

$$x_i = \frac{n_i}{\sum_{i=1}^{N} n_i} \tag{16}$$

where n_i is the number of moles of the i-th component. The preferred reference function for aqueous solutions is molality.

An ideal mixture is one in which a component's absolute activity is proportional to its mole fraction:

$$\lambda_i = \lambda_i^{\ominus} x_i \tag{17}$$

then, the relationship

$$\mu_i = \mu_i^{\ominus} + RT \ln x_i \tag{18}$$

or alternatively $x_i = a_i$, will hold for ideal mixtures. For real mixtures it is often appropriate to define the rational activity coefficient

f_i, a term which accounts for the deviation from the chosen reference behavior; i.e.,

$$f_i \overset{\text{def}}{=} \frac{a_i}{x_i} \tag{19}$$

This yields the usual equation for the chemical potential of a
component:

$$\mu_i = \mu_i^{\ominus} + RT \, \ln(x_i f_i) \tag{20}$$

When dimensionless mole fractions are used for the reference behavior,
activity coefficients are themselves dimensionless.

In the standard state the activity of a component is defined as
unity [Eqs. (2) to (6)] and the most common standard state for a
mixture component is the pure component at a standard pressure $p^{\ominus}$
(currently recommended by IUPAC to be 1 bar or 10^5 Pa) and at a
defined temperature; i.e.,

$$\mu_i^{\ominus}(T) = \mu_i^{*}(T, p^{\ominus}) \tag{21}$$

where μ_i^{*} is the chemical potential of pure component i. In this
standard state both x_i and the activity coefficient are unity.

A "pure" exchange component is a cation exchanger containing a
single exchangeable species. Unfortunately, for cation exchangers
which adsorb water (or salt by mechanisms other than ion-exchange
reactions) the conventional standard state is ambiguous since an exchanger could be in its pure form in the presence of a pure salt
solution of any molality and the amount of adsorbed water is therefore variable. Accordingly, the composition of the liquid phase must
be specified to complete the definition of the solid-phase standard
state. Consequently, the usual solid-phase standard state (and the
one to be adopted in this chapter) for a cation exchange component
is *the homoionic form at equilibrium with an infinitely dilute solution
of the exchanging ion at a defined temperature and pressure of* 0.1
MPa [6]. This is also the reference state for exchangeable cations.
This is convenient since water in the equilibrating solution will also
be in its standard state.

However, the selection of a standard state is arbitrary [7,8]
and other standard states may be appropriate. Babcock and Duckart
[9] recommend a standard state of x_i equal to 0.5 for components in
binary mixtures. In this case the activity coefficient would have a
value of 2 in the standard state. Another choice of standard state
is the hypothetical ideal one molal solution of the exchanger with its
reference state being the infinitely dilute solution. With this convention there is no difference between the standard states for any solid-
phase component and the exchange equilibrium constant approaches
unity. The measured solid phase activity coefficients become the
measures of the selectivities of the exchanger for given ions.

D. Solid-Phase Activity Coefficients

Regardless of the standard state chosen, exchange equilibrium constants have the algebraic form

$$
{}_j^i K_V = \left(\frac{a_{j(aq)}^{z_i} (x_{V_i} f_{V_i})^{z_j}}{a_{i(aq)}^{z_j} (x_{V_j} f_{V_j})^{z_i}} \right)
\tag{22}
$$

and

$$
{}_j^i K_G = \left(\frac{a_{j(aq)}^{1/z_j} x_{G_i} f_{G_i}}{a_{i(aq)}^{1/z_i} x_{G_j} f_{G_j}} \right)
\tag{23}
$$

Argersinger et al. [3] developed an ingenious mathematical procedure
by which solid-phase activity coefficients and exchange equilibrium
constants can be calculated from experimental data of cation-exchange
equilibrium experiments. The working equations for their procedure
are derived from the definitions of the exchange equilibrium constants,
selectivity coefficients, and solid-phase chemical potentials via the
Gibbs-Duhem equation. The following development will derive mathematical procedure for calculating the exchange equilibrium constant
for the type V reaction stoichiometry. A parallel derivation would
yield the allied evaluation procedure for the type G reaction
stoichiometry.

An experimentally determinable mass-action quotient, the Vanselow selectivity coefficient, can be defined as

$$
{}_j^i k_V^a \overset{\text{def}}{=} \left(\frac{a_{j(aq)}^{z_i} x_{V_i}^{z_j}}{a_{i(aq)}^{z_j} x_{V_j}^{z_i}} \right) \tag{24}
$$

The superscript a in ${}_j^i k_V^a$ indicates that the cations in solution are represented by their single-ion activities. It is clear from Eqs. (22) and (24) that

$$
{}_j^i K_V = {}_j^i k_V^a \frac{f_{V_i}^{z_j}}{f_{V_j}^{z_i}} \tag{25}
$$

Equation (25) can be transformed into an equation of constraint by forming the differential of $\ln {}_j^i k_V^a$; i.e.,

$$
d \ln {}_j^i K_V = d \ln {}_j^i k_V^a + z_j\, d \ln f_{V_i} - z_i\, d \ln f_{V_j} = 0 \tag{26}
$$

A second equation of constraint is the Gibbs-Duhem equation which for any closed system is

$$
\Sigma_k\, n_k\, d\mu_k = 0 \tag{27}
$$

where n_k is the number of moles of the k-th component in the system. Three components are associated with the solid phase in a binary cation-exchange system: the two cation species and the water imbibed by the cation exchanger. Accordingly,

$$
n_i\, d \ln(x_{V_i} f_{V_i}) + n_j\, d \ln(z_{V_j} f_{V_j}) + n_w\, d \ln a_w = 0 \tag{28}
$$

Equations (26) and (28) constitute two equations with two unknowns $(f_{V_i}$ and $f_{V_j})$. Dividing Eq. (28) by $n_i + n_j$ gives

$$
x_{V_i}\, d \ln(x_{V_i} f_{V_i}) + z_{V_j}\, d \ln(x_{V_j} f_{V_j}) + \frac{n_w}{n_i + n_j}\, d \ln a_w = 0 \tag{29}
$$

Since $d \ln x_{V_i} = dx_{V_i}/x_{V_i}$,

$$x_{V_i} d \ln f_{V_i} + x_{V_j} d \ln f_{V_j} + \frac{n_w}{n_i + n_j} d \ln a_w = 0 \qquad (30)$$

Then

$$d \ln f_{V_i} = -\frac{1}{x_{V_i}} \left[x_{V_j} d \ln f_{V_j} + \frac{n_w}{n_i + n_j} d \ln a_w \right] \qquad (31)$$

and

$$d \ln f_{V_j} = -\frac{1}{x_{V_j}} \left[x_{V_i} d \ln f_{V_i} + \frac{n_w}{n_i + n_j} d \ln a_w \right] \qquad (32)$$

Substituting for $d \ln f_{V_i}$ from (31) into (26) and simplifying gives

$$d \ln f_{V_i}^{z_j} = E_j d \ln \,_j^i k_V^a - z_i z_j \hat{n}_w \, d \ln a_w \qquad (33)$$

where $\hat{n}_w$ the number of moles of water per mole of positive charge borne by the exchangeable cations; i.e.,

$$\hat{n}_w = \frac{n_w}{z_i n_i + z_j n_j} \qquad (34)$$

and E_i is the solid-phase equivalent fraction

$$E_i = \frac{z_i x_{V_i}}{z_i x_{V_i} + z_j x_{V_j}} \qquad (35)$$

Similarly for $d \ln f_{V_j}$,

$$d \ln f_{V_j}^{z_i} = E_i d \ln \,_i^j k_V^a - z_i z_j \hat{n}_w \, d \ln a_w \qquad (36)$$

In order to evaluate the activity coefficient f_{V_i} it is necessary to integrate Eq. (33) from the equilibrium composition of the mixture to the standard state for the component A_i. This can be over any arbitrary path and perhaps the most convenient path (and the one most commonly followed) is a two-stage integration. Step (1) is from the standard state to a mixture characterized by $z_{V_i} = 1$ and water activity equal to that of a pure M_i salt solution at the anion molality or ionic strength at which the experiment is performed (state designated by ϕ_i). Step (2) is from the final state of step (1) to a state characterized by the chosen solid-phase composition with a water activity equal to that in the mixed electrolyte solution. Performing the integrations (see Appendix 1) gives

$$\ln f_{V_i}^{z_j} = \int_{0(E_i=1)}^{E_j} \ln {}_j^ik_V^a \, dE_j - E_j \ln {}_j^ik_V^a$$

$$- z_i z_j \left[\int_{0(E_i=1)}^{\ln a_{\hat{w}}(\phi_i)} \hat{n}_w^{\phi_i} \, d \ln a_w + \int_{\ln a_w(\phi_i)}^{\ln a_w(E_j)} \hat{n}_w \, d \ln a_w \right]$$

$$(37)$$

Similarly for f_{V_j},

$$\ln f_{V_j}^{z_i} = \int_{0(E_j=1)}^{E_j} \ln {}_i^jk_V^a \, dE_i - E_i \ln {}_i^jk_V^a$$

$$- z_i z_j \left[\int_{0(E_j=1)}^{\ln a_w(\phi_j)} \hat{n}_w^{\phi_j} \, d \ln a_w + \int_{\ln a_w(\phi_j)}^{\ln a_w(E_j)} \hat{n}_w \, d \ln a_w \right]$$

$$(38)$$

The exchange equilibrium constant is obtained by eliminating the exchanger activity coefficients from Eq. (26) using Eqs. (37) and (38); i.e.,

$$\ln {}_j^iK_A = \int_0^1 \ln {}_j^ik_V^a \, dE_j$$

$$+ z_i z_j \left[\int_{0(E_i=1)}^{\ln a_w(\phi_i)} \hat{n}_w \, d \ln a_w - \int_{0(E_j=1)}^{\ln a_w(\phi_j)} \hat{n}_w \, d \ln a_w + \int_{\ln a_w(\phi_i)}^{\ln a_w(\phi_i)} \hat{n}_w \, d \ln a_w \right]$$

$$(39)$$

This equation dictates the experimental measurements necessary to evaluate the thermodynamic properties of a binary cation exchange system. The requirement is a series of equilibrium measurements at a constant sum of cationic charges in solution or ionic strength with solid-phase compositions varying over the complete binary range. At each point the compositions of both ions in both phases should be measured and values of $_j^i k_V^a$ determined. Given the standard states chosen, the activity of water in both phases is identical and directly calculable from the solution composition. Thus, all integrals in Eqs. (37)–(39) can be evaluated graphically or numerically. It is common to neglect water activity terms. Barrer and Klinowski [10] showed that for zeolites the effect on the measured value of the exchange equilibrium constant will be slight because of a the substantial self-compensation of the water terms. This is not the case for the activity coefficients themselves which may vary significantly with ionic strength [11]. For the most precise measurements it may be desirable to perform the water activity corrections for exchangers which show significant water uptake such as clay minerals and synthetic cation-exchange resins. Recently, Fletcher and Sposito [12] on the basis of their analysis of data compiled from cation-exchange reactions on montmorillonite concluded that many cation-exchange reactions of this mineral are ideal. The exceptions were reactions between ions with markedly different hydration characteristics (e.g., between sodium and potassium or cesium). It is well established that large monovalent cations such as potassium and cesium ions can exclude water from montmorillonite particles and in these cases the progressive change in water content accompanying ion exchange with sodium may require explicit chemical-thermodynamic treatment even at low-electrolyte molalities. Equations to predict water activities from solution compositions are given in Section 3.

E. Chemical-Thermodynamic Consistency

In order to be consistent, equations for the value of any specific activity coefficient must tend to unity as the component approaches its reference state. Additionally, for any mixture to be ideal the activities of all components must equal their mole fractions. Equations (37) and (38) will conform to both these requirements for all reactions regardless of ion charge since unit values for activity coefficients are consistent with compositional independence of $_j^i k_V^a$.

Högfeldt [13] has demonstrated that similar equations for activity coefficients and exchange equilibrium constants can be generated for ion exchange using the type G components described earlier in this

chapter. In this case the experimental mass-action coefficient, sometimes called the Gapon selectivity coefficient, as noted earlier, has
the form

$$
_j^i k_G^a = \left(\frac{a_{j(aq)}^{1/z_j} \, x_{G_i}}{a_{i(aq)}^{1/z_i} \, x_{G_j}} \right) \tag{40}
$$

It should be noted that *the mole fraction of a type G component is
equal to the equivalent fraction of its associated cation* regardless of
charge and the two are completely interchangeable in this convention.
Neglecting water activity terms the equations for the activity coefficients are

$$
\ln f_{G_i} = \int_{0(E_i=1)}^{E_j} \ln {_j^i k_G^a} \, dE_j - E_j \ln {_j^i k_G^a} \tag{41}
$$

and

$$
\ln f_{G_j} = \int_{0(E_j=1)}^{E_j} \ln {_i^j k_G^a} \, dE_i - E_i \ln {_i^j k_G^a} \tag{42}
$$

For the exchange equilibrium constant the following relationship holds:

$$
\ln {_j^i K_G} = \int_0^1 \ln {_j^i k_G^a} \, dE_j \tag{43}
$$

These equations, like those for type V components, are completely
self-consistent.

An alternative convention was adopted by Gaines and Thomas [6]
who used Eqs. (1) and (10) for the reaction between type V components but adopted equivalent fractions for expressing their composition.
This is equivalent to using concentration units for type G components
with a type V reaction. The experimentally determinable corrected
selectivity coefficient has the form

$$
_j^i k_{GT}^a = \left(\frac{a_{j(aq)}^{z_i} \, E_i^{z_j}}{a_{i(aq)}^{z_j} \, E_j^{z_i}} \right) \tag{44}
$$

and the activity coefficients f_{GT_i} and f_{GT_j} can be evaluated from

$$\ln f_{GT_i}^{z_j} = -(z_i - z_j)E_j + \int_{0(E_i=1)}^{E_j} \ln {}_j^i k_{GT}^a \, dE_j - E_j \ln {}_j^i k_{GT}^a \quad (45)$$

and

$$\ln f_{GT_j}^{z_i} = (z_i - z_j)E_i + \int_{0(E_j=1)}^{E_j} \ln {}_i^j k_{GT}^a \, dE_i - E_i \ln {}_i^j k_{GT}^a \quad (46)$$

The exchange equilibrium constant can be evaluated from

$$\ln {}_j^i K_{GT} = (z_i - z_j) + \int_0^1 \ln {}_j^i k_{GT}^a \, dE_j \quad (47)$$

As required by the definition of reference functions [Eq. (15)], the value of a Gaines-Thomas activity coefficient for any one component tends to unity as the component approaches its reference state. However, a consequence of this mixed convention is that *for hetero-valent exchange it is not possible for the Gaines-Thomas activity coefficients to be simultaneously equal to unity over the whole iso-therm range.* This can be seen from Eq. (45) by imposing the restriction that ${}_j^i k_{GT}^a$ is a constant (i.e., have unit activity coefficients), in which case f_{GT_i} will show a compositional dependence rather than have unit value. This is not a mathematical inconsistency. Barrer and Townsend [14] have demonstrated that the treatment is correct mathematically and *the activities and the value of the exchange equilibrium constant will be exactly those predicted by the equations of Argersinger et al.* [3]. Thus for heterovalent cation-exchange reactions, it is impossible for both exchangeable cations to behave "ideally" since their solid-phase activity coefficients cannot simultaneously be equal to unity. However, the compositional variation of the Gaines-Thomas activity coefficients is considered by some authors to be unacceptable since "ideal" behavior is possible when adopting either type V or type G components [15,16].

F. Multicomponent Ion-Exchange Equilibria

The chemical-thermodynamic description of multicomponent cation exchange has a greater algebraic complexity than the binary treatment but uses no new principles. The general expression for the reaction, Eq. (1), applies to any binary ion combination in a multispecies cation-exchange system. For a system containing N exchangeable-cation species there are $\frac{1}{2}(N^2 - N)$ independent reactions. A total of $N^2 - N$ reactions can be formulated of which half will be the reverse of the other half.

The Vanselow selectivity coefficients and the general exchange equilibrium constant for a type V component have the same form as the binary constant; i.e.,

$$\,_j^i K_V = \left(\frac{a_{j(aq)}^{z_i} x_i^{z_j}}{a_{i(aq)}^{z_j} x_j^{z_i}} \; \frac{f_{V_i}^{z_j}}{f_{V_j}^{z_i}} \right) \tag{48}$$

where the mole-fraction terms x_ℓ are constrained by

$$\sum_{\ell=1}^{N} x_\ell = 1 \tag{49}$$

These are sometimes called *pseudobinary* exchange equilibrium constants. The value of any single constant will be independent of whether it is determined in a binary or multicomponent system.

The differentials of a minimum of N - 1 type V exchange equilibrium constants (48), combined in the general Gibbs-Duhem equation, formalized as

$$\sum_{\ell=1}^{N} x_\ell \, d \ln f_{V_\ell} = 0 \tag{50}$$

gives a minimum of N equations with N values of f_{V_ℓ} as the basis set. These can be solved in exactly the same way as a binary system. Their solution, a problem first addressed by Soldatov and Bychkova [17] was affected in exactly the same way as it had been for the binary system. For example, a ternary system containing the ions i, j, and k, yields three differential equations derived from the equilibrium constants:

$$d \ln {}_{j}^{i}k_{V}^{a} = z_i \, d \ln f_{V_j} - z_j \, d \ln f_{V_i} \tag{51}$$

$$d \ln {}_{k}^{i}k_{V}^{a} = z_i \, d \ln f_{V_k} - z_k \, d \ln f_{V_i} \tag{52}$$

$$d \ln {}_{j}^{k}k_{V}^{a} = z_k \, d \ln f_{V_j} - z_j \, d \ln f_{V_k} \tag{53}$$

Combining these with Eq. (50) and following the integration procedure described by Chu and Sposito [18] yields a general equation for the k-th activity coefficient in the set; i.e.,

$$z_i z_j \ln f_{V_k} = z_j \left(E_i \ln {}_{i}^{k}k_{V}^{a} - \int_{E_k=1}^{E_j,E_k} \ln {}_{i}^{k}k_{V}^{a} \, dE_i \right)$$

$$+ z_i \left(E_j \ln {}_{j}^{k}k_{V}^{a} - \int_{E_k=1}^{E_j,E_k} \ln {}_{i}^{k}k_{V}^{a} \, dE_j \right) \tag{54}$$

where i, j, and k represent any permutation of the three ions. The general equation for a designated exchange equilibrium constant is

$$\ln {}_{j}^{i}K = \int_{E_i=1}^{E_i,E_j} \left(\ln {}_{j}^{i}k_{V}^{a} \, dE_j + \frac{z_j}{z_k} \ln {}_{k}^{i}k_{V}^{a} \, dE_k \right) \tag{55}$$

which can be evaluated by integration over any chosen path including the binary limit, where the mole fraction of one component leads to zero, in a set of ternary cation-exchange experiments.

A similar set of equations were developed by Fletcher and Townsend [11] who, in extending the Gaines-Thomas approach [6] to ternary cation-exchange reactions, included a water activity correction. This approach has the same features as the Gaines-Thomas formulation for binary exchange since it uses equivalent fractions in the exchange equilibrium constants for type V components. Additionally, Fletcher and Townsend used reaction equations containing all three ions derived from combinations of "pseudobinary" reactions. For homovalent reactions, this approach is formally identical to Eqs. (54), (55) and those produced by Soldatov and Bychkova [17].

Examples of the application of the chemical thermodynamics of ternary cation-exchange reactions [20,21] have shown that internally self-consistent measurements of exchange equilibrium constants and solid-phase activity coefficients can be made. However, a study by Fletcher et al. [22] emphasizes the large amount of experimental data required to characterize completely a ternary system and the difficulty encountered in providing numerical analysis of the data. The amount of data required to characterize a system containing more than three cations is almost prohibitive. The discussion in Section 6 provides statistical techniques to optimize the number of data to collect in equilibrium ion-exchange experiments.

III. CHEMICAL THERMODYNAMICS OF ELECTROLYTE SOLUTIONS

In order to evaluate exchange equilibrium constants and solid phase activity coefficients it is necessary to determine the compositional dependences of $\ln {}_j^i k_V^a$ values over complete isotherm ranges. This includes evaluating the activities of ions in solutions for compositions at which equilibria are established. The variable ${}_j^i k_V^a$ can be formulated as

$$ {}_j^i k_V^a = \frac{x_A^{z_j} m_j^{z_i}}{x_V^{z_i} m_i^{z_j}} \, {}_j^i \Gamma \tag{56} $$

where m_i and m_j are the molalities of the subscripted ions and ${}_j^i \Gamma$ the liquid-phase activity-coefficient ratio defined as

$$ {}_j^i \Gamma \overset{\text{def}}{=} \frac{\gamma_j^{z_i}}{\gamma_i^{z_j}} \tag{57} $$

where γ_i and γ_j are single-ion activity coefficients. This ${}_j^i \Gamma$ term describes nonideal behavior in the liquid phase which, in some cases, can be as large as the nonideality correction for the solid phase [21].

This may be the case, even in dilute solutions, for inorganic ions that exhibit strong ion-ion or ion-solvent interactions. Single-ion activity coefficients cannot be measured independently and it is common to evaluate ${}_j^i \Gamma$ in terms of the activity coefficients of neutral salts.

The following subsections discuss the chemical thermodynamic basis of Eqs. (56) and (57) reviewing selected working equations for the molality dependence of activity coefficients.

A. Standard States, Reference States, and Reference Functions

Molality is the reference function for aqueous electrolyte solutions:

$$\lim_{m_i \to 0} \frac{m_i}{a_{i(aq)}} = 1 \tag{58}$$

The standard state is the hypothetical ideal molal solution whose behavior is extrapolated from infinitely dilute solution (the reference state). The single-ion activity coefficient γ_i can be defined by

$$a_{i(aq)} \overset{\text{def}}{=} \frac{\gamma_i m_i}{m_i^{\ominus}} \tag{59}$$

where $m_i^{\ominus}$ is unit molality of i (not necessarily of its salt). Thus, the activity coefficient so defined, is a dimensionless parameter which tends to unity as $m_i \to 0$. Consequently,

$$\mu_{i(aq)} = \mu_{i(aq)}^{\ominus} + RT \ln \left(\frac{m_i \gamma_i}{m_i^{\ominus}} \right) \tag{60}$$

Consider a neutral salt composed of ν_k moles of anion $M_k^{z_k^-}$ and ν_i moles of $M_i^{z_i^+}$ where z_i represents the charge number of the subscripted ion. For m_{ik} moles of salt which can dissociate into $\nu_i m_{ik}$ moles of cation and $\nu_k m_{ik}$ moles of anion, the chemical potential for the salt μ_{ik} is related to the chemical potential for single ions by

$$m_{ik} \mu_{ik(aq)} = m_i \mu_{k(aq)} + m_k \mu_{i(aq)} \tag{61}$$

Eliminating single-ion molalities and dividing by m_{ik} gives

$$\mu_{ik} = \nu_k \mu_k + \nu_i \mu_k \tag{62}$$

Substituting Eq. (60) for both anion and cation into Eq. (62) gives an equation for the chemical potential of a neutral salt:

$$\mu_{ik} = \nu_i \mu_i^{\ominus} + \nu_k \mu_k^{\ominus} + \nu_i RT \ln(m_i \gamma_i) + \nu_k RT \ln(m_k \gamma_k) \tag{63}$$

Collecting standard chemical potential terms and simplifying yields

$$\mu_{ik} = \mu_{ik}^{\ominus} + (\nu_i + \nu_k)RT \ln [m_{ik}(\nu_k^{\nu_i}\nu_i^{\nu_k})^{1/(\nu_i+\nu_k)}\gamma_{\pm ik}] \tag{64}$$

where

$$\mu_{ik}^{\ominus} = \nu_k\mu_k^{\ominus} + \nu_i\mu_i^{\ominus} \tag{65}$$

This defines the mean-ionic activity coefficient for the neutral salt as

$$\gamma_{\pm ik} \overset{\text{def}}{=} \left[\gamma_i^{\nu_i}\gamma_k^{\nu_k}\right]^{1/(\nu_i+\nu_k)} \tag{66}$$

The liquid-phase activity coefficient ratio ${}^i_j\Gamma$ can now be evaluated by applying Eq. (66) to salts of the ions $M_i^{z_i+}$ and $M_j^{z_j+}$ sharing a common anion $M_k^{z_k-}$. By applying the relation

$$z_i\nu_i = z_k\nu_k \tag{67}$$

for each salt,* the activity coefficient for the anion can be eliminated from the resulting two equations by using Eq. (57) to yield

$$\,{}^i_j\Gamma = \frac{\gamma_{\pm jk}^{z_i(1+z_j/z_k)}}{\gamma_{\pm ik}^{z_j(1+z_i/z_k)}} \tag{68}$$

This equation is valid for aqueous solutions in which any number of salts are dissolved so long as these salts share a unique anionic species. Accurate mean-ionic activity coefficient values of single-salt aqueous solutions are tabulated for most salts of interest to those studying cation-exchange reactions. These experimentally determined values show that $\gamma_{\pm ik}$ vary greatly and nonlinearly with molality. For example, Fig. 1 shows the relationship between mean-ionic activities of electrolytes NaC, HCl, and $CaCl_2$ and molality in simple aqueous solutions at 25°C. Figure 2 shows the same activities, but over a

---

*Please note that the stoichiometric number ν_k may have different values for each of the salts considered (e.g., $\nu_{Cl} = 1$ for NaCl and $\nu_{Cl} = 2$ for $CaCl_2$).

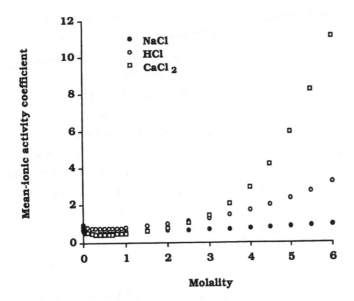

FIG. 1. Mean-ionic activity coefficients of three electrolytes measured in single-electrolyte solutions.

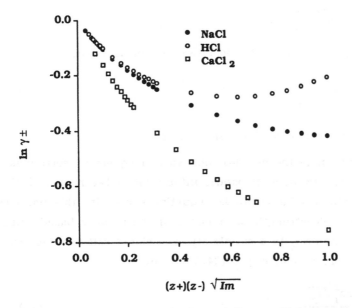

FIG. 2. Measured mean-ionic activity coefficients of three electrolytes plotted against $z_+ z_- \sqrt{I_m}$, showing the Debye-Hückel limiting law.

smaller range of molalities with the units of the axis of ordinates transformed to $z_+ z_- \sqrt{I_m}$, where z_+ and z_- are the charge numbers of the cation and anion, respectively. These curves indicate that mean-ionic activity coefficients at low I_m are directly proportional to $z_+ z_- \sqrt{I_m}$. These data will be returned to in describing the capabilities and shortcomings of several activity coefficient models. The problem confronting researchers of cation-exchange reactions is to predict the activity coefficients of the chosen salts in multicomponent electrolyte solutions from the known properties of single-salt solutions.

Since we are concerned with aqueous solutions, it is appropriate at this stage to define the chemical potential of water μ_w which is related to the activity of water a_w by

$$\mu_w = \mu_w^{\ominus} + RT \ln a_w \qquad (69)$$

For practical purposes the osmotic coefficient of water, ϕ_w, is more convenient than water activity coefficient. The osmotic coefficient of a solvent A is defined by [23]

$$\ln \left(\frac{\lambda_A}{\lambda_A^*} \right) \overset{\text{def}}{=} -\phi_A M_A \sum_B m_B \qquad (70)$$

The value of ϕ_w is linked to the molalities of ions in solution via

$$\phi_w \approx \frac{-1000 \ln a_w}{18.0153 \, \Sigma_j \, m_j} \qquad (71)$$

B. Activity Coefficients in Mixed Electrolytes

A major contribution to the nonideal behavior of aqueous electrolytes is the potential energy of electrostatic interaction between ions of like charge. The extended Debye-Hückel equation deals with this phenomenon giving the ionic-strength dependence of the molality based single-ion activity coefficient γ_i in a completely dissociated salt solution containing a single binary electrolyte [24,25]; i.e.,

$$\ln \gamma_i = \frac{- \alpha z_i^2 \sqrt{I_m}}{1 + \beta d \sqrt{I_m}} \qquad (72)$$

α and β are constants with values of 1.171 $(kg^{1/2} \, mol^{-1/2})$ and 0.328 3 $\times 10^6$ $(kg^{1/2} \, mol^{-1/2})$, respectively, at 25 °C and pressure of 1 atm (101,325 Pa) [4]. The parameter d(m) is the mean diameter of the ions and I_m is the ionic strength [26] defined as

$$I_m \stackrel{def}{=} \frac{1}{2} \Sigma \, m_i z_i^2 \tag{73}$$

Tabulated values of Debye-Hückel equation parameters may be found in standard references of electrolyte-solution thermodynamics [27,28]. Equation (72) is derived from a theory of hard, spherical, unpolarizable, incompressible ions in a homogeneous dielectric continuum. Some of these approximations are invalid for small ions of high valence or solutions of high ionic strength [29,30]. However, comparisons with Monte Carlo simulations of ion-ion interactions [31,32] have suggested that the basic equation may be valid for completely dissociated monovalent electrolyte solutions at molalities up to 1. Despite this, Fig. 3 provides a comparison of mean-ionic activity coefficients measured for three electrolytes with the corresponding Debye-Hückel equation predictions. Generally the equation-based predictions deviate appreciably from the measured values at I_m well below 1 mol kg^{-1},

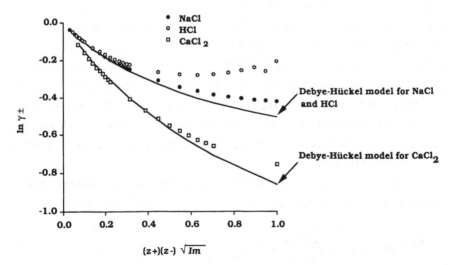

FIG. 3. Measured mean-ionic activity coefficients of three electrolytes plotted with the Debye-Hückel model estimates.

indicating the practical ionic-strength limit for the applicability of
the Debye-Hückel equation.

In real electrolyte solutions ions bind to water molecules and
associate to form complexed species; most improvements in theory
have been concerned with these phenomena. Guggenheim [33] presented an extension to Debye-Hückel theory for multicomponent
electrolytes based on the assumption that all nonelectrostatic contributions to nonideality may be incorporated into a single ion-ion interaction term for each ion-ion combination. For a system containing
N different ionic species, the expression for the single-ion activity
coefficient of the i-th ion is

$$\ln \gamma_i = \frac{-\alpha z_i^2 \sqrt{I_m}}{1 + \sqrt{I_m}} + 2 \sum_{j=1}^{N} \beta_{ij} m_j \qquad (74)$$

where β_{ij} is an ion-ion interaction coefficient term for ions i and j.
To be consistent with Brønsted's principle of specific interaction of
ions [34] these coefficients will be zero for ions of like charge.

Equation (74) omits the product βd from the denominator in the
electrostatic term because d is different for different ion combinations.
Because of this incompatibility with the Gibbs-Duhem equation, it is
assumed that the product βd in Eq. (74) equals 1. (This assumption
may be made also by observing that d for many electrolytes has values
near $1/\beta$.) This limits the validity of the electrostatic term but ensures the equation's chemical-thermodynamic consistency for multicomponent solutions, whereas Eq. (72) is strictly applicable in Eq. (74)
in a single-salt solution. Guggenheim and Turgeon [35] acknowledged
that the interaction terms cannot be attributed to any one single factor and they compensate for all contributions, including inadequacies
in the electrostatic term.

For a neutral salt in a multicomponent electrolyte solution, the
Guggenheim model of the mean-ionic activity coefficient leads to

$$\ln \gamma_{\pm ik} = \frac{-\alpha z_i z_k \sqrt{I_m}}{1 + \sqrt{I_m}} + \frac{2\nu_i}{\nu_i + \nu_k} \sum_{\ell=1}^{M} \beta_{i\ell} m_\ell + \frac{2\nu_k}{\nu_i + \nu_k} \sum_{j=1}^{L} \beta_{jk} m_j \qquad (75)$$

where L is the number of cationic species in the solution and M is

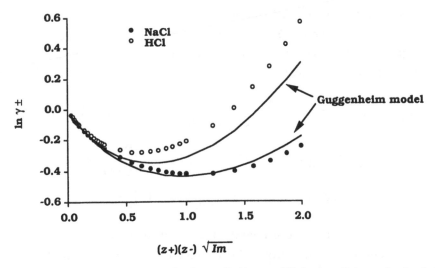

FIG. 4. Measured mean-ionic activity coefficients of two electrolytes plotted with the corresponding Guggenheim-model estimates.

the number of anionic species in the solution. A limited number of tabulated β_{ik} values may be found in Zemaitis et al. [36]. In view of the simplicity of Eq. (74), it may still be the most useful equation available for all electrolyte mixtures at ionic strengths below 0.1 molal. In practice, the procedure would be to evaluate β_{ij} for various ion combinations by analysis of available activity coefficient data for pure electrolytes. The fact that only one interaction parameter is required for each mean-ionic activity coefficient makes this a simple procedure. Figure 4 presents mean-ionic activities of two salts with the Guggenheim-model predictions. For these salts, the results with the Guggenheim-based model represent a modest improvement over those from the Debye-Hückel equation.

For more concentrated solutions it was acknowledged that β_{ij} varied with ionic strength [35]. To circumvent this problem for an aqueous solution of two salts having anionic species, Glueckauf [37] used Eq. (75) to eliminate the ion-ion interaction terms by back-substitution into Eq. (75). This gave an equation for a mean-ionic activity coefficient in terms of the mean-ionic activity coefficients of the two pure salts at the ionic strength of the mixture. Thus, the

compositional dependence of the ion-ion interaction terms was
accounted for implicitly. Fletcher and Townsend [38] generalized
this approach for a salt in a multicomponent electrolyte solution of
known composition to give

$$
\log \gamma_{\pm ik} = \log \bar{\gamma}_{\pm ik} + \frac{1}{4I_m(z_i + z_k)} \left[\sum_{j=1(j \neq i)}^{N} m_j \varepsilon_{kj} \left(\zeta_{1j} \log \bar{\gamma}_{\pm ik} \right. \right.
$$

$$
+ \zeta_{2j} \log \bar{\gamma}_{jk} \zeta_{3j} A_\gamma \Delta - (\zeta_{1j} + \zeta_{2j}) \Gamma \big) + \sum_{j=1(j \neq k)}^{N} m_j \varepsilon_{ij}
$$

$$
\left. \left(\zeta_{4j} \log \bar{\gamma}_{\pm ik} + \zeta_{5j} \log \bar{\gamma}_{ij} \varepsilon_{6j} A_\gamma \Delta - (\zeta_{4j} + \zeta_{5j}) \Gamma \right) \right] \quad (76)
$$

where $\bar{\gamma}_{ij}$ is the activity coefficient of the subscripted salt in the
pure solution and ε_{ij} is unity if ions i and j have charges of different sign and is zero for ionic charges of the same sign. The
terms simply represent algebraic combinations of ionic charge
numbers:

$$
\begin{aligned}
\zeta_{1j} &= z_j(z_i + z_k)(z_i - 2z_j - z_k) \\
\zeta_{2j} &= z_i(z_j + z_k)^2 \\
\zeta_{3j} &= z_i z_j z_k (z_i - z_j)^2 \\
\zeta_{4j} &= z_j(z_i + z_k)(z_k - 2z_j - z_i) \\
\zeta_{5j} &= z_k(z_j + z_i)^2 \\
\zeta_{6j} &= z_i z_j z_k (z_k - z_j)^2
\end{aligned}
\quad (77)
$$

and

$$
\Delta = \left(1 + \frac{1}{\sqrt{I_m}} \right)^{-1} \quad (78)
$$

In principle, Eq. (76) is an equation of mixtures which require
values for the mean-ionic activity coefficients in pure solutions.
Even though this is a limitation, the intrinsic ability of Eq. (76) to
compensate exactly for the compositional dependence of the ion-ion
interactions may render it useful in strongly associated solutions
without recourse to explicit consideration of ion-association equilibria.

There have been many treatments of electrolyte solutions designed to extend the Guggenheim approach or develop similar techniques. A detailed and comprehensive review of these treatments has been assembled recently [36]. One study of particular note is that by Pitzer [39] who accommodated the compositional dependence of the ion-ion interaction terms in the Guggenheim approach with a more generalized expression for excess Gibbs energy of electrolyte solutions. This involved a comprehensive array of ion-ion interaction terms which included terms for ions of like charge. Pitzer acknowledged that many ion-ion interaction terms were not determinable independently. Consequently, interaction terms of similar type were grouped together into a limited set of virial coefficients. Pitzer began by developing an expression for the excess Gibbs energy [to be defined in Eq. (116)] for a multicomponent electrolyte solution:

$$\frac{G^E}{RT} = m_w f(I_m) + \frac{1}{m_w} \sum_i^N \sum_j^N \lambda_{ij}(I_m) n_i n_j$$

$$+ \frac{1}{m_w^2} \sum_i^N \sum_j^N \sum_k^N \mu_{ijk} n_i n_j n_k \qquad (79)$$

where N is the number of ionic species in the solution (1), m_w is the mass of water in the solution (kg), n_i is the amount of ion i (mol), $f(I_m)$ is a function of I_m representing long-range electrostatic forces (kg mol^{-1}), $\gamma_{ij}(I_m)$ is a function of I_m representing short-range forces between ions i and j (kg mol^{-1}) and μ_{ijk} is a parameter representing contribution of short-range interactions between ions i, j, and k (kg^2 mol^{-2}). The single-ion activity coefficient of a cation ε can be derived directly by

$$\ln \gamma_c = \frac{1}{RT} \frac{\partial G^E}{\partial n_c}$$

$$= \frac{z_c^2}{2} f' + 2 \sum_g \lambda_{cg} m_g + \frac{z_c^2}{2} \sum_g^N \sum_h^N \lambda'_{cg} m_g m_h$$

$$+ 3 \sum_g^N \sum_h^N \mu_{cgh} m_g m_h \qquad (80)$$

where $f' = df/dI_m$ and $\lambda'_{ij} = d\lambda_{ij}/dI_m$. For mean ionic activity coefficients in simple, and particularly in mixed, electrolyte solutions. Pitzer and his co-workers found it useful to reformulate (80) with the following substitutions:

$$f' = \frac{1}{2} f'$$

$$B_{ij} = 2\lambda_{ij} + I_m\lambda'_{ij} + \frac{\nu_i}{2\nu_j} (\lambda_{ii} + I_m\lambda'_{ii}) + \frac{\nu_j}{2\nu_i} (\lambda_{jj} + I_m\lambda'_{jj})$$

$$B'_{ij} = \lambda'_{ij} + \frac{z_i}{2z_j} \lambda'_{ii} + \frac{z_i}{2z_j} \lambda'_{jj}$$

$$C^\phi_{ij} = \frac{3}{\sqrt{\nu_i\nu_j}} [\nu_i\mu_{iij} + \nu_j\mu_{ijj}]$$

$$C_{ij} = \frac{C^\phi_{ij}}{2\sqrt{z_iz_j}} \qquad\qquad (81)$$

$$\theta_{ij} = \lambda_{ij} - \frac{z_j}{2z_i} \lambda_{ii} - \frac{z_i}{2z_j} \lambda_{jj}$$

$$\theta'_{ij} = \frac{\partial\theta_{ij}}{\partial I_m}$$

$$\psi_{ijk} = 6\mu_{ijk} - \frac{3z_j}{z_i} \mu_{iik} - \frac{3z_i}{z_j} \mu_{jjk}$$

The single-ion activity coefficient for a cation c in a mixed electrolyte solution becomes

$$\ln \gamma_c = z_c^2 f^\gamma + 2 \sum_{k=1}^{M} m_k \left(B_{ck} + C_{ck} \sum_{g=1}^{N} z_g m_g \right)$$

$$+ 2 \sum_{i=1}^{L} m_i \theta_{ci} + \sum_{i=1}^{L} \sum_{k=1}^{M} m_i m_k (z_c^2 B'_{ik} + z_c C_{ik} + \psi_{cik})$$

$$+ \frac{1}{2} \sum_{k=1}^{M} \sum_{\ell=1}^{M} m_k m_\ell (z_c^2 \theta'_{k\ell} + \psi_{c\ell k}) + \frac{z_c^2}{2} \sum_{i=1}^{L} \sum_{j=1}^{L} m_i m_j \theta'_{ij}$$

$$+ z_c \left[\sum_{i=1}^{L} \frac{m_i \lambda_{ii}}{z_i} - \sum_{k=1}^{M} \frac{m_k \lambda_{kk}}{z_k} \right.$$

$$\left. + \frac{3}{2} \sum_{i=1}^{L} \sum_{k=1}^{M} m_i m_k \left(\frac{\mu_{iik}}{z_i} - \frac{\mu_{ikk}}{z_k} \right) \right] \qquad\qquad (82)$$

where g is the index for any ionic species, i is the index for cationic species, j is the index for cationic species, k is the index for anionic species, ℓ is the index for anionic species, L is the number for cationic species in the solution, M is the number for anionic species in the solution, and N is their sum, L + M. The terms B_{ij}, C_{ij}, θ_{ji}, ψ_{ijk}, and B'_{ij} are virial coefficients for the interactions between the subscripted ions. The terms B_{ij}, $B_{ij}^{\gamma'}$, and C_{ij} have the greatest effect on the values of the equation and represent interactions between ions of different charge; they are zero for ions of like charge. The terms θ_{ij} and ψ_{ijk} are second-order terms for binary and ternary interactions, respectively. The parameters B_{ij} and $B_{ij}^{\gamma'}$ have compositional dependencies given by

$$B_{ij} = \beta_{ij}^{(0)} + \frac{2\beta_{ij}^{(1)}}{\alpha_0^2 I_m}[1 - (1 + \alpha\sqrt{I_m})\exp(-\alpha\sqrt{I_m})]$$

(83)

$$B_{ij}^{\gamma'} = \frac{2\beta_{ij}^{(1)}}{\alpha^2 I_m^2}\left[-1 + \left(1 + \alpha\sqrt{I_m} + \frac{1}{2}\alpha^2 I_m\right)\exp(-\alpha\sqrt{I_m})\right]$$

For 1-1, 2-1, 1-2, 3-1, 4-1, and 5-1 electrolytes α has a value of 2.0. These relations must be expanded for 2-2 electrolytes:

$$B_{ij} = \beta_{ij}^{(0)} + \frac{2\beta_{ij}^{(1)}}{\alpha^2 I_m^2}[1 - (1 + \alpha_1\sqrt{I_m})\exp(-\alpha_1\sqrt{I_m})]$$

$$+ \frac{2\beta_{ij}^{(2)}}{\alpha_2^2 I_m}[1 - (1 - \alpha_2\sqrt{I_m})\exp(-\alpha_2\sqrt{I_m})]$$

(84)

and

$$B_{ij}^{\gamma'} = \frac{2\beta^{(1)}}{\alpha_1 I_m}\left[-1 + \left(1 + \alpha_1\sqrt{I_m} + \frac{1}{2}\alpha_1^2 I_m\right)\exp(-\alpha_1\sqrt{I_m})\right]$$

$$+ \frac{2\beta_{ij}^{(2)}}{\alpha_2 I_m}\left[-1 + \left(1 + \alpha_2\sqrt{I_m} + \frac{1}{2}\alpha_2^2 I_m\right)\exp(-\alpha_2\sqrt{I_m})\right]$$

(85)

where α_1 has the value 1.4 for higher-valence electrolytes and α_2 has the value 12.

The term C_{ij} is related to a compositionally independent parameter C_{ij}^{ϕ} via

$$C_{ij} = \frac{C_{ij}^{\phi}}{2\sqrt{z_i z_j}} \tag{86}$$

The other major term in Eq. (82) is the electrostatic term derived from the pressure equation of statistical mechanics, which has the form

$$f^{\gamma} = -A_{\phi}\left[\frac{\sqrt{I_m}}{1 + b\sqrt{I_m}} + \frac{2}{b}\ln(1 + b\sqrt{I_m})\right] \tag{87}$$

The parameter b represents a mean value for d, having the value 1.2 for all ions, and

$$A_{\phi} = \frac{\alpha}{3}$$

Figure 5 presents a comparison of the mean-ionic activities of the three electrolytes considered in earlier figures with the corresponding Pitzer model-based estimates. The agreement between the model estimates and the measured data is spectacular.

The Pitzer model-based expression for the osmotic coefficient is

$$
(\phi_w - 1) = \frac{1}{\sum_{h=1}^{N} m_h}\left[2I_m f^{\phi}\right.
$$
$$
+ 2\sum_{i=1}^{L}\sum_{k=1}^{M} m_i m_k\left(B_{ik}^{\phi} + C_{ik}^{\phi}\frac{\sum_{h=1}^{N} z_h m_h}{\sqrt{z_i z_k}}\right)
$$
$$
+ \sum_{i=1}^{L}\sum_{j=1}^{L} m_i m_j\left(\theta_{ij} + I_m \theta_{ij}' + \sum_{k=1}^{M} m_k \psi_{ijk}\right)
$$
$$
\left.+ \sum_{k=1}^{M}\sum_{\ell=1}^{M} m_k m_{\ell}\left(\theta_{k\ell} + I_m \theta_{k\ell}' + \sum_{i=1}^{L} m_i \psi_{ik\ell}\right)\right] \tag{88}
$$

where

$$B_{ij}^{\phi} = \beta_{ij}^{(0)} + \beta_{ij}^{(1)}\exp(-\alpha_1\sqrt{I_m}) + \beta_{ij}^{(2)}\exp(-\alpha_2\sqrt{I_m}) \tag{89}$$

and

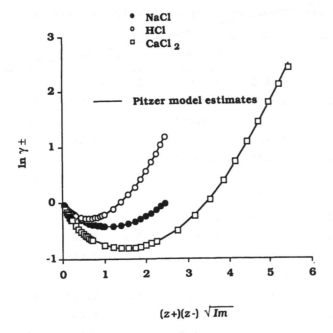

FIG. 5. Measured mean-ionic activity coefficients of three electro-lytes plotted with the corresponding Pitzer-model estimates.

$$f^\phi = \frac{-A_\phi \sqrt{I_m}}{1 + b\sqrt{I_m}}$$

Although algebraically complex, the Pitzer model requires only the parameters $\beta_{ij}^{(0)}$, $\beta_{ij}^{(1)}$, $\beta_{ij}^{(2)}$, and C_{ij}^ϕ for most salts; the inclusion of the higher-order terms is necessary for a small proportion of situations. Values of these parameters for selected inorganic salts are presented in Tables 1-6.

A virtue of this approach is that it is supported by an exten-sive collection of values for the necessary parameters for over 100 salts at 25°C with, in many cases, temperature derivatives valid up to ≈50°C [68]. Computer-based algorithms which calculate solution equilibria with Pitzer's model are available [69].

(text continues on p. 41)

TABLE 1   Pitzer-Model Parameters for Selected Inorganic Type 1-1 Electronics

Solute	$\beta^{(0)}$	$\beta^{(1)}$	$C^\phi = 2C$	Max. m	σ^a	Ref.[b]
HCl	0.1775	0.2945	0.0008	6		
HBr	0.2085	0.3477	0.00152	6.2	0.003	40
HI	0.2211	0.4907	0.00482	6		41
$HClO_4$	0.1747	0.2931	0.00819	5.5	0.002	
HNO_3	0.1168	0.3546	-0.00539	6		41
$H(HSO_4)$	0.2103	0.4711				42
$H(HSO_4)$	0.2065	0.5556				43
LiCl	0.1494	0.3074	0.00359	6	0.001	
LiBr	0.1748	0.2547	0.0053	2.5	0.002	
LiI	0.2104	0.373		1.4	0.006	
LiOH	0.015	0.14		4		
$LiClO_3$	0.1705	0.2294	-0.00524	4.2	0.002	44
$LiClO_4$	0.1973	0.3996	0.0008	3.5	0.002	
$LiBrO_3$	0.0893	0.2157	0.00000	5	0.001	44
$LiNO_2$	0.1336	0.325	-0.0053	6	0.003	
$LiNO_3$	0.142	0.278	-0.00551	6	0.001	
NaF	0.0215	0.2107		1	0.001	
NaCl	0.0765	0.2664	0.00127	6	0.001	
NaBr	0.0973	0.2791	0.00116	4	0.001	
NaI	0.1195	0.3439	0.0018	3.5	0.001	
NaOH	0.0864	0.253	0.0044	6		
$NaClO_3$	0.0249	0.2455	0.004	3.5	0.001	
$NaClO_4$	0.0554	0.2755	-0.00118	6	0.001	
$NaBrO_3$	-0.0205	0.1910	0.0059	2.5	0.001	
NaCNS	0.1005	0.3582	-0.00303	4	0.001	
$NaNO_2$	0.0641	0.1015	-0.0049	5	0.005	
$NaNO_3$	0.0068	0.1783	0.00072	6	0.001	
NaHSe	0.040	(0.253)		2		44
$NaHCO_3$	0.028	0.044				45
$NaHSO_4$	0.0454	0.398				42
NaH_2PO_4	-0.0533	0.0396	0.00795	6	0.003	
NaH_2AsO_4	-0.0442	0.2895		1.2	0.001	

TABLE 1 (continued)

NaB(OH)$_4$	-0.0526	0.1104	0.0154	4.5	0.004	
NaBF$_4$	-0.0252	0.1824	0.0021	6	0.006	
KF	0.08089	0.2021	0.00093	2	0.001	
KCl	0.04835	0.2122	-0.00084	4.8	0.0005	
KBr	0.0569	0.2212	-0.00180	5.5	0.001	
KI	0.0746	0.2517	-0.00414	4.5	0.001	
KOH	0.1298	0.320	0.0041	5.5		
KClO$_3$	-0.0960	0.2481		0.7	0.001	
KBrO$_3$	-0.1290	0.2565		0.5	0.001	
KNO$_2$	0.0151	0.015	0.0007	5	0.003	
KNO$_3$	-0.0816	0.0494	0.00660	3.8	0.001	
KHCO$_3$	-0.0107	0.0478				46
KHSO$_4$	-0.0003	0.1735				42
KH$_2$PO$_4$	-0.0678	-0.1042		1.8	0.003	
KH$_2$AsO$_4$	-0.0584	0.0626		1.2	0.003	
KCNS	0.0416	0.2302	-0.00252	5	0.001	
KCNS	0.0389	0.2536	-0.00192	5	0.001	43
KPF$_6$	-0.163	-0.282		0.5	0.001	
RbF	0.1141	0.2842	-0.0105	3.5	0.002	
RbCl	0.0431	0.1539	-0.00109	7.8	0.003	47
RbBr	0.0396	0.153	-0.00144	5	0.001	
RbI	0.0397	0.133	-0.00108	5	0.001	
RbNO$_2$	0.0269	-0.1553	-0.00366	5	0.002	
RbNO$_3$	-0.0789	-0.0172	0.00529	4.5	0.001	
CsF	0.1306	0.257	-0.0043	3.2	0.002	
CsCl	0.0347	0.0397	0.00049	7.4	0.002	48
CsBr	0.0279	0.0139	0.00004	5	0.002	
CsI	0.0244	0.0262	-0.00365	3	0.001	
CsOH	0.150	0.3				
CsNO$_2$	0.0427	0.06	-0.0051	6	0.004	
CsNO$_3$	-0.0758	-0.0669		1.4	0.002	
AgNO$_3$	-0.0856	0.0025	0.00591	6	0.001	
TlClO$_4$	-0.087	-0.023		0.5	0.001	
TlNO$_3$	-0.105	-0.378		0.4	0.001	

TABLE 1   (continued)

Solute	$\beta^{(0)}$	$\beta^{(1)}$	$C^\phi = 2C$	Max. m	σ^a	Ref.[b]
NH_4Cl	0.0522	0.1918	-0.00301	6	0.001	
NH_4Br	0.0624	0.1947	-0.00436	2.5	0.001	
NH_4I	0.0570	0.3157	-0.00308	7.5	0.002	
NH_4HCO_3	-0.038	0.070			0.7	49
$NH_4H_2PO_4$	-0.0704	-0.4156		3.5	0.003	50
NH_4ClO_4	-0.0103	-0.0194		2	0.004	
NH_4NO_3	-0.0154	0.1120	-0.00003	6	0.001	
$(MgOH)Cl$	-0.10	1.658				42

[a]Estimated maximum standard error of parameter estimates.

[b]Unless noted, all values are from K. S. Pitzer and G. Mayorga, *J. Phys. Chem.* 77:2300 (1973), 78:2698 (1973) or K. S. Pitzer, J. B. Peterson, and L. F. Silvester, *J. Solution Chem.*, 7:45 (1978). Reproduced by permission of American Chemical Society and Plenum Press Corp.

TABLE 2   Pitzer-Model Parameters for Selected Inorganic Type 2-1 Electrolytes

Solute	$\frac{4}{3}\beta^{(0)}$	$\frac{4}{3}\beta^{(1)}$	$\frac{2^{5/2}}{3}C^\phi = \frac{16}{3}C$	Max. m	σ	Ref.[a]
$MgCl_2$	0.4698	2.242	0.00979	4.5	0.003	
$MgCl_2$	0.4679	2.201	0.01227	4	0.003	51
$MgBr_2$	0.5769	2.337	0.00589	5	0.004	
MgI_2	0.6536	2.4055	0.01496	5	0.003	
$Mg(ClO_4)_2$	0.6615	2.678	0.01806	2	0.002	
$Mg(NO_3)_2$	0.4895	2.113	-0.03889	2	0.003	
$Mg(HCO_3)_2$	0.044	1.133				52
$Mg(HSO_4)_2$	0.6328	2.305				42
$CaCl_2$	0.4212	2.152	-0.00064	2.5	0.003	
$CaCl_2$	0.4071	2.278	-0.00406	4.3	0.003	53
$CaBr_2$	0.5088	2.151	-0.00485	2	0.002	
CaI_2	0.5839	2.409	-0.00158	2	0.001	

TABLE 2   (continued)

$Ca(ClO_4)_2$	0.6015	2.342	-0.00943	2	0.005	
$Ca(NO_3)_2$	0.2811	1.879	-0.03798	2	0.002	
$Ca(HCO_3)_2$	0.533	3.97				42
$Ca(HSO_4)_2$	0.286	3.37				42
$SrCl_2$	0.3810	2.223	-0.00246	4	0.003	
$SrCl_2$	0.3779	2.167	-0.00168	3.8	0.002	48
$SrBr_2$	0.4415	2.282	0.00231	2	0.001	
SrI_2	0.5350	2.480	0.00501	2	0.001	
$Sr(ClO_4)_2$	0.5692	2.089	-0.02472	2.5	0.003	
$Sr(NO_3)_2$	0.1795	1.840	-0.03757	2	0.002	
$BaCl_2$	0.3504	1.995	-0.03654	1.8	0.001	
$BaBr_2$	0.4194	2.093	-0.03009	2	0.001	
BaI_2	0.5625	2.249	-0.03286	1.8	0.003	
$Ba(OH)_2$	0.229	1.60		0.1		
$Ba(ClO_4)_2$	0.4819	2.101	-0.05894	2	0.003	
$Ba(NO_3)_2$	-0.043	1.07		0.4	0.001	
$MnCl_2$	0.4429	2.019	-0.04278	4	0.003	47
$FeCl_2$	0.4479	2.043	-0.01623	2	0.002	
$Fe(HSO_4)_2$	0.5697	4.64				54
$CoCl_2$	0.4857	1.967	-0.02869	3	0.004	
$CoBr_2$	0.5693	2.213	-0.00127	2	0.002	
CoI_2	0.695	2.23	-0.0088	2	0.01	
$Co(NO_3)_2$	0.4159	2.254	-0.01436	5.5	0.003	
$NiCl_2$	0.4665	2.040	-0.00888	2.5	0.002	55
$CuCl_2$	0.3955	1.855	-0.06792	2	0.002	
$Cu(NO_3)_2$	0.4224	1.907	-0.04136	2	0.002	
$Cu(NO_3)_2$	0.3743	2.310	-0.01580	8	0.003	56
$ZnCl_2$	0.3043	2.308	-0.1235	1.5	0.007	57
$ZnBr_2$	0.6213	2.179	-0.2035	1.6	0.007	
ZnI_2	0.6428	2.594	-0.0269	0.8	0.002	
$Zn(ClO_4)_2$	0.6747	2.396	0.02134	2	0.003	
$Zn(NO_3)_2$	0.4641	2.255	-0.02955	2	0.001	
$Cd(NO_3)_2$	0.3820	2.224	-0.04836	2.5	0.002	
$Pb(ClO_4)_2$	0.4443	2.296	-0.01667	6	0.004	

GRANT AND FLETCHER

TABLE 2 (continued)

Solute	$\frac{4}{3}\beta^{(0)}$	$\frac{4}{3}\beta^{(1)}$	$\frac{2^{5/2}}{3}C^{\phi}$ $=\frac{16}{3}C$	Max. m	σ	Ref.[a]
$Pb(NO_3)_2$	-0.0482	0.380	0.01005	2	0.002	
UO_2Cl_2	0.5698	2.192	-0.06951	2	0.001	
$UO_2(ClO_4)_2$	0.8151	2.859	0.04089	2.5	0.003	
$UO_2(NO_3)_2$	0.6143	2.151	-0.05948	2	0.002	
Li_2SO_4	0.1817	1.694	-0.00753	3	0.002	
Na_2SO_4	0.0261	1.484	0.00938	4	0.003	
Na_2SO_4	0.0249	1.466	0.010463	4	0.003	58
$Na_2S_2O_3$	0.0882	1.701	0.00705	3.5	0.002	
Na_2CrO_4	0.1250	1.826	-0.00407	2	0.002	
Na_2CO_3	0.0483	2.013	0.0098			41
Na_2HPO_4	-0.0777	1.954	0.0554	1	0.002	
$Na_2HA_5O_4$	0.0407	2.173	0.0034	1	0.001	
K_2SO_4	0.0666	1.039		0.7	0.002	
K_2CrO_4	0.1011	1.652	-0.00147	3.5	0.003	
K_2CO_3	0.1717	1.911	0.00094	1.5	0.001	46
$K_2Pt(CN)_4$	0.0881	3.164	0.0247	1	0.005	
K_2HPO_4	0.0330	1.699	0.0309	1	0.002	
$K_2HA_5O_4$	0.1728	2.198	-0.0336	1	0.001	
Rb_2SO_4	0.0772	1.481	-0.00019	1.8	0.001	
Cs_2SO_4	0.1184	1.481	-0.01131	1.8	0.001	
Cs_2SO_4	0.0954	1.601	0.00549	5		59
$(NH_4)_2SO_4$	0.0545	0.878	-0.00219	5.5	0.004	
$(NH_4)_2SO_4$	0.0521	0.8851	-0.00156	5.8	0.002	50

[a]Estimated maximum standard error of parameter estimates.

TABLE 3 Pitzer–Model Parameters for Selected Inorganic Type 3–1 Electrolytes

Solute	$\frac{3}{2}\beta^{(0)}$	$\frac{3}{2}\beta^{(1)}$	$\frac{3^{3/2}}{2}C^{\phi}$ = 9C	Max. m	σ	Ref.[a]
$AlCl_3$	1.0490	8.767	0.0071	1.6	0.005	
$SrCl_3$	1.0500	7.978	-0.0840	1.8	0.005	
YCl_3	0.939	8.40	-0.040	4.1	0.006	60
$LaCl_3$	0.883	8.40	-0.061	3.9	0.006	
$CeCl_3$	0.907	8.40	-0.074	1.8	0.007	
$PrCl_3$	0.883	8.40	-0.054	3.9	0.006	
$NdCl_3$	0.878	8.40	-0.049	3.9	0.006	
$SmCl_3$	0.900	8.40	-0.053	3.6	0.007	
$EuCl_3$	0.911	8.40	-0.054	3.6	0.006	
$GdCl_3$	0.913	8.40	-0.049	3.6	0.006	
$TbCl_3$	0.922	8.40	-0.046	3.6	0.004	
$DyCl_3$	0.929	8.40	-0.045	3.6	0.005	
$HoCl_3$	0.937	8.40	-0.045	3.7	0.006	
$ErCl_3$	0.928	8.40	-0.038	3.7	0.006	
$TmCl_3$	0.926	8.40	-0.036	3.7	0.005	
$YbCl_3$	0.923	8.40	-0.033	3.7	0.005	
$LuCl_3$	0.922	8.40	-0.033	4.1	0.005	
$CrCl_3$	1.1046	7.883	-0.1172	1.2	0.005	
$Cr(NO_3)_3$	1.0560	7.777	-0.1533	1.4	0.004	
$Ga(ClO_4)_3$	1.2381	9.794	0.0904	2	0.008	
$InCl_3$	-1.68	-3.85		0.01		
$Y(NO_3)_3$	0.915	7.70	-0.189	2.0	0.008	60
$La(NO_3)_3$	0.737	7.70	-0.198	1.5	0.007	61
$Pr(NO_3)_3$	0.724	7.70	-0.173	1.5	0.005	55
$Nd(NO_3)_3$	0.702	7.70	-0.142	2.0	0.008	62
$Sm(NO_3)_3$	0.701	7.70	-0.131	1.5	0.007	
$Eu(NO_3)_3$	0.713	7.70	-0.125	2.0	0.007	60
$Gd(NO_3)_3$	0.776	7.70	-0.170	1.4	0.005	
$Tb(NO_3)_3$	0.838	7.70	-0.202	1.4	0.005	
$Dy(NO_3)_3$	0.848	7.70	-0.1809	2.0	0.008	63

TABLE 3 (continued)

Solute	$\frac{3}{2}\beta(0)$	$\frac{3}{2}\beta(1)$	$\frac{3^{3/2}}{2}C^{\phi}$ $= 9C$	Max. m	σ	Ref.[a]
$Ho(NO_3)_3$	0.876	7.70	-0.1852	2.0	0.009	63
$Er(NO_3)_3$	0.938	7.70	-0.226	1.5	0.006	
$Tm(NO_3)_3$	0.952	7.70	-0.222	1.5	0.006	
$Yb(NO_3)_3$	0.948	7.70	-0.208	1.5	0.006	
$Lu(NO_3)_3$	0.926	7.70	-0.174	2.0	0.008	55
$La(ClO_4)_3$	1.15	9.80	0.001	2.0	0.009	
$Pr(ClO_4)_3$	1.13	9.80	0.016	2.0	0.006	
$Nd(ClO_4)_3$	1.13	9.80	0.019	2.0	0.007	
$Sm(ClO_4)_3$	1.14	9.80	0.014	2.0	0.005	
$Gd(ClO_4)_3$	1.17	9.80	0.014	2.0	0.007	
$Tb(ClO_4)_3$	1.19	9.80	0.012	2.0	0.006	
$Dy(ClO_4)_3$	1.20	9.80	0.014	2.0	0.006	
$Ho(ClO_4)_3$	1.19	9.80	0.013	2.0	0.004	
$Er(ClO_4)_3$	1.20	9.80	0.014	1.8	0.004	
$Tm(ClO_4)_3$	1.19	9.80	0.024	2.0	0.005	
$Yb(ClO_4)_3$	1.20	9.80	0.013	1.8	0.004	
$Lu(ClO_4)_3$	1.18	9.80	0.029	2.0	0.006	
Na_3PO_4	0.2672	5.777	-0.1339	0.7	0.003	
Na_3AsO_4	0.3582	5.895	-0.124	0.7	0.001	
K_3PO_4	0.5594	5.958	-0.2255	0.7	0.001	
$K_3P_3O_9$	0.4867	8.349	-0.0886	0.8	0.004	
K_3AsO_4	0.7491	6.511	-0.3376	0.7	0.001	
$K_3Fe(CN)_6$	0.5033	7.121	-0.1176	1.4	0.003	
$K_3Co(CN)_6$	0.5603	5.815	-0.1603	1.4	0.008	

[a]Estimated maximum standard error of parameter estimates.

TABLE 4 Pitzer-Model Parameters for Selected Inorganic Type
4-1 and 5-1 Electrolytes[a]

4-1 Solute	$\frac{8}{5}\beta^{(0)}$	$\frac{8}{5}\beta^{(1)}$	$\frac{16}{5}C^{\phi}=\frac{64}{5}C$	Max. m	σ
$ThCl_4$	1.622	21.33	-0.3309	1	0.006
$Th(NO_3)_4$	1.546	18.22	-0.5906	1	0.01
$Na_4P_2O_7$	0.699	17.16		0.2	0.01
$K_4P_2O_7$	0.977	17.88	-0.2418	0.5	0.01
$K_4Fe(CN)_6$	1.021	16.23	-0.5579	0.9	0.008
$K_4Mo(CN)_8$	0.854	18.53	-0.3499	0.8	0.01
$K_4W(CN)_8$	1.032	18.49	-0.4937	1	0.005
$(Me_4N)_4Mo(CN)_8$	0.938	15.91	-0.3330	1.4	0.01
5-1 Solute	$\frac{5}{3}\beta^{(0)}$	$\frac{5}{3}\beta^{(1)}$	$\frac{5^{3/2}}{3}C^{\phi}=\frac{50}{3}C$	Max. m	σ
Na_5P_3O10	1.869	36.10	-1.630	0.4	0.01
K_5P_3O10	1.939	39.64	-1.055	0.5	0.015

[a]Estimated maximum standard error of parameter estimates.

TABLE 5 Pitzer-Model Parameters for Selected Inorganic Type 2-2
Electrolytes

Solute	$\beta^{(0)}$	$\beta^{(1)}$	$\beta^{(2)}$	$C^{\phi}=4C$	Range m	σ	Ref.
$MgSO_4$	0.2210	3.343	-37.23	0.0250	0.006-3.0	0.004	64
$MgSO_4$	0.2150	3.365	-37.74	0.0280	0.006-3.6	0.004	58
$NiSO_4$	0.1702	2.907	-40.06	0.0366	0.005-2.5	0.005	64
$MnSO_4$	0.213	2.938	-41.91	0.0155	0.1-5.0	0.005	47
$FeSO_4$	0.2568	3.063	(-42)	0.0209	0.1-2.0	—	54
$CoSO_4$	0.1631	3.346	-30.7	0.03704	0.2-2.4	0.001	65
$CuSO_4$	0.234	2.527	-48.33	0.0044	0.005-1.4	0.003	66
$ZuSO_4$	0.1949	2.883	-32.81	0.0290	0.005-3.5	0.004	64
$CdSO_4$	0.2053	2.617	-48.07	0.0114	0.005-3.5	0.002	64
$CaSO_4$	0.20	3.197	-54.24		0.004-0.011	0.003	42
$SrSO_4$	0.220	2.88	-41.8	0.019	—	—	67
$BeSO_4$	0.317	2.914	?	0.0062	0.1-4.0	0.004	64
UO_2SO_4	0.322	1.827	?	-0.0176	0.1-5.0	0.003	64

TABLE 6   Values of the Virial Ionic Interaction Parameters θ_{ij} and ψ_{ijk} in the Pitzer Model for Selected Ion Combinations [68]

c	c'	$\theta_{cc'}$	$\psi_{cc'Cl}$	$\psi_{cc'SO_4}$	$\psi_{cc'HSO_4}$	$\psi_{cc'NO_3}$	$\psi_{cc'HCO_3}$	$\psi_{cc'CO_3}$
Li	Na	0.0029	—	-0.0039	—	—	—	—
	K	-0.0563	—	-0.0086	—	—	—	—
	Rb	-0.0908	—	0.0024	—	—	—	—
	Cs	-0.1242	—	0.0088	—	—	—	—
Na	K	-0.0029	-0.0018	-0.010	—	—	-0.003	0.003
	Rb	-0.0319	—	0.0048	—	—	—	—
	Cs	-0.0153	—	-0.0035	—	—	—	—
	NH$_4$	0	-0.0003	-0.0013	—	—	—	—
	Ca	0.07	-0.007	-0.055	—	—	—	—
	Mg	0.07	-0.012	-0.015	—	—	—	—
	MgOH	0	—	—	—	—	—	—
	H	0.036	-0.004	—	-0.0016	—	—	—
K	Cs	-0.0049	—	-0.0035	—	—	—	—
	Ca	0.032	-0.025	—	—	—	—	—
	Mg	0.0	-0.022	-0.48	—	—	—	—
	MgOH	0	—	—	—	—	—	—
	H	0.005	-0.011	—	-0.0265	—	—	—
Mg	MgOH	0.0	0.028	—	—	—	—	—
	Cu	-0.0085	—	—	—	-0.0031	—	—
	H	0.005	-0.011	—	-0.0265	—	—	—
Ca	Mg	0.007	-0.012	0.024	—	—	—	—
	MgOH	0	—	—	—	—	—	—
	Cu	-0.0558	—	—	—	0.0026	—	—
	H	0.092	-0.015	—	—	—	—	—
Sr	H	0.0642	-0.0033	—	—	—	—	—
Ba	H	0.0708	0.0018	—	—	—	—	—
Ni	Ca	0.0131	—	—	—	-0.0031	—	—
	H	0.0708	0.0018	—	—	—	—	—

TABLE 6　(continued)

a	a'	$\theta_{aa'}$	$\psi_{aa'Na}$	$\psi_{aa'K}$	$\psi_{aa'Ca}$	$\psi_{aa'Mg}$	$\psi_{aa'}$ MgOH	$\psi_{aa'H}$
Cl	SO_4	0.030	0.000	-0.005	-0.002	-0.008	—	—
	HSO_4	-0.006	-0.006	—	—	—	—	0.13
	OH	-0.050	-0.006	-0.006	-0.025	—	—	—
	HCO_3	-0.03	-0.015	—	-0.096	—	—	0.13
	CO_3	-0.02	0.0085	0.004	-0.096	—	—	—
SO_4	HSO_4	—	-0.0094	-0.0677	—	-0.0425	—	—
	OH	-0.013	-0.009	-0.050	—	—	—	—
	HCO_3	0.01	-0.005	—	-0.161	—	—	0.13
	CO_3	0.02	-0.005	-0.009	—	—	—	—
OH	CO_3	0.10	-0.017	-0.01	—	—	—	—
HCO_3	CO_3	-0.04	0.002	0.012	—	—	—	—

　　　The technique has been tested extensively and shown to be valid
for both major and trace components in concentrated electrolytes.
Indeed, Harvie et al. [70] reported remarkable success in its application to salt evaporites.　The major weakness is in the application to
strongly associating and weak electrolytes where the explicit recogni-
tion of complex-forming equilibria may be required.

C.　Activity Coefficients and Ion-Association Equilibria

The ability of ions in aqueous solutions to associate and form complexed species is well established.　However, the concept of an ion
pair or complex is nebulous.　Bjerrum [71] introduced the idea of an
ion pair to compensate for any ions that may be closer together than
the minimum distance of approach required by the extended Debye-
Hückel equation.　Ion pairs in this class were treated as a single ion
carrying a charge equal to the sum of the charges on the ions from
which they are derived.　The quantity of this ion pair was calculated
from mass-action formalism even though this formalism was not meant
to imply that the ion pair was a stable molecular entity.　However,

for weak electrolytes and many strong electrolytes such as sulfates, carbonates, and salts of transition metals, the formation of a stable molecular complex is apparent from conductivity and freezing-point determinations. In these situations it is necessary to calculate *explicitly* the molalities of the complexed species in order to evaluate ion activities.

In principle there is no conflict between the use of virial coefficients to describe nonideal behavior and the explicit use of ion association equilibria, since an equation of self-consistency exists linking the two approaches. Consider a solution resulting from a neutral salt dissociating into ions $M_i^{z_i}$, $M_k^{z_k}$, and N^c complexes formed from the reactions

$$M_i + jM_k = M_j^c \tag{91}$$

where M_j^c is the j-th complex. Activities can be ascribed to all the solution species; i.e.,

$$a_{i(aq)} = \hat{m}_i \hat{\gamma}_i \tag{92}$$

$$a_{k(aq)} = \hat{m}_k \hat{\gamma}_k \tag{93}$$

and

$$a_{j(aq)}^c = \hat{m}_j^c \hat{\gamma}_j^c \tag{94}$$

where the terms $\hat{m}$ refer to the *true* molalities of the subscripted ions, which for the primary species will be lower than the stoichiometric molalities due to the complex formation. The $\hat{\gamma}$'s are true single-ion activity coefficients as distinct from stoichiometric single-ion activity coefficients.

An equation similar to (61) can describe the chemical potential of the salt but in terms of the molalities and chemical potentials of the ions and complexes; i.e.,

$$\mu_{ik} = \hat{m}_i \mu_i + \hat{m}_k \mu_k + \sum_{j=1}^{N^c} \hat{m}_j^c \mu_j^c \tag{95}$$

Using an approach similar to that for Eqs. (61) to (66) an activity coefficient product can be defined; i.e.,

$$\hat{\gamma}_{\pm ik} = \left(\hat{\gamma}_i^{z_k} \hat{\gamma}_k^{z_i}\right)^{1/(\nu_i + \nu_k)} \tag{96}$$

This function is similar to the mean-ionic activity coefficient defined in Eq. (66), but it refers to the activities of the true ionic species. Since the chemical potential of the salt must be independent of the way the components are formulated, the relationship between the mean-ionic activity coefficient and the true ionic activity coefficient is

$$\ln \gamma_{\pm ik} = \ln \hat{\gamma}_{\pm ik} + \frac{1}{\nu_i + \nu_k} \ln \left(\frac{\hat{m}_i^{z_k} \hat{m}_k^{z_i}}{m_i^{z_k} m_k^{z_i}}\right) \tag{97}$$

This is an equation of constraint. If the species molalities are known then the values of the true single-ion activity coefficients can be constrained by the known values of the mean-ionic activity coefficients. Thus, in any model proposing to describe ion association explicitly the following relationships must hold.*

$$a_i = m_i \gamma_i = \hat{m}_i \hat{\gamma}_i \tag{98}$$

and

$$a_k = m_k \gamma_k = \hat{m}_k \hat{\gamma}_k \tag{99}$$

To perform computations accommodating complex formation requires a procedure for calculating the species molalities from the total ion molalities in solution. For a solution containing N ions and N^c complexes, one can equate the total molality of the i-th ion source, m_i (either anion or cation), to the molality of the free uncomplexed ion plus the sum of the molalities of the ion in each attendant complex. This operation can be described with the following general mass-balance equation:

---

*For simplicity and clarity, the state-standard molality m_i^Θ was not a included as a divisor in this and some of the equations in this discussion [see Eq. (59)], though it should be inserted implicitly into these for dimensional agreement.

$$m_i = \hat{m}_i + \sum_{j=1}^{N^c} \nu_{ij}\hat{m}_j^c \tag{100}$$

where $\hat{m}_j^c$ is the molality of the j-th complex and ν_{ij} is the number of moles of the i-th ion in the j-th complex. If the general reaction to form one mole of the j-th complex is described by the canonical expression

$$\sum_{i=1}^{N} \nu_{ij}M_i + \eta_j H_2O \rightleftharpoons M_j^c \tag{101}$$

and the complementary equilibrium constant is expressed by

$$K_j = \frac{\hat{m}_j^c \hat{\gamma}_j^c}{(a_w)^{\eta_j} \Pi_{i=1}^{N}(\hat{m}_i\hat{\gamma}_i)^{\nu_{ij}}} \tag{102}$$

then, for the situation where the water activity and activity coefficients are known, there are $N + N^c$ solvable equations with $N + N^c$ unknowns (species molalities).

In practice activity coefficients are evaluated using an equation involving *known* species molalities, and the standard procedure is to estimate activity coefficients, solve for species molalities, and then make an improved estimate of activity coefficients from which a better estimate of species molalities can be made. The process is repeated until a specified convergence criterion is met. The equations for species molalities can be solved using multicomponent Newton-Raphson techniques.

In order to perform such calculations a simple equation for single-ion activity coefficients was provided by Davies [72];

$$\ln \gamma_i = -\alpha z_i^2 \left(\frac{\sqrt{\bar{I}_m}}{1 + \sqrt{\bar{I}_m}} - 0.3\bar{I}_m \right) \tag{103}$$

where the true ionic strength, containing terms for complexes, is defined as

$$\bar{I}_m = \frac{1}{2} \left(\sum_{i=1}^{N} \hat{m}_i z_i^2 + \sum_{j=1}^{N^c} \hat{m}_j^c z_j^2 \right) \tag{104}$$

This equation may be useful for ionic strengths less than 0.5 mol kg^{-1} if used with equilibrium constants for complex formation such as those compiled by Smith and Martell [73]. For higher ionic strengths Helgeson et al. developed a similar model [74] (called the HKF model) in which the major contributions to the thermodynamic functions are treated explicitly rather than with nonspecific fitting coefficients. It is argued that the resulting equations have sound molecular basis from which *extrapolations* may be made, rather than being a retrieval equation for *interpolations* over a range of fit, which is the feature of the Pitzer approach. The HKF-model equation for the actual single-ion activity coefficient of cation i is

$$\ln \hat{\gamma}_i = - \frac{\alpha z_i^2 \sqrt{I_m}}{\Lambda} + \Gamma_\gamma + \omega_i^{abs} \sum_{k=1} b_k y_k \bar{I}_m + \sum_{\ell=1} \frac{b_{i\ell} \bar{y}_\ell \bar{I}_m}{\psi_\ell} \quad (105)$$

where ℓ is an index for an anionic species and k is an index for an electrolyte. The dimensionless quantities Λ, Γ_γ, y_k, $\bar{y}_k$, and ψ_k are defined by

$$\Lambda = 1 + \beta d \sqrt{\bar{I}_m} \quad (106)$$

$$\Gamma_\gamma = -\ln(1 + 0.0180153 \, m^*) \quad (107)$$

$$y_k = \frac{z_k^2 m_k}{2 I_m} \quad (108)$$

$$\bar{y}_k = \frac{z_k^2 m_k}{2 \bar{I}_m} \quad (109)$$

$$\psi_k = \frac{\sum_j \nu_{j,k} z_j^2}{2} \quad (110)$$

where J is the index for an ionic species of either polarity (1), m^* is the sum of molalities of all ions in solution (mol kg^{-1}), ν_k is the number of moles of ions in solution from one mole of the k-th thermodynamic component (1), $\nu_{j,k}$ is the number of moles of the j-th ion from one mole of the k-th component (1), ω_i^{abs} is the absolute Born coefficient for the i-th ion (J mol^{-1}), b_k is an ion-solvent interaction term (kg J^{-1}), and b_{ij} is a short-range electrostatic parameter to account for ion-ion interactions not specifically accommodated in an ion-association equilibria (kg mol^{-1}). For single-electrolyte solutions b_k is defined by

$$b_k = \frac{1/\varepsilon_r(k) - 1/\varepsilon_r(w)}{\bar{I}_m RT} \tag{111}$$

where $\varepsilon_r(w)$ is the relative permittivity of pure water (1) and $\varepsilon_r(k)$ is the relative permittivity of the k solution (1). At 25°C and 0.1 MPa pressure the relation

$$b_k = \left(\frac{\omega_k}{\psi_k}\right) 1.75 \times 10^{-10} \tag{112}$$

holds for several electrolytes (HCl and $BaCl_2$ being notable exceptions) where ω_k is the conventional Born coefficient for the k-th electrolyte. The mean distance between ions (d) is estimated from the relative electrostatic radii [r_j^{el} (m)] of the ions by the relation

$$d = \frac{2 \Sigma_k \nu_k (\Sigma_j \nu_{j,k} r_j^{el}/\nu_k)}{\Sigma_k \nu_k} \tag{113}$$

Estimated values of r_j^{el}, ω_i^{abs}, ω_i, b_k, and b_{ij} for several electrolytes and ions (including complexes) are given by Helgeson et al. [74] and are reproduced in Tables 7-9.*

$$\phi_w = -\frac{1}{m^*} \sum_{j=1} m_j \left(\frac{\alpha z_i^2 \sqrt{\bar{I}_m} \sigma}{3} + \frac{\Gamma_\gamma}{0.0180153 \, m^*} \right.$$
$$\left. -\frac{1}{2}\left[\omega_j \sum_{k=1} b_k y_k \bar{I}_m + \sum_{i=1} b_{ij} m_i + \sum_{\ell=1} b_{j\ell} m_\ell \right] \right) \tag{114}$$

where

---

*In the original article, the HFK model for a single-ion activity coefficient was presented as a base-10, rather than natural, logarithm. When calculating activity coefficients with the equations presented here and with tabulated coefficients presented in Helgeson et al., one should change the coefficients by a factor of ln 10.

TABLE 7 Selected Helgeson-Kirkham-Flowers Model Parameters for Calculating Single-Ion Activity Coefficients: The Crystallographic and Effective Electrostatic Ionic Radii (r_j^x and r_j^{el}) and the Absolute and Conventional Born Coefficients (ω_j^{abs} and ω_j)

Species	r_j^x (nm)	r_j^{el} (nm)	ω_j^{abs} (J mol^{-1})	ω_j (J mol^{-1})
H^+	0.214	0.308	8.900	12.895
Li^+	0.068	0.164	2.847	6.866
Na^+	0.097	0.191	4.061	7.997
Rb^+	0.133	0.227	5.568	9.504
K^+	0.147	0.241	6.155	10.090
Cs^+	0.167	0.261	6.992	10.928
NH_4^+	0.137	0.231	5.736	9.672
Tl^+	0.147	0.241	6.155	10.090
Ag^+	0.126	0.220	5.275	9.211
Au^+	0.137	0.231	5.736	9.672
$Cu(I)^+$	0.096	0.190	4.019	7.955
Mg^{2+}	0.066	0.254	2.763	10.634
Sr^{2+}	0.112	0.300	4.089	12.560
Ca^{2+}	0.099	0.287	4.145	12.016
Ba^{2+}	0.134	0.322	5.610	13.481
Pb^{2+}	0.120	0.308	5.024	12.895
Zn^{2+}	0.074	0.262	3.098	10.969
$Cu(II)^{2+}$	0.072	0.260	3.014	10.886
Cd^{2+}	0.097	0.285	4.061	11.932
Hg^{2+}	0.110	0.298	4.605	12.477
$Fe(II)^{2+}$	0.074	0.262	3.098	10.969
Mn^{2+}	0.080	0.268	3.349	11.221
$Fe(III)^{3+}$	0.064	0.346	2.680	14.486
Al^{3+}	0.051	0.333	2.135	13.942
Au^{3+}	0.090	0.372	3.768	15.575
La^{3+}	0.114	0.396	4.773	16.580
Gd^{3+}	0.097	0.379	4.061	15.868

TABLE 7   (continued)

Species	r_j^x (nm)	r_j^{el} (nm)	ω_j^{abs} (J mol^{-1})	ω_j (J mol^{-1})
In^{3+}	0.081	0.363	3.391	15.198
Ga^{3+}	0.062	0.344	2.596	11.403
Tl^{3+}	0.095	0.377	3.977	15.784
F$^-$	0.133	0.133	5.568	5.568
Cl$^-$	0.181	0.181	7.578	7.578
Br$^-$	0.196	0.196	8.206	8.206
I$^-$	0.220	0.220	9.211	9.211
OH$^-$	0.140	0.140	5.862	5.862
HS$^-$	0.184	0.184	7.704	7.704
NO$_3^-$	0.281	0.281	11.765	11.765
HCO$_3^-$	0.210	0.210	8.792	8.792
HSO$_4^-$	0.237	0.237	9.923	9.923
ClO$_4^-$	0.359	0.359	15.031	15.031
ReO$_4^-$	0.423	0.423	17.710	17.710
SO$_4^{2-}$	0.315	0.315	13.188	13.188
CO$_3^{2-}$	0.281	0.281	11.765	11.765

Source: Reprinted from Helgeson et al. [74] by permission of *American Journal of Science*.

TABLE 8   Values of the Helgeson-Kirkham-Flowers-Model Parameter b_k (kg J^{-1}) for Selected Electrolytes

Electrolyte	b_k (kg J^{-1})	Electrolyte	b_k (kg J^{-1})
HCl	0.347 76	LiF	0.542 94
NaCl	0.426 70	KF	0.472 84
LiCl	0.463 86	RbF	0.462 69
KCl	0.393 79	CsF	0.450 08
RbCl	0.383 63	NH$_4$F	0.469 59
CsCl	0.371 02	AgF	0.478 41

TABLE 8 (continued)

NH_4Cl	0.390 54	AuF	0.469 81
AgCl	0.399 35	CuF	0.506 85
AuCl	0.390 75	MgF_2	0.406 92
CuCl	0.427 80	CaF_2	0.382 99
$MgCl_2$	0.354 23	SrF_2	0.375 01
CaCl	0.330 30	BaF_2	0.362 97
$SrCl_2$	0.322 30	FeF_2	0.398 18
$BaCl_2$	0.310 26	MaF_2	0.396 05
$FeCl_2$	0.347 88	PbF_2	0.370 43
$MnCl_2$	0.343 34	ZnF_2	0.400 57
$PbCl_2$	0.317 74	CuF_2	0.402 12
$ZnCl_2$	0.347 88	CdF_2	0.384 28
$CuCl_2$	0.349 43	HgF_2	0.376 21
$CdCl_2$	0.331 59	HBr	0.330 99
$HgCl_2$	0.323 49	NaBr	0.416 38
$AlCl_3$	0.288 17	LiBr	0.453 57
$FeCl_3$	0.281 65	KBr	0.383 47
$GdCl_3$	0.266 50	RbBr	0.373 32
$LaCl_3$	0.259 74	CsBr	0.354 26
HF	0.426 82	NH_4Br	0.373 77
NaF	0.505 76	AgBr	0.382 58
AuBr	0.373 98	FeI_2	0.321 99
CuBr	0.411 03	MnI_2	0.317 45
$MgBr_2$	0.343 05	PbI_2	0.291 85
$CaBr_2$	0.319 12	ZnI_2	0.321 99
$SrBr_2$	0.311 12	CuI_2	0.323 54
$BaBr_2$	0.299 08	CdI_2	0.305 70
$FeBr_2$	0.336 70	HgI_2	0.297 60
$MnBr_2$	0.332 16	KOH	0.457 94
$PbBr_2$	0.306 56	NaOH	0.490 85
$ZnBr_2$	0.336 70	HNO_3	0.269 78
$CuBr_2$	0.338 25	$NaNO_3$	0.348 72
$CdBr_2$	0.320 41	$LiNO_3$	0.385 90
$HgBr_2$	0.312 31	KNO_3	0.315 80

TABLE 8   (continued)

Electrolyte	b_k (kg J^{-1})	Electrolyte	b_k (kg J^{-1})
HI	0.308 02	NH_4ClO_4	0.281 91
NaI	0.387 86	$NaClO_4$	0.318 07
LiI	0.425 05	$NaHCO_3$	0.396 44
KI	0.354 95	NaHS	0.423 12
RbI	0.344 80	H_2SO_4	0.253 63
CsI	0.332 19	Na_2SO_4	0.306 25
NH_4I	0.351 70	K_2SO_4	0.284 32
AgI	0.360 51	Li_2SO_4	0.331 04
AuI	0.351 92	$MgSO_4$	0.282 01
CuI	0.388 96	$ZnSO_4$	0.277 25
MgI_2	0.328 34	$NaHSO_4$	0.374 92
CaI_2	0.304 41	$KHSO_4$	0.342 00
SrI_2	0.296 41	$HReSO_4$	0.222 41
BaI_2	0.284 37	$NaReSO_4$	0.301 35

TABLE 9   Short-Range Electrostatic Interaction
Parameter in the Helgeson-Kirkham-Flowers Model
($b_{j\ell}$, kg mol^{-1}) for Selected Anion-Cation
Combinations

Cation	Anion			
	F^-	Cl^-	Br^-	I^-
H^+	-0.083	0.019	0.056	0.103
Li^+	-0.182	-0.049	-0.024	0.032
Na^+	-0.155	-0.096	-0.075	-0.032
K^+	-0.132	-0.112	-0.101	-0.075
Rb^+	-0.127	-0.119	-0.118	-0.097
Cs^+	-0.119	-0.122	-0.121	-0.108
NH_4^+	-0.131	-0.114	-0.107	-0.087
Ag^+	-0.134	-0.086	-0.067	-0.037
Au^+	-0.131	-0.115	-0.107	-0.087
Cu^+	-0.153	-0.076	-0.047	-0.007

TABLE 9   (continued)

Mg^{2+}	-0.451	-0.256	-0.167	-0.085
Ca^{2+}	-0.493	-0.257	-0.183	-0.100
Sr^{2+}	-0.503	-0.266	-0.204	-0.113
Ba^{2+}	-0.529	-0.292	-0.227	-0.112
Pb^{2+}	-0.554	-0.324	-0.238	-0.140
Zn^{2+}	-0.462	-0.259	-0.186	-0.098
Cu^{2+}	-0.485	-0.277	-0.202	-0.114
Cd^{2+}	-0.463	-0.256	-0.180	-0.092
Hg^{2+}	-0.491	-0.278	-0.198	-0.104
Fe^{2+}	-0.414	-0.282	-0.157	-0.071
Mn^{2+}	-0.499	-0.298	-0.210	-0.118
Al^{3+}		-0.481		

Source: Reprinted from Helgeson et al. [74] by permission of *American Journal of Science*.

$$\sigma = \frac{3}{(d\beta\sqrt{\bar{I}_m})^3}\left[\Lambda - \frac{1}{\Lambda} + 2\ln\Lambda\right] \tag{115}$$

In principle these equations are consistent with the relationship developed in Eq. (97) for cases where significant ion association occurs. In practice, the application of these equations at high molalities require precise knowledge of equilibrium constants and activity coefficients for complexes. Extensive data sources do exist, but there is still no completely self-consistent data set available for this purpose. However, for ions that associate insignificantly, Eqs. (105) and (114) revert to simple equations for single ion activity and osmotic coefficients. In these cases, the equations predict the properties of pure electrolytes at high molalities and at temperatures up to 300 °C on the water-saturation curve. The ability of the ion association technique to predict the properties of trace components in concentrated solutions still requires investigation.

IV. EXCESS GIBBS ENERGY MODELS

Much of the previous section dealt with models for estimating the
activity coefficients for electrolyte solutions. The Pitzer model was
unusually capable of modeling accurately mean-ionic activity coefficients over a wide molality. As was noted, the Pitzer model is
excess Gibbs energy model from which the expressions for individual activity coefficients are derived. In this section, this chapter returns to excess Gibbs energy models, though those used for
estimating the activity coefficients of exchangeable cations in the
solid phase. This section will consider excess Gibbs energy models,
the chemical thermodynamics of these models, the various algebraic
forms these models may take, and the utility they can have for the
determination of solid-phase activity coefficients. The various algebraic forms that the excess Gibbs energy models take and the
methods to evaluate these models experimentally are outlined in Sec.
V. A procedure to develop optimal experimental designs for this
evaluation is presented in Sec. VI.

Experimenters in many disciplines employ chemical thermodynamics to study disparate types of mixtures: electrolyte and nonelectrolyte solutions, minerals, alloys, melts, and mixtures of liquids
[68,75-78]. Many researchers have considered the possibility that
an activity coefficient for every mixture component could be predicted by a single expression, an excess Gibbs energy model.
Excess Gibbs energy models have two useful features. First, they
describe comprehensively and compactly nonideal behavior for all
components in a mixture or solution. Secondly, if an excess Gibbs
energy model is appropriate for a particular type of system, information collected in studies of simple systems may be applied to predict the behavior of systems with many components. The potential
suggested for excess Gibbs energy models is far from realized in
the study of cation-exchange reactions, but most of the analytical
and statistical tools to evaluate them critically are now at hand.

Before accepting the excess Gibbs energy concept for cation-
exchange reactions it must pass two tests, one theoretical and the
other applied. First, it must be demonstrated that the solid-phase

excess Gibbs energy can be defined consistently with the macroscopic and molecular information available for cation-exchange reactions. As will be discussed later, a proposed solid-phase excess
Gibbs energy must be defined relative to standard states of the
solid-phase components in the system. Secondly, it must be
demonstrated that a chosen model works. Several groups of experimenters have shown that excess Gibbs energy models can *describe*
binary cation-exchange equilibria [79,80], but few have shown con-
vincingly that these models can *predict* multicomponent cation exchange equilibria. In the second of these crucial aspects, the
solid-phase excess Gibbs energy concept has yet to be vindicated.

A. Excess Gibbs Energy

The capabilities and limitations of excess Gibbs energy models are
better appreciated if their chemical-thermodynamic foundation is
understood. The excess molar Gibbs energy of mixing $\bar{G}^E$ (sometimes called the molar excess Gibbs energy) is the difference between the Gibbs energy of mixing for a real mixture ($\Delta_{mix}\bar{G}_m$) and
that of an ideal mixture with the same composition [23]; i.e.,

$$\bar{G}^E \overset{def}{=} \Delta_{mix}\bar{G}_m - \Delta_{mix}\bar{G}_m^{id} \tag{116}$$

To make this definition useful, the two terms on the right side of
Eq. (116) must be defined. The molar Gibbs energy of mixing,
$\Delta_{mix}\bar{G}_m$, for a mixture with two components i and j is the molar
Gibbs energy of the mixture ($\bar{G}_m$) minus the weighted sum of the
molar Gibbs energies of the individual components in their respective
standard states ($\bar{G}_i^{\ominus}$, $\bar{G}_j^{\ominus}$); i.e.,

$$\Delta_{mix}\bar{G}_m = \bar{G}_m - x_i\bar{G}_i^{\ominus} - x_j\bar{G}_j^{\ominus} \tag{117}$$

where x_i and x_j are mole fractions of the two mixture components.
The experimenter is free to specify any standard state which can
be defined unambiguously. Agreement by researchers on a single
standard state allows experimental information and theoretical developments to be communicated more easily. In this chapter, we sub-
scribe to the most widely adopted solid-phase standard state—a

homoionic exchanger at equilibrium with an infinitely dilute electrolyte solution. The molar Gibbs energy for the mixture ($\bar{G}_m$) is the
weighted sum of the actual molar Gibbs energies of the two
components

$$\bar{G}_m = x_i \bar{G}_i + x_j \bar{G}_j \tag{118}$$

Combining Eqs. (117) and (118) yields the expression

$$\Delta_{mix} \bar{G}_m = x_i(\bar{G}_i - \bar{G}_i^\ominus) + x_j(\bar{G}_j - \bar{G}_j^\ominus) \tag{119}$$

which shows that the molar Gibbs energy of mixing is the weighted
sum of the differences between the individual component under
Gibbs energies in the mixture and in their respective standard
states.

An ideal mixture can be defined as one in which the absolute
activities of the components are proportional to their mole fractions
in the mixture:

$$\lambda_i^{id} \overset{def}{=} x_i \lambda_i^\ominus \quad \text{and} \quad \lambda_j^{id} \overset{def}{=} x_j \lambda_j^\ominus \tag{120}$$

Since $\bar{G}_i = \mu_i = RT \ln \lambda_i$, then from Eqs. (119) and (120) the molar
Gibbs energy of mixing for an ideal mixture is

$$\Delta_{mix} \bar{G}_m^{id} = RT(x_i \ln x_i + x_j \ln x_j) \tag{121}$$

From Eqs. (116), (119), and (121) the excess molar Gibbs energy
of mixing is therefore

$$\bar{G}^E = x_i(\bar{G}_i - \bar{G}_i^\ominus) + x_j(\bar{G}_j - \bar{G}_j^\ominus) - RT(x_i \ln x_i + x_j \ln x_j) \tag{122}$$

Since

$$\bar{G}_i - \bar{G}_i^\ominus = \mu_i - \mu_i^\ominus = RT \ln \lambda_i - RT \ln \lambda_i^\ominus$$
$$= RT \ln a_i = RT \ln f_i x_i \tag{123}$$

the excess function takes the simple form

$$\bar{G}^E = RT(x_i \ln f_i + x_j \ln f_j) \tag{124}$$

As defined here, molar excess Gibbs energy has meaning only
for solid-phase exchangeable components at equilibrium with solutions
comparable to solid-phase standard-state equilibrating solutions, that
is, infinitely dilute electrolyte solutions. Virtually all cation-exchange
equilibrium experiments are conducted with equilibrating solutions
which cannot be construed to be near infinite dilution. For example,
the aluminum atoms in clay minerals enter solution as the equilibrating electrolyte solution becomes more dilute [81]. For these important classes of cation exchangers, the solid-phase standard state is
hypothetical. At present the excess Gibbs energy approach to solid-
phase activity coefficients cannot adequately bridge the electrolyte
molalities at which cation-exchange experiments are typically conducted and the generally accepted solid-phase standard state. Only
when the differences in solid-phase activities between solutions of
differing electrolyte molalities are described adequately will the theory
of solid-phase activity coefficients be comprehensive.

One way to develop this comprehensive theory is by expanding
the definition of molar excess Gibbs energy. The quantity, which
we designate the total molar excess Gibbs energy, is represented by
$\bar{G}^{E'}$ and by steps similar to those above can be defined by

$$\frac{\bar{G}^{E'}}{RT} = \sum_{i=1}^{N} X_i \ln g_i + \sum_{j=1}^{M} X_j \ln g_j + X_w \ln g_w \qquad (125)$$

where X_i is the mole fraction of the exchangeable ion i, defined by

$$X_i = \frac{n_i}{\sum_{i=1}^{N} n_i + \sum_{j=1}^{M} n_j + n_w} \qquad (126)$$

X_j, the mole fraction of nonexchangeable solute j, X_w, the mole fraction of water associated with the solid exchanger, and g_i, g_j, and
g_w the corresponding activity coefficients. Notice that n, the number of moles which define the mole fractions (X_i, etc.) is the sum
of the exchangeable and nonexchangeable adsorbed solutes combined
with the surface excess of water molecules associated with the solid
exchanger.

The relation in Eq. (125) is of little use in predicting selectivity coefficient values. This problem can be overcome somewhat by multiplying both sides of Eq. (125) by n_T/n, where n_T is the total number of exchangeable cations in the cation-exchange reactions being modeled to yield

$$\frac{n_T}{n} \frac{\bar{G}^{E'}}{RT} = \sum_{i=1}^{N} x_i \ln g_i + \sum_{j=1}^{M} x_j \ln g_i + x_w \ln g_w \qquad (127)$$

While $\sum_{i=1}^{N} x_i$ is equal to 1 the sum of all of the exchangeable-ion-based "mole fractions" is equal to n/n_T. The rational activity coefficient f_i is related to g_i by

$$f_i = \frac{n}{n_T} g_i \qquad (128)$$

B. Excess Gibbs Energy Models

The objective of an excess Gibbs energy model is to equate the excess Gibbs energy to a simple function of composition. This allows the measured activity coefficients to be modeled in terms of compositional dependence via Eq. (124). For ion-exchange reactions all such models are empirical. All have been adopted from reports describing solid and liquid mixtures. Indeed, no one has justified any solid phase excess Gibbs energy model on a molecular level understanding of arrangements and interactions of exchangeable ions, even though the statistical thermodynamics of isomorphic substitution in aluminosilicates has been addressed [82,85]. Given the increasingly clear understanding of the atomic environment of exchangeable ions, development of new or justification of existing solid-phase excess Gibbs energy models can be anticipated.

The particular algebraic form an excess Gibbs energy model may take is constrained by the definition of excess Gibbs energy of mixing and, in principle, by the physical arrangements and interactions exchangeable ions experience. The chemical thermodynamic requirements are explicit; the excess Gibbs energy model is a function of solid-phase mole fractions, and for any solid-phase cation i, if $x_i = 1$ then $\bar{G}^E = 0$. The second set of constraints, those due to the physical interactions of the exchangeable ions themselves, have yet to be resolved adequately, although the advances which have been made in other fields, specifically the chemical thermodynamics of electrolyte solutions, minerals, and melts indicate that similar rapid progress in the study of of exchangeable ions is likely [68,78].

Excess Gibbs energy models can be classified into three algebraic forms: polynomial, fractional, and logarithmic. Currently any of these models are valid candidates, and rigorous experimental tests of these equations are needed so that the list of candidates can be limited to a more manageable number.

1. Polynomial

The regular-solution model is the simplest of the polynomial models [84]. This model is able to correct modest deviations from ideality in systems where the excess Gibbs energy is symmetrical with respect to the mole fractions of the components. In a binary system, the regular-solution model has a single empirical parameter, R_{12}, representing nonspecific interactions between the subscripted ions:

$$\frac{\bar{G}^E}{RT} = R_{12}x_1x_2 \tag{129}$$

For a multicomponent system, this equation can be generalized to

$$\frac{\bar{G}^E}{RT} = \sum_{i=1}^{N-1} \sum_{j>i}^{N} R_{ij}x_ix_j \tag{130}$$

The subregular solution, which follows, was recommended for systems with deviations from ideality greater than can be described by the regular-solution model [85]. For a binary system, this model has two parameters, S_{12} and S_{21}:

$$\frac{\bar{G}^E}{RT} = x_1x_2(S_{12}x_1 + S_{21}x_2) \tag{131}$$

For a multicomponent system, this model becomes

$$\frac{\bar{G}^E}{RT} = \sum_{i=1}^{N-1} \sum_{j>i}^{N} x_ix_j(S_{ij}x_i + S_{ji}x_j) \tag{132}$$

These polynomial models can be expanded indefinitely, yielding the
Margules equation, which has the general form

$$\frac{\bar{G}^E}{RT} = \sum_{i=1}^{N-1} \sum_{j>i}^{N} \left(x_i x_j \sum_{k=1}^{N} \sum_{\ell=0}^{\infty} B_{ijk\ell} x_k^{\ell} \right)$$

$$+ \sum_{i=1}^{N-1} \sum_{j>i}^{N} \sum_{k>j}^{N} \left(x_i x_j x_k \sum_{\ell=1}^{N} \sum_{m=0}^{\infty} T_{ijk\ell m} x_{\ell}^{m} \right) + \cdots \qquad (133)$$

Here, $B_{ijk\ell}$ is a binary coefficient for interations between the i-th
and j-th components where the subscript ℓ defines the power to
which the mole fraction of the k-th component is raised. Similarly,
$T_{ijk\ell m}$ is a ternary coefficient for interactions between the i-th,
j-th, and k-th components.

For the binary case the equation can be truncated to

$$\frac{\bar{G}^E}{RT} = x_1 x_2 (B_{1210} + B_{1220} + B_{1211} x_1 + B_{1221} x_2$$

$$+ B_{1212} x_1^2 + B_{1222} x_2^2 + \cdots) \qquad (134)$$

2. *Fractional*

The fractional equations have been used to describe phase equilibria
of nonelectrolyte solutions [86]. These equations have been used
infrequently to describe cation-exchange equilibria. One example,
in the binary case, is the two-parameter van Laar equation

$$\frac{\bar{G}^E}{RT} = \frac{x_1 x_2 q_1 q_2 2a_{12}}{x_1 q_1 + x_2 q_2} \qquad (135)$$

where q_1 and a_{12} are the model parameters. This can be generalized for multicomponent systems by

$$\frac{\bar{G}^E}{RT} = \frac{\Sigma_{i=1}^{N-1} \Sigma_{j>i}^{N} x_i x_j q_i q_j 2a_{ij}}{\Sigma_{i=1}^{N} x_i q_i} \qquad (136)$$

3. *Logarithmic*

The Wilson equation [87] has been applied successfully to binary
cation-exchange equilibria, though its applicability for multicomponent systems has yet to be demonstrated [88,89]. In binary systems, this equation takes the form

$$\frac{\bar{G}^E}{RT} = -x_1 \ln(1 - x_2 W_{2/1}) - x_2 \ln(1 - x_1 W_{1/2}) \tag{137}$$

where $W_{1/2}$ is a parameter representing differences in interaction energies between pairs of like and unlike components and is related to the molar volumes of the components. It should be noted that in the original paper different forms of Eq. (137) were presented [87]. These contained alternative but equivalent forms for the interaction parameter and should not be confused.

For multicomponent systems, the Wilson equation can be generalized to

$$\frac{\bar{G}^E}{RT} = -\sum_{i=1}^{N} x_i \ln\left(1 - \sum_{j=1}^{N} (1 - \delta_{ij}) x_i W_{j/i}\right) \tag{138}$$

where δ_{ij} is the Kronecker delta symbol.*

C. Derivation of Activity-Coefficient Models

Once one of these excess Gibbs energy models has been adopted, the corresponding activity-coefficient models can be derived. The basic chemical-thermodynamic relationship between a component activity coefficient and the molar excess Gibbs energy of mixing is

$$\ln f_i = \frac{\partial}{\partial n_i} \left(\frac{n\bar{G}^E}{RT}\right) \tag{139}$$

where n is the sum of the number of moles of all components in the phase. Lupis has presented a general equation with which the algebraic representation for the rational activity coefficient can be derived from the parent excess Gibbs energy model [90].

$$RT \ln f_i = \bar{G}^E + \sum_{j=2}^{N} (\delta_{ij} - x_j)\left(\frac{\partial \bar{G}^E}{\partial x_j}\right)_{xk, k \neq i,j} \tag{140}$$

If, on the other hand, the proposed model is for the total molar excess Gibbs energy $\bar{G}^E/RT$ [Eq. (125)], then the solid-phase activity coefficient can be calculated using the relation

---

*$\delta_{ij} = 1$, if $i = j$; $\delta_{ij} = 0$, if $i \neq j$.

$$RT \ln f_i = \bar{G}^{E'} + \sum_{j=2}^{N} (\delta_{ij} - x_j) \left(\frac{\partial \bar{G}^{E'}}{\partial x_j} \right)_{xk, k \neq i, j} + RT \ln \left(\frac{n}{n_T} \right) \quad (141)$$

In this chapter it is assumed conveniently, if not rigorously, that $n = n_T$.

Expressions for individual activity coefficient models can be derived from parent excess Gibbs energy models with Eq. (140). For example, activity-coefficient models corresponding to the regular-solution model for a binary ion-exchange system can be derived by noting that Eq. (140) (recall for a binary system $x_1 = 1 - x_2$) becomes

$$RT \ln f_1 = \bar{G}^E - x_2 \frac{\partial \bar{G}^E}{\partial x_2} \quad (142)$$

and

$$RT \ln f_2 = \bar{G}^E + x_1 \frac{\partial \bar{G}^E}{\partial x_2} \quad (143)$$

The individual solid-phase activity coefficients may then be calculated:

$$\ln f_1 = x_2^2 R_{12} \quad (144)$$

and

$$\ln f_2 = x_1^2 R_{12} \quad (145)$$

By following this procedure the model for an individual solid-phase activity coefficient can be derived directly from any excess Gibbs energy model.

V.   METHODS TO EVALUATE EXCESS GIBBS ENERGY MODELS

The fundamental problem encountered in evaluating solid-phase excess Gibbs energy models arises from the fact that solid-phase excess Gibbs energy cannot be measured directly. Two methods have been proposed to evaluate solid-phase excess Gibbs energy models: both rely on the same general equation:

$$\frac{\bar{G}^E}{RT} = \sum_{i=1}^{N} x_i \ln f_i \quad (146)$$

A. Elprince-Babcock Approach

When following an approach suggested by Elprince and Babcock [79],
individual solid-phase activity coefficients, for interrelated binary
cation-exchange systems, are calculated from the experimental data
by the Argersinger approach, and $\bar{G}^E/RT$ is calculated by Eq. (146).
These $\bar{G}^E/RT$ values are then fitted to a candidate excess Gibbs
energy model. The resulting binary constants are then used to compute the excess Gibbs energy and solid-phase activity coefficients
for systems with more than two exchangeable cations [80].

In order to evaluate the solid-phase activity coefficients it is

This approach presented for the first time a model calculation of
multicomponent ion exchange based on binary cation exchange data.
No data on multicomponent cation-exchange systems are required to
apply the candidate excess Gibbs energy model. The Elprince-
Babcock approach is independent of the method by which solid-phase
activity coefficients are evaluated in the binary cation-exchange systems. Solid-phase activity coefficients and excess Gibbs energy
parameters can be estimated by graphical techniques, although most
recently reported implementations of the Elprince-Babcock approach
have relied on computer-aided regression techniques to estimate
parameters. While regression techniques can be employed with multicomponent cation-exchange equilibria, the binary case will be presented here as an example. The application to systems with more
exchangeable species is straightforward, though there are understandably more integral equations involved in the determination of
activity coefficients.

In order to evaluate the solid-phase activity coefficients it is
usual to fit first, by regression, $\ln {}^i_j k^a_V$ to a function of solid-phase
equivalent fractions. The function is almost always a polynomial of
the form

$$\ln {}^i_j k^a_V = \sum_{\ell=0}^{N} \beta_\ell E^\ell_i \qquad (147)$$

where N is the degree of the polynomial and β_0, β_1, ..., β_N are fitting parameters. Once determined, the polynomial in the right side
of Eq. (147) is substituted for $\ln {}^i_j k^a_V$ in Eqs. (37) and (38) which

are evaluated analytically to estimate $\ln f_{V_i}$ and $\ln f_{V_j}$. (The errors
caused by ignoring changes in the activity of water are usually
assumed to be trivial.)

The choice of the degree of the polynomial is determined by two
factors, one empirical, the other qualitative. First, the degree may
be selected when higher degrees do not appreciably decrease the
mean-square error of the regression. The suitable polynomial degree can be determined by the extra sum of squares procedure,
which is described elsewhere [91]. The second point arises from
the shortage of experimental values for $\ln {}_j^i k_V^a$ that is commonly
encountered in the isotherm limits where E_i tends to 1 or 0. In
this region the polynomial equation may be unconstrained and yield
unreasonable estimates for $\ln {}_j^i k_V^a$ at the limits. Errors of several
log units are not uncommon if high-degree polynomials are selected.
Extrapolation errors such as these may yield significant errors in
the exchange-equilibrium-constant and activity-coefficient values
[92], which result in corresponding errors in the subsequent fitting
of the excess Gibbs energy equation. The careful development of
experimental designs to ensure that measured equilibria will be near
the extreme values of E_i is addressed in Sec. VI.

Fits to data by regression are rarely perfect. The deviation or
error in the regression is attributable to two sources: experimental
error and bias. Experimental error is the variation in the measured
values of the dependent variable, $\ln {}_j^i k_V^a$ in this case, at known
values of the independent variable. As will be shown later, difficulties due to experimental error can be reduced greatly by choosing
carefully equilibrating solution compositions and by replicating treatments sufficiently. Bias is the deviation in the predicted value from
the true value due to misspecification of the regression model [91].
This source of variation is unavoidable when fitting data to empirical
regression equations such as Eq. (147) though the magnitude of the
bias relative to other sources of variation need not be great.

There are two drawbacks to the Elprince-Babcock approach.
First, as noted earlier, the estimated independent variable in the evaluation of a candidate excess Gibbs energy model is biased when fitting

empirical equations such as Eq. (147) by regression techniques. The extent of this bias is not known and no one has developed techniques to reduce its impact, if regression techniques are used, during implementation of the Elprince-Babcock approach. Another severe drawback is that the equations for solid-phase activity coefficients in systems with three or more exchangeable species that correspond to Eq. (54) are much more difficult to evaluate in practice [22]. In fact, no one has successfully estimated parameters for an excess Gibbs energy model from ternary cation-exchange data by following the Elprince-Babcock approach.

B. Rational Approach

The problems of the Elprince-Babcock approach are overcome somewhat by the "rational approach" of Grant and Sparks [93]. In this procedure, excess Gibbs energy model coefficients are estimated directly in a single nonlinear regression analysis of the measured $\ln {}^i_jk^a_V$ data. The procedure eliminates the need to first calculate activity coefficients and values of $\bar{G}^E$. This minimizes problems of bias due to the first regression step, if used, in the Elprince-Babcock approach.

The expression

$$\ln {}^i_jk^a_V = \ln {}^i_jK_V + z_i \ln f_{V_j} - z_j \ln f_{V_i} \tag{148}$$

can be derived from Eq. (25). This is the equation which is fitted by regression techniques after $\ln f_{V_i}$ and $\ln f_{V_j}$ in Eq. (148) are replaced by terms derived from the excess Gibbs energy model to be evaluated. These terms may be derived easily by exploiting the relation in Eq. (140). Derivations of the necessary equations are facilitated by the use of symbolic-manipulation software, such as *Macsyma* and *Mathematica* [94,95].

For systems with more than two exchangeable ions the expression

$$\sum_{i=1}^{N-1} \sum_{j \geq i}^{N} (\ln {}^i_jk^a_V - \ln {}^i_jK_V - z_i \ln f_{V_j} + z_j \ln f_{V_i})^2 \tag{149}$$

can be minimized by nonlinear regression techniques to estimate the excess Gibbs energy model parameters.

This approach simplifies the analysis of ion-exchange data for systems with more than two exchangeable-ion species. This is the major advantage of the rational approach over the conventional chemical-thermodynamic analysis of multicomponent systems which uses involved integral equations for the evaluation of activity coefficients.

A feature of the rational approach is that activity coefficients and exchange-equilibrium constants become dependent on the type of excess Gibbs energy model used and the quality of the cation exchange data (e.g., degree and evenness of exchangeable-ion coverage). This is a drawback if several adsorption isotherms or candidate excess Gibbs energy models are to be evaluated or if both binary and multispecies cation-exchange systems are part of the same investigation. Indeed, $\ln {}_j^i K_V$ can be a fitting parameter in Eq. (149). To make the exchange equilibrium constant independent of composition and the chosen model it must be determined separately and, if Eq. (149) is used, will be subject to the problem of bias which vexes the Elprince-Babcock approach.

Inaccurate estimates of the exchange equilibrium constant will cast doubt on the reliability of excess Gibbs energy model parameters estimated by the rational approach. Even though the magnitude of error is not predeterminable and the accuracy limits of $\ln {}_j^i K_V$ that are needed cannot be readily defined, this problem can be minimized with appropriate experimental designs and adequate replication.

C. Multiple Reactions

The excess Gibbs energy equations, as described above, are applicable to solids which undergo cation-exchange reactions and possibly water adsorption. Many reactive substrates are composite materials, the components of which exhibit a range of reactions in addition to stoichiometric cation-exchange. Soils are a good example exhibiting cation complexation with organics, hydrolysis of clay edges, and ion adsorption on oxides. If multiple reactivity such as this is described as a single-ion exchange process the compositional dependence of nonideality may be incorrectly described. Fletcher and

Sposito [12] evaluated cation-exchange data for Wyoming bentonite
showing that clay hydrolysis and anion adsorption accompanied cation exchange under some circumstances. When these factors were
accounted for explicitly, the remaining contribution from true cation
exchange was found to be ideal with respect to the convention of
Argersinger et al. [3]. If treated solely as ion exchange the data
appeared to exhibit nonideality. Thus, it is essential to establish
the extent of multiple reaction and to show that reversible cation
exchange is occurring before any attempt at modeling the excess
Gibbs energy is made.

VI.   DESIGN OF ION-EXCHANGE EXPERIMENTS
A.   Model Evaluation Formulated as a Regression Problem

There are no generally accepted solid-phase activity coefficient
models for cation-exchange reactions. This is due partly to the absence of theoretical development but also to the lack of experimental
verification. Analyses of data from cation exchange experiments conducted so far have been unable to show that any one of the several
candidate activity coefficient models can describe solid-phase activity
coefficients consistently and accurately. Most studies cannot be
faulted for poor analytical technique, but suffer from deficiencies in
their experimental designs. Since solid-phase excess Gibbs energy
models are evaluated by regression, the design of experiments to
evaluate these models in this way can be viewed as special cases of
problems addressed by the theory of optimal design of regression
experiments. Before discussing the application of the theory of optimal design to cation-exchange experiments, the fundamentals of
regression should be reviewed.

A short example will demonstrate linear regression techniques
and optimal design of regression experiments. For this example,
assume it is believed that $\ln {_2^1}k_V^a$ can be fitted to a linear regression
model

$$\ln {_2^1}k_V^a = \beta_0 + \beta_1 x_{V_1} \tag{150}$$

The estimates $\hat{\beta}_0$ and $\hat{\beta}_1$ of the parameters β_0 and β_1 can be calculated by matrix algebra. Let X be the matrix of values of the independent variable x_1 (the column of ones corresponds to the multiplicands of the parameter β_0),

$$X = \begin{pmatrix} 1 & x_{V_1} & \text{(obs. 1)} \\ 1 & x_{V_1} & \text{(obs. 2)} \\ 1 & x_{V_1} & \text{(obs. 3)} \\ \vdots & \vdots & \end{pmatrix} \tag{151}$$

Y, the matrix of observed dependent variables,

$$Y = \begin{pmatrix} \ln \tfrac{1}{2}k_V^a & \text{(obs. 1)} \\ \ln \tfrac{1}{2}k_V^a & \text{(obs. 2)} \\ \ln \tfrac{1}{2}k_V^a & \text{(obs. 3)} \\ \vdots & \end{pmatrix} \tag{152}$$

and $\hat{\beta}$, the matrix of parameter estimates,

$$\hat{\beta} = \begin{pmatrix} \hat{\beta}_0 \\ \hat{\beta}_1 \end{pmatrix} \tag{153}$$

The matrix of parameter estimates, $\hat{\beta}$, may be calculated from the equation

$$\hat{\beta} = (X'X)^{-1}X'Y \tag{154}$$

where X' is the transpose of X, $X'X$ is the matrix product of X and its transpose, and $(X'X)^{-1}$ is the inverse of this matrix product. Statisticians call $X'X$ the information matrix and $(X'X)^{-1}$ the covariance matrix. The variance of a parameter estimate is

$$\sigma^2_{ij} = c_{ii}\sigma^2_{\ln \tfrac{1}{2}k_V^a} \tag{155}$$

where c_{ii} is the ii-th element (i.e., the i-th diagonal element) of the covariance matrix $(X'X)^{-1}$ [91]. Note that *for a given dependent-variable variance, the variances of the parameter estimates are*

determined completely by the values taken by the independent vari-
ables. Therefore, the experimenter can minimize the parameter estimate variance (i.e., estimate the regression equation parameters
most precisely) by choosing x_1 values so that the c_{ii} elements of
the covariance matrix are minimized. In Sec. VI.C computer programs which can aid the experimenter in selecting the optimal x_1
values will be discussed. Elementary algebra and a moment's reflection will lead one to the optimal design in the linear regression example presented above. The prime objective in linear regression is
to estimate the slope of the line defined by Eq. (150). This will be
accomplished best by selecting experimental data at the extremes of
the independent variable range—near the limits $x_1 = 0$ and $x_1 = 1$
in an ion-exchange experiment.

The design of a cation-exchange equilibrium experiment consists
of establishing the equilibrating solution compositions and number of
replicates necessary to test the hypothesis that a candidate excess
Gibbs energy model describes the chemical behavior of exchangeable
ion species. The hypothesis to be tested is that the equality

$$\frac{\bar{G}^E}{RT} = f(x_{V_1}, x_{V_2}, \ldots, x_{V_{(N-1)}}) \tag{156}$$

where $f(x_{V_1}, x_{V_2}, \ldots, x_{V_{(N-1)}})$ represents the candidate excess
Gibbs energy model, is true. The representation in Eq. (156) shows
the two serious problems confronting the experimenter. First, the
dependent variable in Eq. (156), $\bar{G}^E/RT$, is not measured directly.
Rather the experimenter must interpret trends in the measured
$\ln {}_j^i k_V^a$ values to deduce evaluations about the excess Gibbs energy
behavior. Second, the experimenter cannot predetermine directly
the independent variable values (e.g., x_i) for the experiments being conducted. The experimenter can only establish the equilibrating-solution electrolyte molalities and hope that the resulting solid-
phase mole fractions will cover completely and evenly the range of
solid-phase compositions.

The evaluation of excess Gibbs energy models relies on regression techniques, indirectly with the Elprince-Babcock approach and

directly with the rational approach. Whichever approach is followed, the problem is to fit by regression an equation or set of equations having the form

$$\ln {}^1_2k^a_V = g(x_{V_1}, x_{V_2}, \ldots, x_{V_{(N-1)}}) \tag{157}$$

The expression $g(x_{V_1}, x_{V_2}, \ldots, x_{V_{(N-1)}})$ in Eq. (157) is equal to an arbitrary-degree polynomial in the Elprince-Babcock approach or an expression equivalent to

$$\ln {}^1_2K_V + z_1 \ln f_{V_2} - z_2 \ln f_{V_1} \tag{158}$$

in the rational approach.

To the best of our knowledge, this chapter is the first to provide a comprehensive procedure for the statistical design of cation-exchange experiments. The surprising absence of earlier work in this direction may be due to the fact that many of the statistical tools for such an approach, response surface methodology, and the theories of optimal designs and design of mixture experiments were developed after the most active period of cation-exchange research had come to an end. Similarly, powerful statistical analysis, graphics, and symbolic-manipulation computer software, taken for granted today, have become widely available only recently. The experimental design procedure suggested here takes advantage of the theoretical and computational developments since this earlier ion-exchange research period.

Regardless of the approach, either Elprince-Babcock or rational, the dependent variable to be fitted to a model by regression is $\ln {}^i_jk^a_V$. The entire procedure outlined here rests on the accurate measurement of this quantity. Systematic errors in its determination arise from errors in the measurements of the electrolyte solution molalities and the solid-phase mole fractions, or errors in the estimation of the chosen liquid-phase activity-coefficient correction. Analytical procedures to determine electrolyte solution molalities are well established. Estimation of activity coefficients for ions and salts in solution was discussed extensively earlier.

In order to measure accurately the solid-phase mole fractions, the experimenter must measure accurately the surface excess of each exchangeable species. Some experimenters have neglected determining one of the exchangeable species, relying on the surface excesses of the other exchange ions and the exchange capacity. This can greatly increase error in the measurement of solid-phase mole fractions [19]. First, since one of the ions is measured by difference, the variation in its measured value will be the sum of the variances of the exchange capacity and the variances of the other measured exchangeable-ion specific surface excesses. Secondly, exchangeable H^+ or OH^-, if present, will not be determined. They will instead contribute to the inaccuracy in the measured amount of exchangeable ions being determined by difference.

The procedure suggested here for experimental design consists of four steps:

1. Specification of the regression model
2. Determination of solid-phase mole fractions
3. Determination of replication necessary
4. Calculation of appropriate equilibrating-solution composition

Assuming that systematic errors in the determination of $\ln {}_j^i k_V^a$ can be avoided, an experimenter following these steps can stipulate the equilibrating-solution compositions and number of replicates appropriate to evaluate statistically the candidate excess Gibbs energy model. The procedure can be applied to multicomponent cation-exchange systems, but, for simplicity, the examples presented here will be restricted to binary cation-exchange systems.

B. Specification of the Regression Model

The specification of the regression model depends on which evaluation approach is being followed. If the Elprince-Babcock approach is followed, then an arbitrary-degree polynomial is the regression model. The precise degree to be chosen will depend on the experience of the experimenter and the degree of Eq. (157) needed to describe the behavior of $\ln {}_2^1 k_V^a$ in similar exchange systems.

If the rational approach is followed, the candidate excess Gibbs energy model will specify the corresponding $\ln f_{V_i}$ expressions and thus, by Eq. (158), the regression model to be fitted. Table 10 presents the algebraic form of polynomial models corresponding to regular-solution and sub-regular-solution models for systems with two or three exchangeable ions. Procedures for specifying solid-phase mole fractions are well established for linear equations but much less so for nonlinear equations which arise from the Wilson model and van Laar equations [96]. Accordingly, to simplify the experimental design procedure for the evaluation of a nonlinear excess Gibbs energy model, the procedure should employ a polynomial model of sufficient degree to mimic the nonlinear equation.

C. Determination of Solid-Phase Mole Fractions

With the regression model specified, the exchangeable-cation mole fraction experimental sites can be chosen so as to estimate precisely the model parameters. The fact that, for a given $\sigma^2_{\ln \frac{1}{2}k_V^a}$, the covariance matrix $(X'X)^{-1}$ determines completely variance of parameter estimates is drawn upon to select these sites. Still, a rule is needed by which to choose them. This rule is a *design criterion*. Of the several design criteria, three will be presented here: D-optimal, integrated mean-square error (IMSE), and average mean-square error of the response (AM) [97-99]. There is more than one design criterion because the experimenter rarely knows the correct regression model before the experimental data have been collected nor the consequences an incorrect choice of regression model would have on the parameter estimates for the true model. The many design criteria allow the experimenter to balance his or her expectation that the specified regression model is correct while minimizing the consequences should the specified regression model prove to be incorrect.

Experimental designs which adhere to the D-optimal criterion maximize the determinant of the information matrix $X'X$. For a given dependent-variable variance, this will minimize the sum of

TABLE 10 Polynomial Expressions Corresponding to Regular-Solution and Sub-Regular-Solution Models of $\ln \frac{1}{2}k_V^a$ in Systems with Two and Three Exchangeable Ions

Regular-solution model, binary

$$\ln \tfrac{1}{2}K_V + z_1 R_{12} x_1^2 - z_2 R_{12} x_2^2 \tag{159}$$

Sub-regular-solution model, binary

$$\ln \tfrac{1}{2}K_V + z_1 S_{12} x_1^3 + x_1^2 x_2 (-(z_1 S_{12}) + 2z_1 S_{21})$$

$$- S_{21} x_2^3 z_2 + x_1 x_2^2 (S_{21} z_2 - 2 S_{12} z_2) \tag{160}$$

Regular-solution model, ternary

$$\ln \tfrac{1}{2}K_V + R_{12}(x_1^2 z_1 + x_1 x_2 z_1 - x_2^2 z_1 + x_1 x_3 z_1$$

$$+ x_2 x_3 z_1 - x_2^2 z_2 - x_1 x_3 z_2 - x_2 x_3 z_2)$$

$$+ R_{13}(-(x_1^2 z_1) - x_1 x_2 z_1 - x_1 x_3 z_1 - x_2 x_3 z_1$$

$$+ x_3^2 z_1 - x_1 x_2 z_2 - x_2 x_3 z_2 - x_3^2 z_2)$$

$$+ R_{23}(x_2^2 z_1 + x_1 x_3 z_1 + x_2 x_3 z_1 + x_3^2 z_1$$

$$- x_2^2 z_2 + x_2 x_3 z_2 - x_3^2 z_2) \tag{161}$$

Sub-regular-solution model, ternary

$$\ln \tfrac{1}{2}K_V + S_{13}(-(x_1^2 z_1) + x_1^3 z_1 - 2x_1 x_2 z_1 - 2x_1^2 x_2 z_1$$

$$+ 2x_1 x_2^2 z_1 - 2x_1 x_2 + 4x_1^2 z_2 + 2x_1^3 z_2 + 2x_1 x_2 z_2$$

$$- 3x_1^2 x_2 z_2) + S_{12}(2x_1^2 z_1 - 2x_1^3 z_1 - 2x_1 x_2^2 z_1$$

$$- x_1^2 z_2 + x_1^3 z_2 - 2x_1 x_2 z_2 + 3x_1^2 x_2 z_2)$$

$$+ S_{21}(4x_1 x_2 z_1 - 4x_1^2 x_2 z_1 - x_1 x_2^2 z_1 - x_2^3 z_1$$

$$- 2x_1 x_2 z_2 + 2x_1^2 x_2 z_2 - x_2^2 z_2 + 4x_1 x_2^2 z_2)$$

$$+ S_{31}(-3x_1 z_1 + 6x_1^2 z_1 - 3x_1^3 z_1 - x_2 z_1 + 6x_1 x_2 z_1$$

$$- 5x_1^2 x_2 z_1 + 2x_2^2 z_1 - 3x_1 x_2^2 z_1 - x_2^3 z_1 - z_2 + 4x_1 z_2$$

TABLE 10   (continued)

$$- 5x_1^2z_2 + 2x_1^3z_2 - 2x_2z_2 - 8x_1x_2z_2 + 6x_1^2x_2z_2$$

$$- x_2^2z_2 + 4x_1x_2^2z_2)$$

$$+ S_{23}(4x_2z_1 - 8x_1x_2z_1 + 4x_1^2x_2z_1 + 7x_2^2z_1$$

$$+ 7x_1x_2^2z_1 + 2x_2^3z_1 - 2x_2z_2 + 4x_1x_2z_2 - 2x_1^2x_2z_2$$

$$+ 6x_2^2z_2 - 6x_1x_2^2z_2 - 5x_2^3z_2)$$

$$+ S_{32}(2z_1 - 6x_1z_1 + 6x_1^2z_1 - 2x_1^3z_1 - 8x_2z_1$$

$$+ 16x_1x_2z_1 - 8x_1^2x_2z_1 + 8x_2^2z_1 - 8x_1x_2^2z_1 - 2x_2^3z_1 - z_2$$

$$+ 3x_1z_2 - 3x_1^2x_2 + x_1^3z_2 + 5x_2z_2 - 10x_1x_2z_2$$

$$+ 5x_1^2x_2z_2 - 9x_2^2z_2 + 9x_1x_2^2z_2 + 5x_2^3z_2) \tag{162}$$

the parameter estimate variances and covariances. There are several computerized algorithms for calculating mixture component mole fractions based on a given linear regression model and the D-optimal design criterion. The D-optimal design criterion assumes that bias in the parameter estimates for a misspecified selected regression model is negligible. Many statisticians think that this assumption is unrealistic. Several other design criteria have been developed which, when applied, attempt to minimize the variances and covariances of the parameter estimates while minimizing simultaneously the bias due to potential inadequacies of the chosen regression model.

The first of these criteria to be studied extensively was the IMSE criterion suggested by Box and Draper [98]. Assuming that the true model is of higher degree than the regression model, the criterion will minimize the MSE between the model prediction and the true behavior as predicted by the true model. A related optimal design criterion is the average mean-square error of the response (AM) of Welch [99]. This attempts to minimize the sum of variance error and the maximum bias due to inadequate model selection. This criterion is popular, in part, because of a widely available FORTRAN program, developed by Welch, for calculating mixture experimental designs which offers this criterion as an option.

The following example shows the effect the D-optimal and IMSE criteria have upon the specified experimental locations. Returning to the experimenter who conveniently continues to assume that Eq. (150) is appropriate for this regression model, the D-optimal criterion will specify points at x_i = 0.0 and 1.0. (Since the selectivity coefficients are undefined in the complete absence of one of the exchangeable ions, in practice, this would mean that an experimenter would select nearby points, for example, at x_i = 0.01 and 0.99.) If Eq. (150) is indeed the correct one, data points at 0.0 and 1.0 will yield minimum variance parameter estimates. Rarely are experimenters so prescient. If Eq. (150) is not the correct relation, but rather

$$\ln {}_2^1 k_V^a = \beta_0 + \beta_1 x_{V_1} + \beta_2 x_{V_1}^2 \tag{163}$$

or

$$\ln {}_2^1 k_V^a = \beta_0 + \beta_1 x_{V_1} + \beta_2 x_{V_1}^2 + \beta_3 x_{V_1}^3 \tag{164}$$

is appropriate, then sampling points at x_1 = 0.0 and 1.0 would not yield parameter estimates for β_2 or β_3. Similar concerns would be valid for regression models of higher degree. For this reason, the experimenter should consider either the IMSE or AM design criteria. For example, the IMSE would specify x_1 values of 0.1185, 0.5, and 0.8815. Application of the AM criterion would yield comparable values. For Eq. (150), specification at those values suggested by the IMSE criterion would yield acceptable variances of the parameter estimates while protecting the experimental objectives should a higher-degree polynomial provide the true model [100]. Once the model and design criteria have been selected, the appropriate solid-phase mole fractions can be specified. These can be calculated by computer programs such as DETMAX [101] and ACED [102]. Tables 11 and 12

TABLE 11a Solid-Phase Mole Fractions Suggested by the D-Optimal
and Average Mean-Square Error Design Criteria to Evaluate the
Regular-Solution Model in Binary Cation-Exchange Experiments

D-optimal criterion			Average mean-square error criterion		
Ionic charge numbers	Solid-phase mole fractions		Ionic charge numbers	Solid-phase mole fractions	
z_1, z_2	x_1	x_2	z_1, z_2	x_1	x_2
1.1	0.01	0.99	1,1	0.01	0.99
	0.99	0.01		0.02	0.98
1,2	0.01	0.99		0.98	0.02
	0.99	0.01		0.99	0.01
1,3	0.01	0.99	1,2	0.01	0.99
	0.99	0.01		0.02	0.98
2,2	0.01	0.99		0.98	0.02
	0.99	0.01		0.99	0.01
2,3	0.01	0.99	1,3	0.01	0.99
	0.99	0.01		0.02	0.98
3,3	0.01	0.99		0.98	0.02
	0.99	0.01		0.99	0.01
			2,2	0.01	0.99
				0.02	0.98
				0.98	0.02
				0.98	0.01
			2,3	0.01	0.99
				0.02	0.98
				0.98	0.02
				0.99	0.01
			3,3	0.01	0.99
				0.02	0.98
				0.98	0.02
				0.99	0.01

TABLE 11b   Solid-Phase Mole Fractions Suggested by the D-Optimal and Average Mean-Square Error Design Criteria to Evaluate the Sub-Regular-Solution Model in Binary Cation-Exchange Experiments

D-optimal criterion			Average mean-square error criterion		
Ionic charge numbers	Solid-phase mole fractions		Ionic charge numbers	Solid-phase mole fractions	
z_1, z_2	x_1	x_2	z_1, z_2	x_1	x_2
1,1	0.01	0.99	1,1	0.01	0.99
	0.50	0.50		0.02	0.98
	0.99	0.01		0.51	0.48
1,2	0.01	0.99		0.52	0.48
	0.45	0.55		0.54	0.46
	0.99	0.01		0.99	0.01
1,3	0.01	0.99	1,2	0.01	0.99
	0.42	0.58		0.43	0.57
	0.99	0.01		0.44	0.56
2,2	0.01	0.99		0.45	0.55
	0.50	0.50		0.98	0.02
	0.99	0.01		0.99	0.01
2,3	0.01	0.90	1,3	0.01	0.99
	0.47	0.53		0.41	0.59
	0.99	0.01		0.42	0.58
3,3	0.01	0.99		0.43	0.57
	0.50	0.50		0.98	0.02
	0.99	0.01		0.99	0.01
			2,2	0.02	0.98
				0.51	0.49
				0.98	0.02
				0.99	0.01
			2,3	0.01	0.99
				0.41	0.59
				0.42	0.58
				0.43	0.57
				0.98	0.02
				0.99	0.01
			3,3	0.01	0.99
				0.02	0.98
				0.51	0.49
				0.52	0.48
				0.54	0.46
				0.99	0.01

TABLE 12a Solid-Phase Mole Fractions Suggested by the D-Optimal and Average Mean-Square Error Design Criteria to Evaluate the Regular-Solution Model in Ternary Cation-Exchange Experiments

D-optimal criterion

Ionic charge numbers z_1, z_2	Solid-phase mole fractions x_1	x_2	x_3
1,1	0.01	0.01	0.98
	0.01	0.49	0.5
	0.01	0.98	0.01
	0.98	0.01	0.01
1,2	0.01	0.01	0.98
	0.01	0.49	0.5
	0.01	0.98	0.01
	0.98	0.01	0.01
1,3	0.01	0.01	0.98
	0.01	0.49	0.5
	0.01	0.98	0.01
	0.98	0.01	0.01
2,2	0.01	0.01	0.98
	0.01	0.49	0.5
	0.01	0.98	0.01
	0.98	0.01	0.01
2,3	0.01	0.01	0.98
	0.01	0.49	0.5
	0.01	0.98	0.01
	0.66	0.33	0.01
	0.33	0.01	0.66
	0.98	0.01	0.01
3,3	0.01	0.01	0.98
	0.01	0.49	0.5
	0.01	0.98	0.01
	0.98	0.01	0.01

Average mean-square error criterion

Ionic charge numbers z_1, z_2	Solid-phase mole fractions x_1	x_2	x_3
1,1	0.01	0.01	0.98
	0.01	0.49	0.50
	0.01	0.98	0.01
	0.66	0.33	0.01
	0.98	0.01	0.01
	0.04	0.44	0.52
	0.04	0.48	0.48
	0.92	0.04	0.04
1,2	0.01	0.01	0.98
	0.01	0.49	0.5
	0.01	0.33	0.66
	0.01	0.98	0.01
	0.66	0.01	0.33
	0.98	0.01	0.01
	0.04	0.44	0.52
	0.04	0.48	0.48
1,3	0.01	0.01	0.98
	0.01	0.49	0.5
	0.01	0.98	0.01
	0.98	0.01	0.01
	0.04	0.48	0.48
	0.76	0.04	0.2
	0.8	0.04	0.16
2,2	0.01	0.01	0.98
	0.01	0.49	0.5
	0.01	0.98	0.01
	0.66	0.33	0.01
	0.98	0.01	0.01
	0.04	0.4	0.56
	0.04	0.44	0.52
	0.92	0.04	0.04
2,3	0.01	0.49	0.5
	0.01	0.98	0.01
	0.66	0.33	0.01
	0.33	0.01	0.66
	0.98	0.01	0.01
	0.04	0.48	0.48

TABLE 12a   (continued)

	z_1,z_1	x_1	x_2	x_3
3,3		0.01	0.01	0.98
		0.01	0.49	0.5
		0.01	0.49	0.01
		0.66	0.33	0.01
		0.98	0.01	0.01
		0.04	0.4	0.56
		0.04	0.44	0.52
		0.92	0.04	0.04

TABLE 12b   Solid-Phase Mole Fractions Suggested by the D-Optimal and Average Mean-Square Error Design Criteria to Evaluate the Sub-Regular-Solution Model in Ternary Cation-Exchange Experiments

D-optimal criterion				Average mean-square error criterion			
Ionic charge numbers	Solid-phase mole fractions			Ionic charge numbers	Solid-phase mole fractions		
z_1,z_1	x_1	x_2	x_3	z_1,z_2	x_1	x_2	x_3
1,1	0.01	0.01	0.98	1,1	0.01	0.01	0.98
	0.01	0.49	0.5		0.01	0.49	0.5
	0.01	0.98	0.01		0.01	0.98	0.01
	0.49	0.5	0.01		0.49	0.5	0.01
	0.49	0.01	0.5		0.49	0.01	0.5
	0.98	0.01	0.01		0.98	0.01	0.01
1,2	0.01	0.01	0.98		0.04	0.48	0.48
	0.01	0.49	0.5		0.04	0.52	0.44
	0.01	0.98	0.01		0.36	0.32	0.32
	0.49	0.5	0.01		0.44	0.52	0.04
	0.49	0.01	0.5		0.44	0.52	0.04
	0.98	0.01	0.01		0.44	0.04	0.52
1,3	0.01	0.01	0.98		0.48	0.48	0.04
	0.01	0.49	0.5	1,2	0.01	0.01	0.98
	0.01	0.98	0.01		0.01	0.49	0.5
	0.49	0.5	0.01		0.01	0.98	0.01
	0.49	0.01	0.5		0.49	0.5	0.01
	0.98	0.01	0.01		0.49	0.01	0.5
2,2	0.01	0.01	0.98		0.98	0.01	0.01
	0.01	0.49	0.5		0.04	0.48	0.48
	0.01	0.98	0.01		0.04	0.52	0.44
	0.49	0.5	0.01		0.36	0.32	0.32
	0.49	0.01	0.5		0.44	0.52	0.04
	0.98	0.01	0.01		0.44	0.52	0.04
					0.44	0.04	0.52
					0.48	0.48	0.04

TABLE 12b (continued)

D-optimal criterion				Average mean-square error criterion			
Ionic charge numbers	Solid-phase mole fractions			Ionic charge numbers	Solid-phase mole fractions		
z_1, z_2	x_1	x_2	x_3	z_1, z_2	x_1	x_2	x_3
2,3	0.01	0.01	0.98	1,3	0.01	0.01	0.98
	0.01	0.49	0.5		0.01	0.49	0.5
	0.01	0.98	0.01		0.01	0.98	0.01
	0.49	0.5	0.01		0.49	0.5	0.01
	0.49	0.01	0.5		0.49	0.01	0.5
	0.98	0.01	0.01		0.98	0.01	0.01
3,3	0.01	0.01	0.98		0.04	0.48	0.48
	0.01	0.49	0.5		0.04	0.52	0.44
	0.01	0.98	0.01		0.36	0.32	0.32
	0.49	0.5	0.01		0.44	0.52	0.04
	0.49	0.01	0.5		0.44	0.52	0.04
	0.98	0.01	0.01		0.44	0.04	0.52
					0.48	0.48	0.04
				2,2	0.01	0.01	0.98
					0.01	0.49	0.5
					0.01	0.98	0.01
					0.49	0.5	0.01
					0.49	0.01	0.5
					0.98	0.01	0.01
					0.04	0.48	0.48
					0.04	0.52	0.44
					0.36	0.32	0.32
					0.44	0.52	0.04
					0.44	0.52	0.04
					0.44	0.04	0.52
					0.48	0.48	0.04
				2,3	0.01	0.01	0.98
					0.01	0.49	0.5
					0.01	0.98	0.01
					0.49	0.5	0.01
					0.49	0.01	0.5
					0.98	0.01	0.01
					0.04	0.48	0.48
					0.04	0.52	0.44
					0.36	0.32	0.32
					0.44	0.52	0.04
					0.44	0.52	0.04
					0.44	0.04	0.52
					0.48	0.48	0.04

TABLE 12b   (continued)

	3,3	0.01	0.01	0.98
		0.01	0.49	0.5
		0.01	0.98	0.01
		0.49	0.5	0.01
		0.49	0.01	0.5
		0.98	0.01	0.01
		0.04	0.48	0.48
		0.04	0.52	0.44
		0.36	0.32	0.32
		0.44	0.52	0.04
		0.44	0.52	0.04
		0.44	0.04	0.52
		0.48	0.48	0.04

present the solid-phase mole fractions suggested by the D-optimal
and AM design criteria for cation-exchange experiments with two and
three exchangeable ions, respectively.

The data in suggested mole fractions in Table 11 may be used
directly as solid-phase mole-fraction recommendations in cation-exchange experiments. For ternary systems, two sets of solid-phase
mole-fraction recommendations should be combined when developing
an experimental design. For example, an experimenter planning an
experiment with three cations, with charge numbers 1, 2, and 3,
could combine the recommended solid-phase mole fractions for charge
numbers 1 and 2 with those for charge numbers 2 and 3.

D. Determination of Number of Replicates

The next step in this procedure is to calculate number of replicates
necessary to reduce the standard error of the parameter estimate to
specified level. After the experimental data have been analyzed,
the experimenter will calculate the confidence interval of the parameter β_1 with

$$\hat{\beta}_1 - t \sqrt{\sigma^2_{\hat{\beta}_1}} \leq \beta_1 \leq \hat{\beta}_1 + t \sqrt{\sigma^2_{\hat{\beta}_1}} \qquad (165)$$

where t is the value obtained from tabulated Student's t distribution
for the appropriate degrees of freedom and desired confidence coefficient. The variance of the parameter estimate is

$$\sigma^2_{\hat{\beta}_1} \approx \frac{c_{11} \sigma^2_{\ln \, {}^1_2 k^a_V}}{r} \tag{166}$$

where r is the number of replicates of solid-phase composition. To determine $\hat{\beta}_1$ to within δ of β_1, one must choose r so that

$$r \geq \frac{t^2 c_{11} \sigma^2_{\ln \, {}^1_2 k^a_V}}{\delta^2} \tag{167}$$

To calculate the number of replicates from Eq. (167), the experimenter must determine or estimate

1. An appropriate Student's t distribution value
2. A desired confidence interval
3. An appropriate diagonal element of the covariance matrix
4. A variance of the logarithm of the measured selectivity coefficient

The objective of the cation-exchange experiment will be to determine which parameter estimates dictate the number of replicates, to discriminate, for example, between regular-solution and sub-regular-solution models. Generally, the confidence limits will be set for the most variable of all the parameter estimates. In this case the number of replicates will be dictated by the largest c_{ii} terms in the covariance matrix. As noted above, the diagonal members of the covariance matrix specify the variances of the corresponding parameter estimates. These covariance matrix diagonal elements, corresponding to the solid-phase mole fractions suggested in Tables 11 and 12, are presented in Tables 13 and 14.

The expected variance of experimentally determined $\ln \, {}^i_j k^a_V$ values depends, understandably, on the procedure by which the solid-phase mole fractions are estimated. Only one procedure will be discussed here: the extraction of the cation exchanger which is at equilibrium with a small volume of supernatant equilibrating solution. The reader should be able to extend this example to calculate $\sigma^2_{\ln \, {}^1_2 k^a_V}$ when other procedures are followed in the determination of x_i. If the precision of measured cation-solution molalities is known, the variance of

TABLE 13   Diagonal Elements (c_{ii}) of the Covariance Matrice Corresponding to the Suggested Solid-Phase Mole Fractions in Table 11

Parameter	1,1	1,2	1,3	2,2	2,3	3,3
	\multicolumn{6}{c}{z_1, z_2}					

Parameter	1,1	1,2	1,3	2,2	2,3	3,3
\multicolumn{7}{c}{D-optimal criterion}						
\multicolumn{7}{c}{Regular-solution model}						
$\ln \,{}_2^1 K_V$	1.000	1.111	1.250	1.000	1.040	1.000
R_{12}	1.041	0.463	0.260	0.260	0.167	0.116
\multicolumn{7}{c}{Sub-regular-solution model}						
$\ln \,{}_2^1 K_V$	1.509	1.476	1.491	1.509	1.494	1.509
S_{12}	3.746	2.556	1.798	0.936	0.772	0.416
S_{21}	3.746	1.120	0.528	0.936	0.471	0.416
\multicolumn{7}{c}{Average mean-square error criterion}						
\multicolumn{7}{c}{Regular-solution model}						
$\ln \,{}_2^1 K_V$	1.000	1.111	1.250	1.000	1.040	1.000
R_{12}	1.063	0.472	0.266	0.266	0.170	0.118
\multicolumn{7}{c}{Sub-regular-solution model}						
$\ln \,{}_2^1 K_V$	1.107	1.102	1.204	1.107	1.076	1.110
S_{12}	5.635	2.176	1.518	1.409	0.671	0.373
S_{21}	3.325	1.762	0.847	0.831	0.721	0.620

$\ln \,{}_2^1 k_V^a$ may be estimated easily. The variance of a random variable y, which is a function of N other random variables u_i,

$$y = f(u_i, \, u_j, \, \ldots, \, u_N) \tag{168}$$

may be approximated by

$$\sigma_y^2 = \sum_{i=1}^N \left[\sigma_{u_i}^2 \left(\frac{\partial y}{\partial u_i}\right)^2 + \sum_{j \neq i} \left(\sigma_{u_i u_j} \left(\frac{\partial y}{\partial u_i}\right)\left(\frac{\partial y}{\partial u_j}\right) \right) \right]$$

where σ_y^2 is the variance of y and $\sigma_{u_i u_j}$, the covariance between u_i and u_j [103]. Recalling Eq. (24), we have

TABLE 14 Diagonal Elements (c_{ii}) of the Covariance Matrices Corresponding to the Solid-Phase Mole Fractions Suggested in Table 12 for Evaluating the Regular-Solution Model

Parameter	z_1, z_2					
	1,1	1,2	1,3	2,2	2,3	3,3
D-optimal criterion *Regular-solution model*						
$\ln \frac{1}{2}K_V$	4.235	35.44	10.21	4.235	29.05	4.235
R_{12}	6.818	3.028	1.171	1.704	1.345	0.7575
R_{23}	15.59	34.26	3.756	3.896	10.9	1.732
R_{31}	6.641	4.80	1.351	1.66	1.244	0.7379
Average mean-square error criterion *Regular-solution model*						
$\ln \frac{1}{2}K_V$	8.728	36.27	9.546	8.728	33.41	8.728
R_{12}	6.918	2.793	1.215	1.729	1.169	0.7687
R_{23}	24.41	35.33	4.417	3.896	12.59	2.713
R_{31}	5.571	6.204	1.591	1.66	1.268	0.6189

The corresponding elements could not be calculated for the sub-regular-solution model.

$$\ln \frac{1}{2}k_V^a = z_2 \ln x_{V_1} - z_1 \ln x_{V_1} + z_1 \ln m_{2(aq)}$$
$$- z_2 \ln m_{1(aq)} + z_1 \ln \gamma_2 - z_2 \ln \gamma_1 \qquad (170)$$

Estimating $\sigma^2_{\ln \frac{1}{2}k_V^a}$ with Eqs. (168) and (170) is easier if it is reformulated so that the covariances of all the independent random variables [the x_{V_i} in Eq. (170)] are zero. Further it can be assumed that any errors due to the single-ion activity coefficients are systematic, so that $\sigma^2_{\gamma_{1(aq)}} = 0$. For a binary cation-exchange system,

$$x_{V_1} = \frac{n_{1(ads)}}{\sum_{i=1}^{N} n_{i(ads)}} \qquad (171)$$

where

$$n_{i(ads)} = m_{(extr)} \times m_{i(aq,extractant)} - m_{(equil)}$$

$$\times m_{i(aq,equilibrating)} \tag{172}$$

and $m_{(equil)}$ and $m_{(extr)}$ are the masses of the equilibrating and extractant solutions. For the purpose of calculating $\sigma^2_{\ln \frac{1}{2}k^a_V}$ by applying Eq. (24) assume

$$n_{i(ads)} \approx m_{(extr)} \times m_{i(aq,extractant)} \tag{173}$$

$$a_{i(aq)} \approx m_{1(aq,equilibrating)} \tag{174}$$

Then

$$\ln \tfrac{1}{2}k^a_V = z_1 \ln m_{2(aq,equilibrating)} - z_2 \ln m_{1(aq,equilibrating)}$$

$$- z_1 \ln m_{2 (aq,extractant)} + z_2 \ln m_{1(aq,extractant)}$$

$$+ (z_1 - z_2)\ln(\Sigma^N_{i=1} m_{i(aq,extractant)}) \tag{175}$$

and

$$\sigma^2_u \left(\frac{d \ln u}{du}\right)^2 = \frac{\sigma^2_u}{u^2} \approx (CV(u))^2 \tag{176}$$

In a binary cation-exchange system, therefore

$$\sigma^2_{\ln \frac{1}{2}k^a_V} \approx [z_1 CV(m_{2(aq,equilibrating)})]^2$$

$$+ [z_2 CV(m_{1(aq,equilibrating)})]^2$$

$$+ [CV(m_{1(aq,extractant)})^2 CV(m_{2(aq,extractant)})^2]$$

$$\times \left(\frac{z_2 m_{1(aq,extractant)} + z_1 m_{2(aq,extractant)}}{m_{1(aq,extractant)} + m_{2(aq,extractant)}} \right) \tag{177}$$

Similar equations can be generated for Vanselow selectivity coefficients calculated in systems with more than two exchangeable ions. By assuming that the molalities of the ions in the extracting solution are comparable and that the coefficient of variation for the ions are approximately equal—both reasonable assumptions—$\sigma^2_{\ln \frac{1}{2}k_V}$ can be calculated as a function of the CV of the measured ionic solution molalities and the charges of the exchangeable ions. The variance

TABLE 15   Population Standard-Deviation Values of Measured $\ln {}_2^1k_V^a$ for Several Coefficients of Variation for Measured Solution Molalities of Ions in the Equilibrating and Extracting Solutions

Molality CV (%)	z_1, z_2					
	1,1	1,2	1,3	2,2	2,3	3,3
	Population standard deviation					
0.500	0.010	0.015	0.021	0.020	0.025	0.030
1.000	0.020	0.031	0.042	0.040	0.050	0.060
2.500	0.050	0.077	0.106	0.100	0.126	0.150
5.000	0.100	0.154	0.212	0.200	0.252	0.300
7.500	0.150	0.231	0.318	0.300	0.379	0.450
10.000	0.200	0.308	0.424	0.400	0.505	0.600

and population standard deviations for ionic-solution molalities CV are presented in Table 15.

E.   Calculation of Equilibrating-Solution Composition

The final step in the design of cation-exchange experiments is to calculate the equilibrating-solution molalities. The experimenter needs an estimate of the exchange-equilibrium constant for the cation exchanger to be studied. This estimate can be obtained from measured exchange-equilibrium constants or Vanselow selectivity coefficients of similar materials or from the best judgment of the experimenter.

If the Vanselow selectivity coefficient is the dependent variable to be determined, then the variable Y_{ij} is defined as

$$Y_{ij} = {}_j^i k_V^a \frac{x_{V_j}^{z_i} \gamma_i^{z_j}}{\gamma_j^{z_i} x_{V_i}^{z_j}} \tag{178}$$

On the other hand, if the Gapon selectivity coefficient is the dependent variable to be determined, then the variable Y_{ij} is defined as

$$Y_{ij} = \left({}_j^i{}_k^a{}_G \frac{x_{G_j(ads)}}{x_{G_i(ads)}} \right)^{z_i z_j} \frac{\gamma_j^{z_i}}{\gamma_i^{z_j}} \tag{179}$$

In either case, the relationship

$$Y_{ij} m_{i(aq)}^{z_j} = m_{j(aq)}^{z_i} \tag{180}$$

can be exploited. For a series cation-exchange experiments for which the target solid-phase mole fractions have been calculated previously, molalities of the ions in solution can be calculated by solving simultaneously the following set of equations:

$$Y_{12} m_{1(aq)}^{z_2} = m_{2(aq)}^{z_1}$$

$$Y_{23} m_{2(aq)}^{z_3} = m_{3(aq)}^{z_2}$$

$$\vdots$$

$$Y_{(N-1)N} m_{(N-1)(aq)}^{z_N} = m_{N(aq)}^{z_{(N-1)}}$$

$$\sum_{i=1}^{N} z_i m_i = z_A m_A \tag{181}$$

where z_A and m_A are the valence and molality of the ion having charge of opposite sign from the exchangeable ions (e.g., the anions for a cation-exchange equilibrium). The last relation in Eq. (181) represents the charge-balance requirement for the equilibrating electrolyte solution. The formulae in Eq. (181) were evaluated for homovalent and heterovalent cation-exchange equilibria with two and three exchangeable ions. They are presented in Tables 16 and 17.

F. An Example

An application of this theoretical development is demonstrated in the practical example that is considered next. In this example, an experimenter wishes to test the hypothesis that the Ca-Na cation-exchange reaction on montmorillonite is ideal. Since the regular-solution

TABLE 16   Formulae to Calculate Equilibrating-Solution Ion Molalities from Target Solid-Phase Mole Fractions in Binary Cation-Exchange Experiments

$z_1 = 1$ $z_2 = 1$ $m_1 = \dfrac{z_A m_A}{1 + Y_{12}}$ 　　　　　　　　(182)

$z_1 = 1$ $z_2 = 2$ $m_1 = \dfrac{-1/Y_{12} + \sqrt{Y_{12}^{-2} + (8 z_A m_A)/Y_{12}}}{4}$ 　　(183)

$z_1 = 1$ $z_2 = 3$ $m_1 = \dfrac{-1}{9 Y_{12}\left((z_A m_A)6 Y_{12} + \sqrt{(4 + 81 Y_{12}(z_A m_A)^2)}/54 Y_{12}^{3/2}\right)}$

$$+ \frac{z_A m_A}{6 Y_{12}} + \frac{\sqrt{4 + 81 Y_{12}(z_A m_A)^2}}{54 Y_{12}^{3/2}} \qquad (184)$$

$z_1 = 2$ $z_2 = 2$ $m_1 = \dfrac{z_A m_A}{2 + 2\sqrt{Y_{12}}}$ 　　　　　　　(185)

$z_1 = 3$ $z_2 = 2$ $m_1 = \dfrac{3}{4 Y_{12}^{2/3}}$

$$- \frac{4 Y_{12}^{2/3}\left(-9/(16 Y_{12}^{4/3}) + z_A m_A/(2 Y_{12}^{2/3})\right)}{\left(27 - 36 Y_{12}^{2/3}(z_A m_A) + 8 Y_{12}^{4/3}(z_A m_A)^2 + 8 Y_{12}\sqrt{-(z_A m_A)^3 + Y_{12}^{2/3}(z_A m_A)^4}\right)^{1/3}}$$

$$+ \frac{\left(27 - 36 Y_{12}^{2/3}(z_A m_A) + 8 Y_{12}^{4/3}(z_A m_A)^2 + 8 Y_{12}\sqrt{-(z_A m_A)^3 + Y_{12}^{2/3}(z_A m_A)^4}\right)^{1/3}}{4 Y_{12}^{2/3}}$$

$$(186)$$

$z_1 = 3$ $z_2 = 3$ $m_1 = \dfrac{z_A m_A}{3 + 3 Y_{12}^{1/3}}$ 　　　　　　　(187)

TABLE 17 Formulae to Calculate Equilibrating-Solution Ion Molalities from Target Solid-Phase Mole Fractions in Ternary Cation-Exchange Experiments

---

$z_1 = 1 \qquad z_2 = 1 \qquad z_3 = 1 \qquad m_1 = \dfrac{z_A m_A}{1 + Y_{12} + Y_{12} Y_{23}}$

$$m_2 = \dfrac{Y_{12}(z_A m_A)}{1 + Y_{12} + Y_{12} Y_{23}} \qquad (188)$$

$z_1 = 1 \qquad z_2 = 1 \qquad z_3 = 2$

$$m_1 = \dfrac{-1 - Y_{12} + Y_{12} Y_{23} \sqrt{(1 + Y_{12})^2 / (Y_{12}^2 Y_{23}^2) + 8(z_A m_A)/Y_{23}}}{4 Y_{12}^2 Y_{23}}$$

$$m_2 = \dfrac{-(1 + Y_{12})/Y_{12} Y_{23} + \sqrt{(1 + Y_{12})^2 / (Y_{12}^2 Y_{23}^2) + 8(z_A m_A)/Y_{23}}}{4} \qquad (189)$$

$z_1 = 1 \qquad z_2 = 1 \qquad z_3 = 3$

$$m_1 = \dfrac{-1 - Y_{12} + 9 Y_{12} Y_{23} \left[\dfrac{z_A m_A}{6 Y_{23}} + \sqrt{\dfrac{(1 + Y_{12})^3}{729 Y_{12}^3 Y_{23}^3} + \dfrac{(z_A m_A)^2}{36 Y_{23}^2}} \right]^{2/3}}{9 Y_{12}^2 Y_{23} \left[\dfrac{z_A m_A}{6 Y_{23}} + \sqrt{\dfrac{(1 + Y_{12})^3}{729 Y_{12}^3 Y_{23}^3} + \dfrac{(z_A m_A)^2}{36 Y_{23}^2}} \right]^{1/3}}$$

$$m_2 = \left[\dfrac{z_A m_A}{6 Y_{23}} + \sqrt{\dfrac{(1 + Y_{12})^3}{729 Y_{12}^3 Y_{23}^3} + \dfrac{(z_A m_A)^2}{36 Y_{23}^2}} \right]^{1/3}$$

$$+ 9 Y_{12} Y_{23} \left[\dfrac{z_A m_A}{6 Y_{23}} + \sqrt{\dfrac{(1 + Y_{12})^3}{729 Y_{12}^3 Y_{23}^3} + \dfrac{(z_A m_A)^2}{36 Y_{23}^2}} \right]^{1/3} \qquad (190)$$

$z_1 = 2 \qquad z_2 = 2 \qquad z_3 = 2 \qquad\qquad\qquad\qquad (191)$

$$m_1 = \dfrac{z_A m_A}{2 + 2\sqrt{Y_{12}} + 2\sqrt{Y_{12}}\sqrt{Y_{23}}} \qquad m_2 = \dfrac{\sqrt{Y_{12}}(z_A m_A)}{2 + 2\sqrt{Y_{12}} + 2\sqrt{Y_{12}}\sqrt{Y_{23}}}$$

$z_1 = 3 \qquad z_2 = 3 \qquad z_3 = 3 \qquad\qquad\qquad\qquad (192)$

$$m_1 = \dfrac{z_A m_A}{3 + 3 Y_{12}^{1/3} + 3 Y_{12}^{1/3} Y_{23}^{1/3}} \qquad m_2 = \dfrac{Y_{12}^{1/3}(z_A m_A)}{3 + 3 Y_{12}^{1/3} + 3 Y_{12}^{1/3} Y_{23}^{1/3}}$$

---

model is the most elementary excess Gibbs energy model, the experimenter can test the hypothesis that the regular-solution model parameter R_{12} is equal to zero (1 and 2 are the indices for Na and Ca, respectively).

As stated in Sec. VI.A, the first step in this process is to formulate the regression problem appropriate for the test of this hypothesis. To do this the experimenter must first select an evaluation procedure choosing between the Elprince-Babcock and rational approaches. In this example the rational approach is selected. The regression equation to be fitted is therefore

$$\ln \, {}_2^1 k_V^a = \ln \, {}_2^1 K_V + R_{12}(x_1^2 - 2x_2^2) \tag{193}$$

The second step is to determine the target exchangeable-cation mole fractions. To use the tables presented in this chapter, the experimenter must decide whether to select D-optimal or average mean-square error design criterion. In this example, the AM criterion is selected, so by consulting Table 11 the target exchangeable-ion mole fractions are $x_1 = 0.01, 0.01, 0.98, 0.99$.

The third step is determine the number of replicates. The experimenter decides that resolving $R_{12} \pm 0.1$ with a Student's t value of 1.96 will provide a fair test of the hypothesis. Assume that the CV for measured cation solution molalities is 1% which corresponds to a $\sigma^2_{\ln \, {}_2^1 k_V^a}$ of $0.031^2 = 0.000961$. From Table 13, the value of c_{ii} for R_{12} is 0.472. With Eq. (163), then, the number of replicates will be

$$r \geq \frac{(1.96)^2(0.472)(0.000961)}{(0.1)^2} = 0.174 \tag{194}$$

Accordingly, one replicated determination at each experimental site is determined to be adequate. (Most experimenters would prudently choose two replicates per experimental site.)

The final step is calculating the appropriate equilibrating-solution compositions. The experimenter decides to conduct the experiments with a chloride background medium of 0.01 mol kg^{-1}. From tabulated exchange-equilibrium-constant values of $\ln \, {}_2^1 K_V$ [104-106] a $\ln \, {}_2^1 k_V^a$ value of 0.06 is adopted. With the equations presented in Table 16, the experimenter quickly establishes Na solution molalities of the five equilibrating-solution compositions to be 0.000667, 0.001295, 0.009996, and 0.009998 mol kg^{-1}. Application of this procedure to ternary cation-exchange systems is identical, except that the experimental sites are recommended in Table 12 for pairs of cation charge numbers. The sites chosen must therefore be combinations for sets of recommended experimental sites for two cation-charge-number pairs.

G. Conclusion

In many situations, cation-exchange equilibria can be anticipated with acceptable precision from the exchange equilibrium constant. For more precise estimates, changes in corrected selectivity coefficient values with exchangeable-cation composition must be made available. Solid-phase excess Gibbs energy models provide an internally consistent, theoretically satisfying foundation to represent these changes. Despite the large body of experimental cation-exchange equilibrium data available, it cannot be determined if these models, despite their intellectual appeal, are any more useful than simpler empirical models. It is hoped that the chemical thermodynamics and experimental and statistical information presented here will stimulate a conclusive evaluation of solid-phase excess Gibbs energy models.

ACKNOWLEDGMENTS

We thank with warm appreciation J. A. Cornell, A. M. Elprince, S. V. Mattigod, R. D. Rhue, N. C. Scrivner, and one of the series editors, J. A. Marinsky, who carefully reviewed this chapter in manuscript form. K. S. Pitzer kindly shared galley proofs of a new tabulation of parameter values for his model. Part of the work done by S. A. Grant was supported by USDA-CSRS, project no. 89-COOP-1-473.

VII. APPENDIX 1. DERIVATION OF EQUATIONS FOR SOLID-PHASE ACTIVITY COEFFICIENTS

The exchange-equilibrium constant has the algebraic form

$$
{}_j^i K_V = \left(\frac{a_{j(aq)}^{z_i}(x_{V_i} f_{V_i})^{z_j}}{a_{i(aq)}^{z_j}(x_{V_j} f_{V_j})^{z_i}} \right)
\tag{195}
$$

An experimentally determinable mass-action quotient is

$$
{}_j^i k_V^a = \left(\frac{a_{i(aq)}^{z_j} x_j^{z_i}}{a_{j(aq)}^{z_i} x_i^{z_j}} \right)
\tag{196}
$$

Whereby,

$$
{}_j^i K_V = {}_j^i k_V^a \ \frac{f_{V_i}^{z_j}}{f_{V_j}^{z_i}}
\tag{197}
$$

Forming the differential of Eq. (197) yields

$$
d \ln {}_j^i K_V = d \ln {}_j^i k_V^a + z_j \, d \ln f_{V_i} - z_i \, d \ln f_{V_j} = 0
\tag{198}
$$

A second equation of constraint is the Gibbs-Duhem equation

$$
\Sigma_k \, n_k \, d\mu_k = 0
\tag{199}
$$

where n_k is the number of moles of the k-th component in the system. For the binary mixture under consideration,

$$
n_i \, d \ln(x_{V_i} f_{V_i}) + n_j \, d \ln(x_{V_j} f_{V_j}) + n_w \, d \ln a_w = 0
\tag{200}
$$

Equations (198) and (200) constitute two equations with two unknowns (f_{V_i} and f_{V_j}). Dividing Eq. (200) by $n_i + n_j$ gives

$$
x_{V_i} \, d \ln(x_{V_i} f_{V_i}) + x_{V_j} \, d \ln(x_{V_j} f_{V_j}) + \frac{n_w}{n_i + n_j} \, d \ln a_w = 0
\tag{201}
$$

Since $d \ln x_{A_i} = (1/x_{A_i}) \, dx_{A_i}$

$$x_{V_i} \, d \ln f_{V_i} + x_{V_j} \, d \ln f_{V_j} + \frac{n_w}{n_i + n_j} \, d \ln a_w = 0 \qquad (202)$$

Then

$$d \ln f_{V_i} = -\frac{1}{x_{V_i}} \left[x_{V_j} \, d \ln f_{V_j} + \frac{n_w}{n_i + n_j} \, d \ln a_w \right] \qquad (203)$$

and

$$d \ln f_{V_j} = -\frac{1}{x_{V_j}} \left[x_{V_i} \, d \ln f_{V_i} + \frac{n_w}{n_i + n_j} \, d \ln a_w \right] \qquad (204)$$

Substituting for $d \ln f_{V_j}$ into Eq. (198) gives

$$d \ln {}_j^i k_V^a + z_j \, d \ln f_{V_i} + \frac{z_i}{x_{V_j}} \left[x_{V_i} \, d \ln f_{V_i} + \frac{n_w}{n_i + n_j} \, d \ln a_w \right] = 0$$
$$(205)$$

Multiplying by $z_j x_{V_j}/(z_i x_{V_i} + z_j x_{V_j})$ and noting that the equivalent fractions of ions in the exchanger E_i and E_j are

$$E_i = \frac{z_i x_{V_i}}{z_i x_{V_i} + z_j x_{V_j}} \qquad (206)$$

and

$$E_j = \frac{z_j x_{V_j}}{z_i x_{V_i} + z_j x_{V_j}} \qquad (207)$$

then

$$(208)$$
$$E_j \, d \ln {}_j^i k_V^a + d \ln f_{V_i}^{z_j} + \frac{z_i z_j}{z_i x_{V_i} + z_j x_{V_j}} \frac{n_w}{n_i + n_j} \, d \ln a_w = 0$$

Defining $\hat{n}_w$ as the number of moles of water per exchange equivalent, i.e.,

$$\hat{n}_w = \frac{n_w}{z_i n_i + z_j n_j} \qquad (209)$$

and noting $n_w = \eta_w/(n_i + n_j)$ and $E_i + E_j = 1$, then

$$E_j \, d \ln {}_j^i k_V^a + E_j \, d \ln f_{V_i}^{z_j} + E_i \, d \ln f_{V_i}^{z_j} + z_i z_j \hat{n}_w \, d \ln a_w = 0 \tag{210}$$

Rearranging Eq. (210) gives

$$d \ln f_{V_i}^{z_j} = -E_j \, d \ln {}_j^i k_V^a - z_i z_j n_w \, d \ln a_w \tag{211}$$

The integration steps are

1. From the standard state to a mixture characterized by $x_{A_i} = 1$ and water activity equal to that of a pure M_i salt solution at the normality at which the experiment is performed (state designated by ϕ_i)

2. From the final state of step 1 to a state characterized by the chosen exchanger composition with water activity being that in the attendant mixed-electrolyte solution

Integrating the r.h.s. of (211) for step 1 gives

$$\int_0^{\ln f_{V_i}(\phi_i)} d \ln f_{V_i}^{z_j} = \ln f_{V_i}^{z_j}(\phi_i) \tag{212}$$

Integrating the l.h.s. of (211) for step 1 gives

$$\ln f_{V_i}^{z_j}(\phi_i) = -z_i z_j \int_0^{\ln a_w(\phi_i)} n_w \, d \ln a_w \tag{213}$$

Integrating r.h.s. of (211) for step 2 gives

$$\int_{\ln f_{V_i}(\phi_i)}^{\ln f_{V_i}} d \ln f_{V_i}^{z_j} = \ln f_{V_i}^{z_j} - \ln f_{V_i}^{z_j}(\phi_i) \tag{214}$$

Integrating the l.h.s. of (211) for step 2 gives

$$\int_{\ln k_V^a(\phi_j)}^{\ln k_V^a} E_j \, d \ln {}_j^i k_V^a - z_i z_j n_w \, d \ln a_w$$

$$= \int_{E_i=1}^{E_j} \ln {}_j^i k_V^a \, dE_j - E_j \ln {}_j^i k_V^a - z_i z_j \int_{\ln a_w(\phi_i)}^{\ln n_w} \hat{n}_w \, d \ln a_w \tag{215}$$

Combining Eqs. (212) and (215) gives

$$\ln f_{V_i}^{z_j} = \int_{0(E_i=1)}^{E_j} \ln {_j^i}k_V^a \, dE_j - E_j \ln {_j^i}k_V^a$$

$$- z_i z_j \left[\int_{0(E_i=1)}^{\ln a_w(\phi_i)} \hat{n}_w^{\phi_i} \, d \ln a_w + \int_{\ln a_w(\phi_i)}^{\ln a_w} \hat{n}_w \, d \ln a_w \right]$$

(216)

Similarly for f_{V_j},

$$\ln f_{V_j}^{z_i} = \int_{0(E_j=1)}^{E_j} \ln {_j^i}k_V^a \, dE_i - E_i \ln {_i^j}k_V^a$$

$$- z_i z_j \left[\int_{0(E_j=1)}^{\ln a_w(\phi_j)} \hat{n}_w^{\phi_j} \, d \ln a_w + \int_{\ln a_w(\phi_j)}^{\ln a_w} \hat{n}_w \, d \ln a_w \right]$$

(217)

The equilibrium constant is obtained by eliminating the exchanger activity coefficients from Eq. (197) using Eqs. (216) and (217), i.e.,

$$\ln {_j^i}K_A = \int_0^1 \ln {_j^i}k_V^a \, dE_j + z_i z_j \left[\int_{0(E_i=0)}^{\ln a_w(\phi_i)} \hat{n}_w \, d \ln a_w \right.$$

$$\left. - \int_{0(E_j=1)}^{\ln a_w(\phi_j)} \hat{n}_w \, d \ln a_w + \int_{\ln a_w(\phi_i)}^{\ln a_w(\phi_i)} \hat{n}_w \, d \ln a_w \right]$$

(218)

VIII. APPENDIX 2. ADJUSTING EXCHANGE-EQUILIBRIUM-CONSTANT VALUES FOR DIFFERENT CHOICES OF IONIC-SOLUTE ACTIVITY

Exchange equilibrium constants are properly calculated from corrected selectivity-coefficient values in which cations in solution are represented by the mean-ionic activities of their salts. For the type V cation-exchange-reaction stoichiometry, this corrected selectivity coefficient may be given a unique symbol, ${_j^i}k_V^\pm$, defined by

$${_j^i}k_V^\pm = \frac{x_{V_i}^{z_j} a_{\pm jk}^{z_i}}{x_{V_j}^{z_i} a_{\pm ik}^{z_j}}$$

(219)

A similar mean-ionic-activity-based corrected selectivity coefficient
may be defined for type G cation-exchange-reaction stoichiometry.
Please note that in this appendix it is assumed, reasonably, that the
electrolytes in the equilibrating solutions share a common anionic
species k.

In practice, exchange-equilibrium-constant values calculated from
corrected-selectivity-coefficient values in which cations in solution
are represented by their single-ion activities are more useful when
predicting cation-exchange reactions in aqueous electrolyte solutions
in which more than one anionic species are present. This more mer-
cenary corrected selectivity coefficient has been defined by

$$_j^i k_V^a = \frac{x_{V_i}^{z_j} a_{j(aq)}^{z_i}}{x_{V_j}^{z_i} a_{i(aq)}^{z_j}} \tag{220}$$

Its utility has led to the adoption in this chapter of corrected selectivity coefficients with single-ion (rather than mean-ionic) activity
coefficients. Still, many of the tabulated exchange-equilibrium-constant values were determined from mean-ionic-activity-based corrected
selectivity coefficients. If the experimental conditions in which these
values were determined, these exchange-equilibrium-constant values
can be adjusted to the corresponding single-ion activity based values.
If the corrected selectivity coefficients were determined in aqueous
solutions of equal ionic strengths, then the exchange-equilibrium-constant values should be largely unaffected by the choice of ionic-solute
activity. If the ionic strengths of the equilibrating solutions were not
equal, then the resulting exchange-equilibrium-constant values should
be adjusted to take account of that fact.

To begin, the two Vanselow selectivity coefficients can be shown
to be equal to

$$_j^i k_V^\pm = {_j^i k_V} \, _j^i \Gamma^\pm \tag{221}$$

and

$$^i_j k^a_V = {}^i_j k_V \; {}^i_j \Gamma \tag{222}$$

where ${}^i_j k_V$ is the uncorrected Vanselow selectivity coefficient defined by

$$^i_j k_V \; \overset{\text{def}}{=} \; \frac{x_{V_i}^{z_j} m_j^{z_i}}{x_{V_j}^{z_i} m_i^{z_j}} \tag{223}$$

${}^i_j \Gamma$ is the single-ion activity-based liquid-phase activity coefficient ratio (57)

$$^i_j \Gamma \; \overset{\text{def}}{=} \; \frac{\gamma_j^{z_i}}{\gamma_i^{z_j}}$$

and ${}^i_j \Gamma^{\pm}$ is based on mean-ionic activities:

$$^i_j \Gamma^{\pm} \; \overset{\text{def}}{=} \; \frac{\gamma_{\pm jk}^{z_i}}{\gamma_{\pm ik}^{z_i}} \tag{224}$$

Since it is assumed in this appendix that electrolytes in the equilibrating solutions share a common anionic species k, ${}^1_j \Gamma$ has been shown (68) to equal

$$^i_j \Gamma = \frac{\gamma_{\pm jk}^{z_i(1 + z_j/z_k)}}{\gamma_{\pm ik}^{z_j(1 + z_i/z_k)}} \tag{225}$$

Let us consider exchange equilibrium constants which are determined (for clarity's sake, neglecting the activity of water) from mean-ionic activity-based corrected-selectivity-coefficient values:

$$\ln \; {}^i_j K^{\pm}_V = \int_0^1 \ln \; {}^i_j k^{\pm}_V \; dE_j \tag{226}$$

and from single-ion-activity-based corrected-selectivity-coefficient
values

$$\ln {}_j^i K_V = \int_0^1 \ln {}_j^i k_V^a \, dE_j \tag{227}$$

By Eqs. (221) and (222) these integrals are equivalent to

$$\ln {}_j^i K_V^{\pm} = \int_0^1 \ln {}_j^i k_V \, dE_j + \int_0^1 \ln {}_j^i \Gamma^{\pm} \, dE_j \tag{228}$$

and

$$\ln {}_j^i K_V = \int_0^1 \ln {}_j^i k_V \, dE_j + \int_0^1 \ln {}_j^i \Gamma \, dE_j \tag{229}$$

These can be combined to yield

$$\ln {}_j^i K_V = \ln {}_j^i K_V^{\pm} + \int_0^1 \ln {}_j^i \Gamma \, dE_j - \int_0^1 \ln {}_j^i \Gamma^{\pm} \, dE_j \tag{230}$$

which by Eq. (225) is equivalent to

$$\ln {}_j^i K_V = \ln {}_j^i K_V^{\pm}$$
$$+ \frac{z_i z_j}{z_k} \left(\int_0^1 \ln \gamma_{\pm ik} \, dE_j - \int_0^1 \ln \gamma_{\pm jk} \, dE_j \right) \tag{231}$$

For equilibrating solutions with ionic strengths less than 0.02
mol kg^{-1}, the mean-ionic activity coefficients may be estimated with
good accuracy by the Debye-Hückel equation

$$\ln \gamma_{\pm ik} = - \frac{\alpha |z_i z_k| \sqrt{I_m}}{1 + \beta d \sqrt{I_m}} \tag{232}$$

If it assumed, as it may with small error, that equilibrating-solution
ion strength varies linearly with solid-phase equivalent fraction, then
I_m may be written as a function E_j:

$$I_m = I_m(E_j = 0) + [I_m(E_j = 1) - I_m(E_j = 0)]E_j \tag{233}$$

where $I_m(E_j = 0)$ is the ionic strength of the equilibrating solution in which $(M_i)_{\nu_i}(M_k)_{\nu_k}$ is the only electrolyte present and $I_m(E = 1)$ is the ionic strength of the equilibrating solution in which $(M_j)_{\nu_j}$ $(M_k)_{\nu_k}$ is the only electrolyte present. The integrals in Eq. (231) can then be approximated by

$$\int_0^1 \ln \gamma_{\pm ik} \; dE_j = \frac{2\alpha|z_i z_k|}{\beta^3 d^3 (I_m(E_j = 1) - I_m(E_j = 0))}$$

$$\times \; [\beta d \; (\sqrt{I_m(E_j = 0)} - \sqrt{I_m(E_j = 1)})$$

$$+ \frac{\beta^2 d^2}{2} \; (I_m(E_j = 1) - I_m(E_j = 0))$$

$$- \beta d \; \ln(1 + \sqrt{I_m(E_j = 0)})$$

$$+ \ln(1 + \beta d \; \sqrt{I_m(E_j = 1)})] \tag{234}$$

More accurate adjustments could be made by estimating mean-ionic activity coefficients with more sophisticated activity-coefficient models or by applying an empirically fitted polynomial function to describe changes in ionic strength with solid-phase equivalent fraction.

LIST OF SYMBOLS

$a_{i(aq)}$	Single-ion activity for cation i (1)
a_{G_i}	Type G component activity (1)
a_{V_i}	Type V component activity (1)
$a^c_{j(aq)}$	Activity of the j-th complex species (1)
a_i	Activity of i (1)
a_w	Activity of water associated with a cation exchanger (1)
A_ϕ	$= \alpha/3$ (kg$^{1/2}$ mol$^{-1/2}$)
b	$= \beta d$ (kg$^{1/2}$ mol$^{-1/2}$)
b_k	Solvent interaction term in HKF model (kg J^{-1})
$b_{i\ell}$	Short-range ionic interaction parameter (kg mol^{-1})
B^ϕ_{ij}	Pitzer-model coefficient (kg mol^{-1})
$B_{ijk\ell}$	Binary Margules coefficient (1)

B_{ij} Virial coefficient for ionic interaction in Pitzer model ($kg\ mol^{-1}$)

B'_{ij} Virial coefficient for ionic interaction in Pitzer model ($kg^2\ mol^{-2}$)

c_{ii} i-th diagonal element of the covariance matrix (1)

C_{ij} Virial coefficient for ionic interaction in Pitzer model ($kg^2\ mol^{-2}$)

C^{ϕ}_{ij} Compositionally independent parameter ($kg^2\ mol^{-2}$)

$CV(u)$ Coefficient of variation (1)

d Mean distance between ions in solution (m)

E_i Solid-phase equivalent fraction (1)

$f(I_m)$ Function representing long range electrostatic forces ($mol\ kg^{-1}$)

f' $= df/dI_m$ ($mol^2\ kg^{-2}$)

f^{ϕ} Pitzer-model coefficient (1)

f_i Rational activity coefficient of i (1)

f_{GT_i} Gaines-Thomas activity coefficient of cation i (1)

f_{G_i} Type G rational activity coefficient of cation i (1)

f_{V_i} Type V rational activity coefficient of cation i (1)

g_i Rational activity coefficient for an exchangeable solute (1)

g_j Rational activity coefficient for a nonexchangeable solute (1)

g_w Rational activity coefficient for water (1)

G^E Excess Gibbs energy (J)

$\bar{G}^E$ Molar excess Gibbs energy of mixing ($J\ mol^{-1}$)

$\bar{G}^{E'}$ Total molar excess Gibbs energy ($J\ mol^{-1}$)

$\bar{G}_m$ Molar Gibbs energy of a real mixture ($J\ mol^{-1}$)

$\bar{G}^{\ominus}_i$ Standard-state molar Gibbs energy of mixture component $\overset{def}{=} M_{i(ads)1/z_i}X$ (1)

i Index for a cation (1)

I_m Molality-based ionic strength ($mol\ kg^{-1}$)

$I_m(E_j = 0)$ Equilibrating-solution ionic strength for $E_j = 0$ ($mol\ kg^{-1}$)

$I_m(E_j = 1)$ Equilibrating-solution ionic strength for $E_j = 1$ ($mol\ kg^{-1}$)

$\bar{I}_m$	True ionic strength (mol kg^{-1})
j	Index for a cation (1)
k	Index for an anion (1)
$_j^i k_G^a$	Gapon selectivity coefficient (1)
$_j^i k_{GT}^a$	Gaines-Thomas selectivity coefficient (1)
$_j^i k_V$	Uncorrected Vanselow selectivity coefficient (1)
$_j^i k_V^a$	Vanselow selectivity coefficient (1)
$_j^i k_V^\pm$	Mean-ionic-activity-based Vanselow selectivity coefficient (1)
$_j^i K_{GT}$	Type GT exchange equilibrium constant (1)
$_j^i K_G$	Type G exchange equilibrium constant (1)
$_j^i K_V$	Type V exchange equilibrium constant (1)
$_j^i K_V^\pm$	Mean-ionic-activity-based exchange equilibrium constant (1)
ℓ	Index for an anion (1)
ln	Base-e, or natural, logarithm
log	Base-10 logarithm
L	Number of cationic species in solution (1)
$m_{(extr.)}$	Mass of extractant solution (kg)
$m_{(equil)}$	Mass of the entrained equilibrating solution (kg)
m_w	Mass of water of a solution (kg)
$\hat{m}_i$	Actual single-ion molality of ion i (mol kg^{-1})
m_i	Molality of ion i in aqueous solution (mol kg^{-1})
$m_i^{\ominus}$	Standard state of molality of ion i in aqueous solution (mol kg^{-1})
$m_{i(aq,equil-ibrating)}$	Molality of cation i in the equilibration solution (mol kg^{-1})
$m_{i(aq,ex-tractant)}$	Molality of cation i in the extractant solution (mol kg^{-1})

$\hat{m}_j^c$ Single-ion molality of the j-th complex species (mol kg^{-1})

m^* Total molality of an electrolyte solution (mol kg^{-1})

$M_i(ads)$ Exchangeable cation i (1)

$M_i(aq)$ Cationic solute i (1)

M_j^c The j-th complex species formed in solution by a salt (1)

M Number of anionic species in a solution (1)

M_A Molecular mass of solvent A (g mol^{-1})

$\hat{n}_w$ $= n_w/(z_i n_i + z_j n_j)$

n_i Quantity of i (mol)

n_T Total number of exchangeable cations (mol)

n_w Number of water molecules associated with a cation exchanger (mol)

N Number of components in a mixture (1)

N Number of ions in a solution (1)

N Number of independent random variables (1)

N^c Number of complex species formed (1) in solution by a salt

$p^{\ominus}$ Standard-state pressure (Pa)

q_i van Laar model parameter (1)

r Number of replicates (1)

$r(\xi)$ A reference function (1)

r_j^{el} Relative electrostatic radius (m)

R Molar gas constant (J K^{-1} mol^{-1})

R_{ij} Regular-solution-model parameter (1)

S_{ij} Sub-regular-solution model parameter (1)

t Student's t (1)

T Thermodynamic temperature (K)

T_{ijklm} Ternary Margules coefficient (1)

u_i An independent random variable (1)

V_i Type V component $= M_i(ads)X_{z_i}$ (1)

$W_{j/i}$ Wilson equation parameter (1)

x_i Mole fraction of mixture component i (1)

x_{G_i} Type G solid-phase mole fraction of cation i (1)

x_{v_i} Type V solid-phase mole fraction of cation i (1)

X One mole negative charge of cation exchanger (1)

X_i Mole fraction of exchangeable solute i (1)

X_j Mole fraction of nonexchangeable solute j (1)

X Matrix of independent variable values (1)

X' Transpose of the matrix X (1)

X^{-1} Inverse of the matrix X (1)

y A dependent random variable (1)

y_k $= z_k^2 m_k / (2I_m)$ (1)

$\bar{y}_k$ $= z_k^2 m_k / (2\bar{I}_m)$ (1)

Y_{ij} Variable for calculating equilibrating-solution
compositions (1)

z_i Charge number of cation i (1)

Greek Symbols

α_1 Parameter in Pitzer model ($kg^{1/2}$ $mol^{-1/2}$)

α_2 Parameter in Pitzer model ($kg^{1/2}$ $mol^{-1/2}$)

α Constant in Debye-Hückel equation ($kg^{1/2}$ $mol^{-1/2}$)

β Constant in Debye-Hückel equation ($kg^{1/2}$ $mol^{-1/2}$ m^{-1})

$\beta_{ij}^{(0)}$ Parameter in Pitzer model (kg mol^{-1})

$\beta_{ij}^{(1)}$ Parameter in Pitzer model (kg mol^{-1})

$\beta_{ij}^{(2)}$ Parameter in Pitzer model (kg mol^{-1})

β_ℓ Polynomial fitting parameters (1)

β_{ij} Ion-ion interaction term in Guggenheim's equation
(kg mol^{-1})

$\hat{\beta}_i$ Estimate of the parameter β_i (1)

$\hat{\gamma}_{\pm ik}$ True ionic activity coefficient (1)

$\hat{\gamma}_j^c$ Single-ion activity coefficient of the j-th complex
species (1)

γ_i Single-ion activity coefficient of cation i (1)

$\gamma_{\pm ik}$ Mean-ionic activity coefficient of salt ik (1)

$\bar{\gamma}_{\pm ik}$ True mean-ionic activity coefficient of salt ik (1)

$\hat{\gamma}_i$ Actual single-ion activity coefficient of ion i (1)

Γ Mole fraction to molality conversion factor (1)

Γ_γ Mole fraction to molality conversion factor (1)

$= -\ln(1 + 0.018015\ 3m^*)$

${}^i_j\Gamma$ Single-ionic activity-coefficient ratio (1)

${}^i_j\Gamma^\pm$ Mean-ionic activity-coefficient ratio (1)

δ Desired absolute difference between a parameter and its estimated value (1)

δ_{ij} Kronecker delta symbol (1)

${}^i_j\Delta\bar{G}^\ominus_G$ Standard molar Gibbs energy change for type G reaction $(J\ mol^{-1})$

${}^i_j\Delta\bar{G}^\ominus_V$ Standard molar Gibbs energy change for type V reaction $(J\ mol^{-1})$

$\Delta_{mix}\bar{G}_m$ Molar Gibbs energy of mixing $(J\ mol^{-1})$

$\Delta_{mix}\bar{G}^{id}_m$ Molar Gibbs energy of mixing for ideal mixture $(J\ mol^{-1})$

ε_{ij} $= 1$, if $sgn(z_i) \neq sgn(z_j)$; $= 0$, if $sgn(z_i) = sgn(z_j)$ (1)

$\varepsilon_r(k)$ Relative permittivity of a aqueous solution of k (1)

$\varepsilon_r(w)$ Relative permittivity of a pure water (1)

ζ_{ij} Ion-charge term in Fletcher-Townsend model

η_j Stoichiometric number of eater in complex for formation reaction (1)

θ_{ij} Virial coefficient for ionic interaction in Pitzer model $(kg\ mol^{-1})$

θ'_{ij} $= d\theta_{ij}/dI_m$ $(kg^2\ mol^{-2})$

λ^{id}_i Ideal mixture absolute activity of i (1)

λ_i Absolute activity of i (1)

$\lambda^\ominus_i$ Standard state absolute activity of i (1)

$\lambda_{ij}(I_m)$ Function representing short range electrostatic forces between ions i and j $(kg\ mol^{-1})$

λ'_{ij} $= d\lambda_{ij}(I_m)/dI_m$ $(mol^2\ kg^{-2})$

Λ HKF model parameter $= 1 + \beta d\ \sqrt{I_m}$ (1)

μ_i Chemical potential of i $(J\ mol^{-1})$

μ_{ijk} — Parameter representing short-range electrostatic forces between ions i, j, and k ($kg^2 mol^{-2}$)

μ_i^* — Chemical potential of pure mixture component i ($J mol^{-1}$)

$\mu_i^\ominus$ — Standard-state chemical potential of i ($J mol^{-1}$)

μ_j^c — Chemical potential of the j-th complex species ($J mol^{-1}$)

μ_w — Chemical potential of water ($J mol^{-1}$)

$\mu_w^\ominus$ — Standard-state chemical potential of water ($J mol^{-1}$)

$\mu_{ik(aq)}$ — Chemical potential of salt ik in aqueous solution ($J mol^{-1}$)

$\mu_{i(aq)}$ — Chemical potential of ion i in aqueous solution ($J mol^{-1}$)

$\mu_{V_i}^\ominus$ — Type V component std. state chemical potential ($J mol^{-1}$)

$\mu_{i(aq)}^\ominus$ — Single-ion std. state chemical potential for cation i ($J mol^{-1}$)

ν_i — Stoichiometric number of i (1)

ν_k — Number of moles of ions generated by the dissolution of k (1)

$\nu_{j,k}$ — Number of moles of ion j generated by the dissolution of k (1)

ξ — Exemplary independent variable for reference function

ξ^ρ — Reference state

σ — $= 3(\Lambda - 1/\Lambda + 2 \ln \Lambda)/(d\beta \sqrt{I_m})^{-3}$

$\sigma_{\hat{\beta}i}^2$ — Variance of the estimate of β_i (1)

$\sigma_{\ln \frac{1}{2}k_V^a}^2$ — Variance of Vanselow selectivity coefficient (1)

σ_x^2 — Variance of a random variable x (1)

σ_{xy} — Covariance between the random variables x and y (1)

ϕ_i — Intermediate step in exchange-equilibrium constant calculation (1)

ϕ_w — Osmotic coefficient (1)

ψ_k — $= \frac{1}{2} \Sigma_j \nu_{j,k} z_j^2$

ψ_{ijk} — Virial coefficient for ionic interaction in Pitzer model ($kg^2 mol^{-2}$)

ω_k — Absolute Born coefficient of the k-th electrolyte ($J mol^{-1}$)

ω_j^{abs} — Absolute Born coefficient of the j-th ion ($J mol^{-1}$)

REFERENCES

1. G. Sposito, *Thermodynamics of Soil Solutions*, Oxford University Press, Oxford, 1981.

2. V. S. Soldatov, in *Ion Exchangers* (K. Dorfner, ed.), Walter de Gruyter, Berlin, 1991.

3. W. J. Argersinger, A. W. Davidson, and O. D. Bonner, *Trans. Kansas Acad. Sci.*, *53*:404 (1950).

4. E. A. Guggenheim, *Thermodynamics*, North-Holland, New York 1967.

5. F. T. Wall, *Chemical Thermodynamics*, 3rd ed., W. H. Freeman, San Francisco, 1974.

6. G. L. Gaines and H. C. Thomas, *J. Chem. Phys.*, *21*:714 (1953).

7. G. N. Lewis, *J. Am. Chem. Soc.*, *35*:1 (1913).

8. K. L. Babcock, *Hilgardia*, *34*:1 (1963).

9. K. L. Babcock and E. C. Duckart, *Soil Sci.*, *130*:64 (1980).

10. R. M. Barrer and J. Klinowski, *J. Chem. Soc. Farad. Trans. I*, *70*:2080 (1974).

11. P. Fletcher and R. P. Townsend, *J. Chem. Soc. Farad. Trans. II*, *77*:965 (1981).

12. P. Fletcher and G. Sposito, *Clay Minerals*, *24*:375 (1989).

13. E. Högfeldt, *Ark. Kemi.*, *5*:147 (1953).

14. R. M. Barrer and R. P. Townsend, *J. Chem. Soc. Farad. Trans. II*, *80*:629 (1984).

15. G. Sposito, *Chemistry in the Soil Environment* (R. H. Dowdy et al., eds.), American Society of Agronomy, Madison, WI, 1981.

16. G. Sposito and S. V. Mattigod, *Clays Clay Miner*, *27*:125 (1979).

17. V. S. Soldatov and V. A. Bychkova, *Russ. J. Phys. Chem.*, *44*:1297 (1970).

18. S.-Y. Chu and G. Sposito, *Soil Sci. Soc. Am. J.*, *45*:1084 (1981).

19. G. Sposito, *Soil Physical Chemistry* (D. L. Sparks, ed.), CRC Press, Boca Raton, Florida, 1986.

20. G. Sposito, C. Jouany, K. Holtzclaw, and C. L. Levesque, *Soil Sci. Soc. Am. J.*, *47*:1081 (1983).

21. P. Fletcher and R. P. Townsend, *J. Chem. Soc. Farad. Trans. I*, *81*:1731 (1985). For a criticism, see Z. Tao and G. Yang, *Kexue Tongbao*, *32*:1240 (1987).

22. P. Fletcher, K. Franklin, and R. P. Townsend, *Phil. Trans. Roy. Soc. A*, *312*:141 (1984).

23. M. L. McGlashan, *Chemical Thermodynamics*, Academic Press, New York, 1979.

24.   P. Debye and E. Hückel, *Phys. Z.*, *24*:185 (1923).

25.   P. Debye, *Phys. Z.*, *25*:97 (1924).

26.   G. N. Lewis and M. Randall, *J. Amer. Chem. Soc.*, *43*:1141 (1921).

27.   R. A. Robinson and R. H. Stokes, *Electrolyte Solutions*, 2nd ed., Butterworths, London, 1970.

28.   H. S. Harned and B. B. Owen, *Physical Chemistry of Electrolytic Solutions*, 3rd ed., Reinhold, New York, 1958.

29.   L. Onsager, *Chem. Rev.*, *13*:73 (1933).

30.   J. Kirkwood, *J. Chem. Phys.*, *2*:767 (1934).

31.   D. N. Card and J. P. Valleau, *J. Chem. Phys.*, *52*:6232 (1970).

32.   J. C. Rassiah, D. N. Card, and J. P. Valleau, *J. Chem. Phys.*, *56*:248 (1972).

33.   E. A. Guggenheim, *Philos. Mag.*, *19*:588 (1935).

34.   J. N. Brønsted, *J. Am. Chem. Soc.*, *44*:877 (1922).

35.   E. A. Guggenheim and J. C. Turgeon, *Trans. Farad. Soc.*, *51*:747 (1955).

36.   J. F. Zemaitis, D. M. Clark, M. Rafal, and N. C. Scrivner, *Handbook of Aqueous Electrolyte Thermodynamics*, American Institute of Chemical Engineers, New York, 1986. (This work serves as an exceptionally useful reference.)

37.   E. Glueckauf, *Nature*, *163*:414 (1949).

38.   P. Fletcher and R. P. Townsend, *J. Chem. Soc. Farad. Trans. II*, *77*:2077 (1981).

39.   K. S. Pitzer, *J. Phys. Chem.*, *77*:286 (1973).

39a. J. B. Macaskill and R. G. Bates, *J. Solution Chem.*, *12*:607 (1983).

40.   S. L. Clegg, *J. Phys. Chem. Ref. Data*, *1*:1047 (1972).

41.   K. S. Pitzer, R. N. Roy, and I. F. Silvester, *J. Am. Chem. Soc.*, *99*:4930 (1977).

42.   C. E. Harvie, H. P. Eugster, and J. H. Weare, *Geochim. Cosmochim. Acta*, *48*:723 (1984).

43.   H.-T. Kim and W. J. Frederick, Jr., *J. Chem. Eng. Data*, *33*:177 (1988).

44.   D. E. Levy and R. J. Myers, *J. Phys. Chem.*, *94*:7842 (1990).

45.   J. C. Peiper and K. S. Pitzer, *J. Chem. Thermodyn.*, *14*:613 (1982).

46.   R. N. Roy, J. J. Gibbons, R. Williams, G. Baker, J. M. Simmonson, and K. S. Pitzer, *J. Chem. Thermodyn.*, *16*:303 (1984).

47.   J. A. Rard, *J. Chem. Eng. Data*, *29*:443 (1984).

48. J. A. Rard and D. G. Miller, *J. Chem. Eng. Data, 27*:169 (1982).

49. R. N. Roy, K. Hufford, P. J. Lord, D. R. Mrad, L. N. Roy, and D. A. Johnson, *J. Chem. Thermodyn., 20*:63 (1988).

50. V. K. Filippov, M. V. Charykova, and Yu. M. Trofimov, *J. Appl. Chem. U.S.S.R., 58*:1807 (1985).

51. J. A. Rard and D. G. Miller, *J. Chem. Eng. Data, 26*:38 (1981).

52. K. S. Pitzer, J. Olsen, J. M. Simonson, R. N. Roy, J. J. Gibbons, and L. A. Rowe, *J. Chem. Eng. Data, 30*:14 (1985).

53. R. C. Phutela and K. S. Pitzer, *J. Solution Chem., 12*:201 (1983).

54. E. J. Reardon and R. D. Beckie, *Geochim. Cosmochim. Acta, 51*:2355 (1987).

55. J. A. Rard, *J. Chem. Eng. Data, 32*:334 (1987).

56. V. K. Filippov, D. S. Barkov, and Ju. A. Federov, *Z. Phys. Chem. Leipzig, 266*:129 (1985).

57. J. A. Rard and D. G. Miller, *J. Chem. Thermodyn., 21*:463 (1989).

58. J. A. Rard and D. G. Miller, *J. Chem. Eng. Data, 26*:33 (1981).

59. V. K. Filippov, A. M. Kalinkin, and S. K. Vasin, *J. Chem. Thermodyn., 19*:185 (1987).

60. J. A. Rard and F. H. Spedding, *J. Chem. Eng. Data, 27*:454 (1982).

61. J. A. Rard and D. G. Miller, *J. Chem. Eng. Data, 32*:92 (1987).

62. J. A. Rard, D. G. Miller, and F. H. Spedding, *J. Chem. Eng. Data, 24*:948 (1979).

63. J. A. Rard and F. H. Spedding, *J. Chem. Eng. Data, 26*:391 (1981).

64. K. S. Pitzer and G. Mayorga, *J. Solution Chem., 3*:539 (1974).

65. V. K. Filippov, G. V. Dmitriev, and S. I. Yakovieva, *Dokl. Akad. Nauk. SSSR Fiz. Khim., 252*:156 (1980). In English translation *252*:359 (1980).

66. C. J. Downes and K. S. Pitzer, *J. Solution Chem., 5*:389 (1976).

67. E. J. Reardon and D. K. Armstrong, *Geochim. Cosmochim. Acta, 51*:63 (1987).

68. K. S. Pitzer, *Activity Coefficients in Electrolyte Solutions*, 2nd ed. (K. S. Pitzer, ed.), CRC Press, Boca Raton, FL, 1992.

69. L. N. Plummer, D. L. Parkhurst, G. W. Flemming, and S. A. Dunkle. *A Computer Program Incorporating Pitzer's Equation for Calculation of Geochemical Reactions in Brines*, U.S. Geological Survey, Water Resources Investigations Report 88-4152, Reston, VA, 1988.

70. C. Harvie, H. P. Eugster, and P. Wear, *Geochim. Cosmochim. Acta, 46*:1603 (1982).

71. N. Bjerrum, *Kgl. Danske Vid. Selsk. Mat-Fys. Medd.* 7: (1926).

72. C. W. Davies, *Ion Association*, Butterworths, London, 1962.

73. R. M. Smith and A. E. Martell, *Inorganic Complexes*, Vol. 4, Plenum Press, 1976.

74. H. C. Helgeson, D. H. Kirkham, and G. C. Flowers, *Am. J. Sci.*, *281*:1249 (1981).

75. N. A. Gocken, *Statistical Thermodynamics of Alloys*, Plenum Press, New York, 1986.

76. J. Grover, in *Thermodynamics in Geology* (D. G. Fraser, ed.), Reidel, Dordrecht, The Netherlands, 1977.

77. D. S. Abrams and J. M. Prausnitz, *AlChE J.*, *21*:116 (1975).

78. A. Navrotsky, in *Macroscopic to Microscopic: Atomic Environments to Mineral Thermodynamics* (S. W. Kieffer and A. Navrotsky, eds.), Mineralogical Society of America, Washington, DC, 1985.

79. A. M. Elprince and K. L. Babcock, *Soil Sci.*, *120*:332 (1975).

80. A. M. Elprince, A. P. Vanselow, and G. Sposito, *Soil Sci. Soc. Am. J.*, *44*:964 (1980).

81. L. L. Schramm and J. C. T. Kwak, *Soil Sci.*, *137*:1 (1984).

82. R. M. Barrer and J. Klinowski, *Phil. Trans. Roy. Soc.*, *A*, *285*:637 (1977).

83. R. M. Barrer and J. Klinowski, *Geochim. Cosmochim. Acta*, *43*:755 (1979).

84. J. H. Hildebrand, *J. Am. Chem. Soc.*, *51*:66 (1929).

85. H. K. Hardy, *Acta Metall.*, *1*:202 (1953).

86. K. Wohl, *Trans. Am. Inst. Chem. Engrs.*, *42*:215 (1946).

87. G. M. Wilson, *J. Am. Chem. Soc.*, *86*:127 (1964).

88. R. P. Wiedenfeld and L. R. Hossner, *Soil Sci. Soc. Am. J.*, *42*:709 (1978).

89. A. M. Elprince and K. L. Babcock, *Soil Sci.*, *120*:332 (1975).

90. C. H. P. Lupis, *Chemical Thermodynamics of Materials*, North-Holland, New York, 1983.

91. N. R. Draper and H. Smith, *Applied Regression Analysis*, 2nd ed., Wiley, New York, 1981.

92. P. Fletcher and R. P. Townsend, *J. Chromatogr.*, *201*:93 (1980).

93. S. A. Grant and D. L. Sparks, *J. Phys. Chem.*, *93*:6265 (1989).

94. Symbolics, Inc., *VAX UNIX MACSYMA Reference Manual Version 11*, Symbolics, Inc., Cambridge, MA, 1985.

95. S. A. Wolfram, *Mathematica: A System for Doing Mathematics by Computer*, Addison-Wesley, Redwood City, CA, 1988.

96. I. Ford, D. M. Titterington, and C. P. Kitsos, *Technometrics*, *31*:49 (1989).

97. A. Wald, *Ann. Math. Stat.*, *14*:134 (1943).

98. G. E. P. Box and N. R. Draper, *J. Am. Stat. Assoc.*, *54*:622 (1959).

99. W. J. Welch, *Biometrika*, *70*:205 (1983).

100. A. I. Khuri and J. A. Cornell, *Response Surfaces: Designs and Analyses*, Marcel Dekker, New York, 1987.

101. T. J. Mitchell, *Technometrics*, *16*:67 (1974).

102. W. J. Welch, *Technometrics*, *26*:217 (1984).

103. P. R. Bevington, *Data Reduction and Error Analysis for the Physical Sciences*, McGraw-Hill, New York, 1969.

104. M. G. M. Bruggenwert and A. Kamphorst, in *Soil Chemistry B: Physico-Chemical Models* (G. H. Bolt, ed.), Elsevier, Amsterdam, 1979.

105. Y. Marcus and D. G. Howery, *Ion Exchange Equilibrium Constants*, IUPAC Chemical Data Series No. 10, Butterworths for IUPAC, London, 1975.

106. E. Högfeldt, *Ion Exchange Data: Thermodynamic and Other Equilibrium Data*, Marcel Dekker, New York (in press).

2

A Three-Parameter Model for Summarizing Data in Ion Exchange

ERIK HÖGFELDT   The Royal Institute of Technology, Stockholm, Sweden

I. INTRODUCTION

About 10 years ago a simple three-parameter model based on the Guggenheim zeroth approximation was introduced [1,2]. Recently, the experience with this model was reviewed [3]. In this chapter the model is applied to many different systems to show its versatility.

II. THE MODEL

Consider the following ion-exchange reaction:

$$\begin{array}{cc} -R^-A^+ & -R^-A^+ \\ -R^-A^+ + B^+ \rightleftharpoons -R^-B^+ + A^+ \\ -R^-A^+ & -R^-A^+ \end{array}$$

$$(I)$$

or

$$\begin{array}{cc} -R^-B^+ & -R^-B^+ \\ -R^-A^+ + B^+ \rightleftharpoons -R^-B^+ + A^+ \\ -R^-A^+ & -R^-A^+ \end{array}$$

$$(II)$$

109

or

$$-R^-B^+ \qquad\qquad -R^-B^+$$
$$-R^-A^+ + B^+ \rightleftharpoons -R^-B^+ + A^+$$
$$-R^-B^+ \qquad\qquad -R^-B^+$$

(III)

Observe, that even though the ion-exchange reaction is the same, three different site environments prevail. In reaction (I) the exchanging site is surrounded by $-R^-A^+$, in reaction (II) by both $-R^+A^-$ and $-R^-B^+$, and in reaction (III) by $-R^-B^+$ only. If the neighboring sites interact in the same way with the exchanging site, the reaction can be expected to be similarly affected and the equilibrium quotient of the three reactions should be the same. If, on the other hand, the sites interact with each other, one should expect three different ion-exchange constants. If the ions are randomly distributed over the sites, the number of each kind is provided by Guggenheim's zeroth approximation [4], giving for the equilibrium quotient κ the expression

$$\log \kappa = \log \kappa(1)\bar{x}^2 + \log \kappa(0)(1 - \bar{x})^2 + 2 \log \kappa_m \bar{x}(1 - \bar{x})$$

Here, $\bar{x}$ = stoichiometric mole fraction of $-R^-B^+$ in the ion exchanger.

For any molar property Y, the expression above becomes

$$Y = y_B\bar{x}^2 + y_A(1 - \bar{x})^2 + 2y_m\bar{x}(1 - \bar{x}) \tag{1}$$

In the ion-exchange literature the property under study is most often plotted versus $\bar{x}$, not $\bar{x}^2$. To facilitate this approach it has been convenient to substitute Eq. (2) for Eq. (1).

$$Y = y_B\bar{x} + y_A(1 - \bar{x}) + B\bar{x}(1 - \bar{x}) \tag{2}$$

The fitting of experimental data to Eq. (2) by least-squares procedures leads to resolution of the parameters y_A, y_B, and the empirical constant B introduced by the substitution. The third parameter, y_m, is then accessible with

$$y_m = \frac{1}{2}(y_A + y_B + B) \tag{3}$$

Equation (3) is obtained by equating Eqs. (1) and (2).

III. APPLICATION TO ION EXCHANGE

A. The Activity Scale

The above model refers to binary mixtures. Its application to three-component ion-exchange systems is considered justified by the pseudo-binary approach first outlined by Sillén [5-7] as well as Argersinger et al. [8]. In this approach an ion exchanger is assumed to contain a mixture of the two ionic forms. Since the activity of water, the third component, often is kept practically constant, it can, to a first approximation, be neglected and the system treated as a binary one [9,10]. The three-component treatment of this system by Thomas [11] has been used by many workers, who often have disregarded the water component and ended up with essentially the pseudobinary approach as well.

B. The Ion-Exchange Reaction

1. Monovalent Ions

Let us first consider ion exchange between monovalent ions. The ion-exchange reaction can be written

$$M_2^z + M_1X \rightleftharpoons M_2X + M_1^z \tag{4}$$

where X = anion or cation framework of exchanger. For cations z = +1, for anions z = -1.

The equilibrium quotient of reaction (4) is given by

$$\kappa = \frac{[M_2X]\{M_1^z\}}{[M_1X]\{M_2^z\}} = \frac{\bar{x}[M_1^z]}{(1 - \bar{x})[M_2^z]} \left(\frac{y_1}{y_2}\right) \tag{5}$$

where $\bar{x}$ = mole fraction of M_2X, { } denotes activity, and [ ] denotes concentration. The activity-coefficient ratio in the aqueous phase, y_1/y_2, can be kept constant by controlling the ionic strength of the medium and need not be included in κ. Otherwise it can be measured or estimated by a suitable formula. For example, the specific ionic interaction theory of Guggenheim and Scatchard can be used to estimate activity-coefficient changes with ionic strength.

On the molal scale the expression for the activity coefficient is given by [12]

$$\log \gamma_i = -\frac{A z_i^2 \sqrt{I}}{1 + B\mathring{a}\sqrt{I}} + \Sigma\, b_{ik} m_k \tag{6}$$

where I = ionic strength, A and B are constants in the Debye-Huckel equation, $\mathring{a}$ = distance of closest approach, and b_{ik} = ionic interaction coefficient. For ions of the same charge $b_{ik} = 0$. For ions of opposite charge b_{ik} can be estimated from experimental data. A table of such coefficients for $B\mathring{a} = 1.5$ $\mathring{A}$ has been published [12].

2. Multivalent Ions

Let us consider cation exchange. For anion exchange analogous expressions can be found. For the reaction

$$z_1 M_2^{z_2} + z_2 M_1 X_{z_1} \rightleftharpoons z_2 M_1^{z_1} + z_1 M_2 X_{z_2} \tag{7}$$

The equilibrium quotient λ is given by

$$\lambda = \frac{\bar{x}^{z_1} \{M_1^{z_1^+}\}^{z_2}}{(1-\bar{x})^{z_2} \{M_2^{z_2^+}\}^{z_1}} = \frac{\bar{x}^{z_1} [M_1^{z_1^+}]^{z_2}}{(1-\bar{x})^{z_2} [M_2^{z_2^+}]^{z_1}} \left(\frac{y_1^{z_2}}{y_2^{z_1}} \right) \tag{8}$$

where $\bar{x}$ = mole fraction of $M_2 X_{z_2}$.

As before, the activity coefficient ratio in the aqueous phase $(y_1^{z_2}/y_2^{z_1})$ can be measured or estimated by suitable methods. If equivalent fractions are used to express concentrations in the ion exchanger, the equilibrium quotient q may be introduced:

$$q = \frac{\beta^{z_1} \{M_1^{z_1^+}\}^{z_2}}{(1-\beta)^{z_2} \{M_2^{z_2^+}\}^{z_1}} \tag{9}$$

where β = equivalent fraction of $M_2 X_{z_2}$.

The ion-exchange reaction can also be written

$$z_1 M_2^{z_2^+} + z_1 z_2 (M_1)_{1/z_1} X \rightleftharpoons z_2 M_1^{z_1^+} + z_1 z_2 (M_2)_{1/z_2} X \tag{10}$$

with equilibrium quotient κ' given by

$$\kappa' = \left(\frac{\beta}{1-\beta}\right)^{z_1 z_2} \frac{\{M_1^{z_1+z_2}\}}{\{M_2^{z_2+z_1}\}}$$ (11)

C. Free Energy, Enthalpy, and Entropy

The logarithm of the equilibrium quotient is related to the free energy by

$$\Delta G = -RT \ln \kappa \quad (\lambda, q, \text{etc.})$$ (12)

Very few direct enthalpy measurements have been reported. In most instances ΔH has been estimated from the temperature dependence of the equilibrium quotient. Entropy estimates have then been obtained by employing Eq. (13).

$$\Delta S^\circ = \frac{\Delta H^\circ - \Delta G^\circ}{T}$$ (13)

The procedure for employment of $\log \kappa$ or $\ln \kappa$ (λ, q, etc.) for fitting ion-exchange data to the model is presented next. From Eq. (2),

$$\log \kappa = \log \kappa(1)\bar{x} + \log \kappa(0)(1 - \bar{x}) + B\bar{x}(1 - \bar{x})$$ (14)

where $\bar{x} = \bar{x}_{M_2 X}$. Similar expressions are obtained for other equilibrium quotients.

The parameter $\log \kappa(0)$ is the limiting value of $\log \kappa$ for $\bar{x} = 0$, and $\log \kappa(1)$ is that for $\bar{x} = 1$. These parameters are not available by direct experiment. They can be estimated, however, by tracer techniques.

The third parameter, $\log \kappa_m$, is computed with Eq. (3):

$$\log \kappa_m = \frac{1}{2} [\log \kappa(0) + \log \kappa(1) + B]$$ (15)

The integral free energy of reaction (4) is accessible through the equilibrium constant that is computed with Eq. (16):

$$\log K = \int_0^1 \log \kappa(\bar{x}) \, d\bar{x} = \frac{1}{3}[\log \kappa(0) + \log \kappa(1) + \log \kappa_m]$$ (16)

Equations (14)-(16) are applicable to all equilibrium quotients defined in Eqs. (5), (8), (9), and (11).

Observe that [9,10]

$$\log K = \int_0^1 \log \lambda(\bar{x}) \, d\bar{x} = \int_0^1 \log \kappa'(\beta) \, d\beta \qquad (17a)$$

$$\log Q = \int_0^1 \log q(\beta) \, d\beta = \log K + \frac{z_2 - z_1}{\ln 10} \qquad (17b)$$

Equation (16) is obtained by introducing Eq. (14) into the equation

$$\log K = \int_0^1 \log \kappa(\bar{x}) \, d\bar{x} \qquad (18)$$

given by Argersinger et al. [8], by integrating the resulting expression termwise, and by substituting B from Eq. (15).

When treating protonation equilibria of weak-acid and chelating resins, the independent variable is chosen as the fraction of metal ion in the resin, α, when titrating the acid form with the metal hydroxide, MeOH; i.e.,

$$\alpha = \frac{n_{MeX}}{s_0} = \bar{x}_{RMe} \qquad (19)$$

where n_{MeX} = number of millimoles of MeX, and s_0 = capacity of sample in milliequivalents. For titrations of polyelectrolytes the same independent variable is used when estimating the acidity constant, i.e., the equilibrium constant of the reaction

$$-RH \rightleftharpoons -R^- + H^+(aq) \qquad (20)$$

D. Activity Coefficients

1. *Monovalent Ions*

Equation (14) can be written

$$\log \kappa = \alpha + b\bar{x} + c\bar{x}^2 \qquad (21)$$

From Eq. (5) and the definition of K,

$$K = \kappa\left(\frac{f_2}{f_1}\right) \tag{22}$$

From the Gibbs-Duhem equation

$$\bar{x} \, d \ln f_2 + (1 - \bar{x}) \, d \ln f_1 = 0 \quad (p, \, T \, const.) \tag{23}$$

By using the pseudobinary approach with the pure components as standard and reference states and Eqs. (21)-(23), integration gives

$$\log f_1 = \frac{1}{2} b\bar{x}^2 + \frac{2}{3} c\bar{x}^3 \tag{24a}$$

$$\log f_2 = \frac{1}{2} (b + 2c)(1 - \bar{x})^2 - \frac{2}{3} c(1 - \bar{x})^3 \tag{24b}$$

The constants b and c are related to the parameters of the model by

$$b = 2[\log \kappa_m - \log \kappa(0)] \tag{25a}$$

$$c = \log \kappa(0) + \log \kappa(1) - 2 \log \kappa_m \tag{25b}$$

Observe that when $c = 0$ a straight-line relationship, $\log \kappa = f(\bar{x})$, is the result, and the expressions for the activity coefficients reduce to those of a regular solution:

$$\log f_1 = \frac{b\bar{x}^2}{2}$$

$$\log f_2 = \frac{b(1 - \bar{x})^2}{2}$$

2. Multivalent Ions

Treatment of the functions $\kappa' = f(\beta)$ and $\lambda = f(\beta)$ lead to simple activity expressions while other possibilities with $\bar{x}$ as independent variable or $q = f(\beta)$ do not. The first two cases are considered next.

   a. $\kappa' = f(\beta)$. The equilibrium constant, K, of reaction (10) is related to κ' by

$$K = \kappa'\left(\frac{\gamma_2}{\gamma_3}\right)^{z_1 z_2} \tag{26}$$

The Gibbs-Duhem equation gives

$$\beta_1 \; d \ln \gamma_1 + \beta_2 \; d \ln \gamma_2 = 0 \qquad (p, \; T \; const.) \tag{27}$$

According to the model

$$\log \kappa' = a + b\beta + c\beta^2 \tag{28}$$

with b and c related to the parameters of the model by Eqs. (25a,b). From Eqs. (26)-(28) with $\beta = \beta_2$,

$$\log \gamma_1 = \frac{1}{z_1 z_2} \left[\frac{1}{2} b\beta^2 + \frac{2}{3} c\beta^3 \right] \tag{29a}$$

$$\log \gamma_2 = \frac{1}{z_1 z_2} \left[\frac{1}{2} (b + 2c)(1 - \beta)^2 - \frac{2}{3} c(1 - \beta)^3 \right] \tag{29b}$$

b. $\lambda = f(\beta)$. It has been shown elsewhere [9,10] that it is convenient to use λ as a function of β when evaluating activity coefficients. The equilibrium constant, K, of reaction (7), with mole fraction as concentration unit, is related to λ by

$$K = \lambda \left(\frac{g_2^{z_1}}{g_1^{z_2}} \right) \tag{30}$$

From the Gibbs-Duhem equation

$$z_2 \beta_1 \; d \log g_1 + z_1 \beta_2 \; d \log g_2 = 0 \qquad (p, \; T \; const.)$$

This equation, together with Eq. (28) with λ substituted for κ' in Eq. (30), gives

$$\log g_1 = \frac{1}{z_2} \left[\frac{1}{2} b\beta^2 + \frac{2}{3} c\beta^2 \right] \tag{31a}$$

$$\log g_2 = \frac{1}{z_1} \left[\frac{1}{2} (b + 2c)(1 - \beta)^2 - \frac{2}{3} c(1 - \beta)^3 \right] \tag{31b}$$

Equations (29a,b) and (31a,b) reduce to Eqs. (24a,b) for $z_1 = z_2 = 1$.

The model requires that $\log \kappa'$ and $\log \lambda$ have finite limits. It is shown elsewhere [9,10] that this may not always be true. Selection of one or the other needs to be based on plots of the respective functions. The limited experience of this author indicates that $\lambda = f(\beta)$ has the desired properties.

If one of the curves can be fitted to the model, the thermodynamic properties of the other can be computed from [9,10]

$$g_1 \bar{x}_1 = (\gamma_1 \beta_1)^{z_1} \qquad (32a)$$

$$g_2 \bar{x}_2 = (\gamma_2 \beta_2)^{z_2} \qquad (32b)$$

The two concentration units are related by [9,10]

$$\frac{1 - \beta}{\beta} = \frac{(1 - \bar{x})z_1}{\bar{x}z_2} \qquad (33)$$

E. Water Uptake

The number of water molecules per equivalent of exchanger, W, is given by

$$W = \frac{n_{H_2O}}{s_0} \qquad (34)$$

As with Eq. (19), s_0 = capacity of sample in milliequivalents. For liquid ion exchangers

$$n_{H_2O} = n_{tot} - n_{solv}$$

where n_{solv} is the number of millimoles of water extracted by the solvent (diluent) containing the liquid ion exchanger. For solid resins

$$n_{H_2O} = n_{tot}$$

From the model

$$W = w(1)\bar{x} + w(0)(1 - \bar{x}) + B\bar{x}(1 - \bar{x}) \qquad (35)$$

The water uptake of the mixed-ionic form is estimated from Eq. (3):

$$w_m = \frac{1}{2} [w(0) + w(1) + B] \qquad (36)$$

Observe that $w(0)$ and $w(1)$, the water extracted by the two pure ionic forms, can be determined by separate experiments, leaving only one unknown parameter to be determined. The water uptake can thus be used to test the validity of the model.

IV. APPLICATIONS

A. Some Statistical Criteria

To examine the applicability of the model, the goodness of fit that such treatment of the experimental data provides will be determined by the following criteria obtained from statistical analysis of the residuals.

Residual squares sum

$$U = \sum_1^n (y_{exp} - y_{calc})^2 = \Sigma\, r_i^2 \tag{37}$$

where n = number of experimental points, and r_i = i-th residual.

Residual mean

$$|r| = \frac{\Sigma_1^n |r_i|}{n} \tag{38}$$

Mean residual

$$\bar{r} = \frac{\Sigma_1^n r_i}{n} \tag{39}$$

Variance

$$m_{r,2} = \frac{\Sigma\, r_i^2}{n} - \left[\frac{\Sigma\, r_i}{n}\right]^2 \tag{40}$$

Standard deviation

$$s(Y) = \sqrt{\frac{U}{n-k}} \tag{41}$$

where k = number of parameters estimated.   Sometimes

$$s'(Y) = \sqrt{\frac{U}{n-1}} \tag{42}$$

is used to express the standard deviation.

Hamilton R-factor in percent

$$R(\%) = 100 \sqrt{\frac{\Sigma\, r_i^2}{\Sigma\, Y_{exp}^2}} \tag{43}$$

The summations in Eqs. (40)-(43) are taken from 1 to n as in Eqs. (37)-(39). The weighting factors are taken as unity.

B. Liquid Ion Exchangers

1. *Free Energy*

The model was first applied to the liquid ion exchanger dinonylnaphthalene sulfonic acid (HD) with general structure

(I)

For HD, $R = C_9H_{19}$ and for didodecylnaphthalene sulfonic acid (HDDNS), also studied by Soldatov et al., $R = C_{12}H_{25}$. The commercial products of HD and HDDNS are isomer mixtures. In all examples in this chapter the concentration of HD and HDDNS in the organic phase is 0.100 mol dm^{-3}. In the aqueous phase the ionic strength is kept at 0.100 mol dm^{-3}.

   a. The Reaction $CH_3NH_3^+ + HD(org) \rightleftharpoons CH_3NH_3D(org) + H^+$.

Mikulich [13] studied the equilibrium distribution of various organic ammonium and hydrogen ions between aqueous medium and HD in various solvents and at different temperatures. Application of the model to these data has been published elsewhere [14]. Among the systems studied is the methylammonium-hydrogen exchange in heptane at 0°C, 25°C, and 63°C. In Fig. 1 log κ is plotted versus $\bar{x} = \bar{x}_{MeNH_3D}$. The data can be fitted to straight lines. As indicated in Fig. 1, log $\kappa(1)$ seems to be practically independent of temperature, giving

$$\log \kappa(1) = 0.00 \; (\mp 0.02) \tag{44a}$$

Data for the other parameters were fitted to $1/T$ by linear regression, giving

$$\log \kappa(0) = -0.04 + \frac{464}{T} \tag{44b}$$

In Table 1 experimental and computed log κ values are compared. The statistical analysis indicates an acceptable fit. Thus, both composition and temperature dependence of the free energy can be satisfactorily

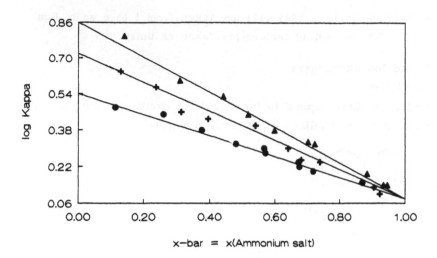

FIG. 1.   Log κ plotted versus $\bar{x}_{CH_3NH_3D}$ for the system $CH_3NH_3^+$ -H^+ on HD in heptane at various temperatures: $\blacktriangle$, 0°C; +, 25°C; $\bullet$, 63°C.   Ionic strength = 0.1 mol dm^{-3} (CH_3NH_3,H)Cl.   The curves have been computed from Eqs. (44a,b).   (Reproduced with permission from *Chemica Scripta*.)

TABLE 1   Comparison between Experimental and Computed log κ Values for the Exchange of $CH_3NH_3^+$ and H^+ Ions on HD in Heptane at 0°C, 25°C, and 63°C

$\bar{x}$	log κ exp	log κ calc	Residual analysis	
			0°C	
0.140	0.80	0.751	Residual-squares sum	4.90×10^{-3}
0.310	0.60	0.618	Mean residual	1.92×10^{-2}
0.442	0.53	0.515	Residual mean	1.14×10^{-2}
0.517	0.45	0.457	Variance	3.60×10^{-4}
0.598	0.38	0.394	Standard deviation	0.025
0.702	0.33	0.312	Hamilton R-factor (%)	5.06
0.723	0.32	0.296		
0.884	0.19	0.170		
0.934	0.14	0.131		
0.946	0.14	0.122		

TABLE 1   (continued)

			25°C		
0.129	0.64	0.637	Residual-squares sum	7.81×10^{-3}	
0.237	0.57	0.568	Mean residual	2.18×10^{-2}	
0.314	0.46	0.519	Residual mean	-1.56×10^{-2}	
0.395	0.43	0.467	Variance	5.37×10^{-4}	
0.540	0.40	0.374	Standard deviation	0.031	
0.640	0.30	0.310	Hamilton R-factor (%)	7.15	
0.681	0.25	0.284			
0.739	0.24	0.247			
0.904	0.13	0.141			
0.923	0.10	0.129			
			63°C		All data[a]
0.113	0.48	0.488	Residual-squares sum	1.92×10^{-3}	1.46×10^{-2}
0.260	0.45	0.420	Mean residual	1.13×10^{-2}	1.74×10^{-2}
0.376	0.38	0.367	Residual mean	5.70×10^{-3}	5.00×10^{-4}
0.480	0.32	0.319	Variance	1.59×10^{-4}	4.87×10^{-4}
0.566	0.30	0.280	Standard deviation	0.015	0.023
0.570	0.28	0.278	Hamilton R-factor (%)	4.34	5.73
0.672	0.24	0.231			
0.675	0.22	0.230			
0.718	0.20	0.210			
0.869	0.15	0.140			

[a]Here all data at the three temperatures are treated together.
Source: Data from Soldatov et al. [14].

described with the three constants in Eqs. (44a,b) over the temperature studied.

b. *The Reaction* $K^+ + HX(org) \rightleftharpoons KX(org) + H^+$. Kuvaeva et al. [15] studied the potassium-hydrogen exchange of HD and HDDNS at 298 K with heptane as solvent. By least-squares fitting to Eq. (14),

$$\log \kappa = -0.89\bar{x} + 0.12(1 - \bar{x}) + 0.79\bar{x}(1 - \bar{x}) \quad \text{(HD)} \quad (45)$$
$$\log \kappa = -0.69\bar{x} + 0.12(1 - \bar{x}) \quad \text{(HDDNS)} \quad (46)$$

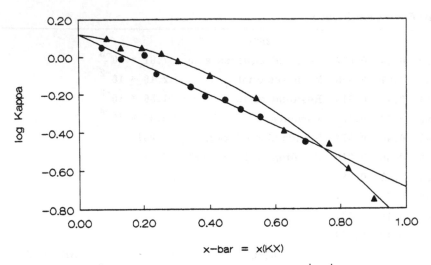

FIG. 2. Log κ plotted versus $\bar{x}_{KX}$ for the K^+-H^+ system on HD and HDDNS in heptane. Temperature = 298 K. Ionic strength = 0.10 mol dm^{-3} (K,H)Cl: ▲ , HD; ●, HDDNS. The curves have been computed from Eqs. (45) and (46). (Reproduced with permission from *Solvent Extraction and Ion Exchange.*)

In Fig. 2, log κ is plotted versus $\bar{x} = \bar{x}_{KX}$. The curves are obtained using Eqs. (45) and (46). In Table 2, experimental and computed log κ values are compared. An acceptable fit is found.

   c.   *The Reaction* $(1/2)Co^{2+} + HDDNS(org) \rightleftharpoons CoDDNS\frac{1}{2}(org) + H^+$. Kuvaeva et al. [16] studied the exchange of Co(II) and H^+ ions in the HDDNS, heptane water system at 298 K. The reaction may be written

$$\frac{1}{2} Co^{2+} + HDDNS(org) \rightleftharpoons Co_{1/2}DDNS(org) + H^+ \qquad (47)$$

For the equilibrium quotient the following expression was used:

$$\kappa_2 = \frac{\beta^{1/2}\{H^+\}}{(1-\beta)\{Co^{2+}\}^{1/2}} = \kappa^* \left(\frac{y_{H^+}}{y_{Co^{2+}}^{1/2}} \right) \qquad (48)$$

In Fig. 3, log κ^* is plotted versus β. The straight line, obtained by linear regression, is given by

$$\log \kappa^* = -0.05\beta + 0.01(1 - \beta) \qquad (49)$$

TABLE 2 Comparison between Experimental and Computed log κ Values for the Exchange of K^+ and $-H^+$ on HD and HDDNS in Heptane at 298 K

x̄	log κ exp	log κ calc	Residual analysis	
			HD	
0.088	0.10	0.095	Residual-squares sum	8.40×10^{-3}
0.130	0.05	0.078	Mean residual	1.89×10^{-2}
0.195	0.05	0.047	Residual mean	-4.00×10^{-3}
0.255	0.02	0.013	Variance	7.48×10^{-4}
0.305	-0.02	-0.021	Standard deviation	0.032
0.400	-0.10	-0.094	Hamilton R-factor (%)	7.89
0.540	-0.22	-0.229		
0.624	-0.39	-0.325		
0.764	-0.46	-0.509		
0.824	-0.59	-0.598		
0.903	-0.75	-0.723		
			HDDNS	
0.072	0.05	0.062	Residual-squares sum	4.49×10^{-3}
0.131	-0.01	0.014	Mean residual	1.56×10^{-2}
0.203	0.01	-0.044	Residual mean	-1.00×10^{-3}
0.239	-0.09	-0.074	Variance	4.48×10^{-4}
0.343	-0.16	-0.158	Standard deviation	0.024
0.386	-0.21	-0.193	Hamilton R-factor (%)	9.32
0.446	-0.23	-0.241		
0.493	-0.28	-0.279		
0.553	-0.32	-0.328		
0.690	-0.45	-0.439		

Source: Data from Kuvaeva et al. [15].

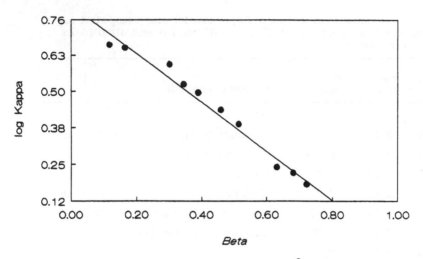

FIG. 3. Log κ^* plotted versus β for the $Co^{2+}-H^+$ system on HDDNS in heptane. Temperature = 298 K. Ionic strength = 0.10 equiv dm^{-3} (Co,H)Cl. The straight line is that of Eq. (49). (Reproduced with permission from *Hydrometallurgy*.)

TABLE 3 Comparison between Experimental and Computed log κ^* Values for the Exchange of Co^{2+} and H^+ on HDDNS in Heptane at 298 K

β	log κ^* exp	log κ^* calc	Residual analysis	
0.117	0.67	0.709	Residual-squares sum	6.84×10^{-3}
0.164	0.66	0.669	Mean residual	2.27×10^{-2}
0.301	0.60	0.551	Residual mean	4.50×10^{-3}
0.345	0.53	0.513	Variance	6.64×10^{-4}
0.389	0.50	0.475	Standard deviation	0.029
0.458	0.44	0.416	Hamilton R-factor (%)	5.50
0.513	0.39	0.369		
0.630	0.24	0.268		
0.680	0.22	0.225		
0.721	0.18	0.190		

Source: Data from Kuvaeva et al. [16].

In Table 3 experimental and computed log κ values are compared. A fair fit is found.

In the original paper the second degree polynomial was used. Use of the straight line is acceptable since the standard deviation, $s(\log \kappa) = \mp 0.03$ is only slightly larger than the ∓ 0.02 value associated with the second-degree polynomial.

2. Water Uptake

In the course of ion exchange in liquid ion exchangers water is extracted by the organic diluent. This phenomenon has sometimes been studied together with the ion-exchange equilibria. As mentioned above, for liquid ion exchangers the water extracted by the diluent has to be subtracted from the total.

When applying the model to water extraction only one parameter needs to be determined in cases where $w(0)$ and $w(1)$ have already been measured in separate experiments. Using data in the range $0.1 < \bar{x} < 9$, B is computed as

$$B = \frac{W_{exp} - w(0)(1 - \bar{x}) - w(1)\bar{x}}{\bar{x}(1 - \bar{x})} \tag{50}$$

The average B-value is then used in Eq. (35) to compute W. For systems where the limiting values have not been determined the data are treated in the same way as they were for the free energy.

a. The Reaction $K^+ + HX(org) \rightleftharpoons KX(org) + H^+$. Kuvaeva et al. [15] also measured the water uptake during the exchange of K^+ and H^+ ions with HD and HDDNS in heptane at 298 K. In Fig. 4, W is plotted versus $\bar{x} = \bar{x}_{KX}$ for the two sulfonic acids. By using the average B-value the following equations were obtained:

$$W_I = 7.6\bar{x} + 9.8(1 - \bar{x}) - 8.2\bar{x}(1 - \bar{x}) \quad \text{(HD)} \tag{51}$$

$$W_I = 6.2\bar{x} + 7.5(1 - \bar{x}) - 5.4\bar{x}(1 - \bar{x}) \quad \text{(HDDNS)} \tag{52}$$

By least-squares fitting to all data,

$$W_{III} = 7.5\bar{x} + 9.6(1 - \bar{x}) - 7.5\bar{x}(1 - \bar{x}) \quad \text{(HD)} \tag{53}$$

$$W_{III} = 7.2\bar{x} + 7.4(1 - \bar{x}) - 5.1\bar{x}(1 - \bar{x}) \quad \text{(HDNNS)} \tag{54}$$

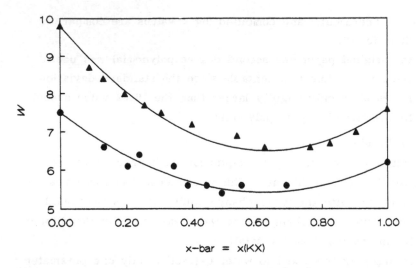

FIG. 4.   W plotted versus $\bar{x}_{KX}$ for the K^+-H^+ system on HD and HDNNS in haptane.   Temperature = 298 K.   Ionic strength = 0.10 mol dm^{-3} (K,H)Cl: ▲, HD; ●, HDDNS.   The curves have been computed from Eqs. (51) and (52).   (Reproduced with permission from *Solvent Extraction and Ion Exchange*.)

In Table 4 experimental and computed W-values are compared. Practically the same fit is obtained with one or three unknown parameters.

   b.   The Reaction $(1/2)Co^{2+}$ + HDDNS(org) $\rightleftharpoons$ $Co_{1/2}$ DDNS(org) + H^+.   Water uptake was also measured in the study of the Co(II)-H^+ system by Kuvaeva et al. [16].   By using the average B-value the following expression was obtained:

$$W_I = 7.63\beta + 6.45(1 - \beta) - 2.22\beta(1 - \beta) \tag{55}$$

and by nonlinear regression of all data

$$W_{III} = 7.60\beta + 6.49(1 - \beta) - 2.28\beta(1 - \beta) \tag{56}$$

Experimental and computed W-values are compared in Table 5.   The fit is about the same for the two equations.

TABLE 4 Comparison between Experimental and Computed W-Values for the Exchange of K^+ and H^+ on HD and HDDNS in Heptane at 298 K

$\bar{x}$	W exp	W_I	W_{III}	Residual analysis	No. of parameters 1	No. of parameters 3
				HD		
0	9.8	—	9.60	Residual-squares sum	0.314	0.197
0.088	8.7	8.95	8.85	Mean residual	0.142	0.109
0.130	8.4	8.59	8.52	Residual mean	3.64×10^{-3}	-3.08×10^{-3}
0.195	8.0	8.08	8.08	Variance	2.85×10^{-2}	1.51×10^{-2}
0.255	7.7	7.68	7.72	Standard deviation	0.177	0.140
0.305	7.5	7.39	7.45	Hamilton R-factor (%)	2.28	1.61
0.400	7.2	6.95	7.06			
0.540	6.9	6.58	6.70			
0.624	6.6	6.50	6.62			
0.764	6.6	6.64	6.72			
0.824	6.7	6.80	6.84			
0.903	7.0	7.10	7.08			
1	7.6	—	7.50			
				HDDNS		
0	7.5	—	7.40	Residual-squares sum	0.255	0.241
0.131	6.6	6.71	6.66	Mean residual	0.151	0.124
0.203	6.1	6.36	6.33	Residual mean	1.78×10^{-2}	2.36×10^{-2}
0.239	6.4	6.21	6.19	Variance	2.80×10^{-2}	2.13×10^{-2}
0.343	6.1	5.84	5.84	Standard deviation	0.179	0.173
0.386	5.6	5.72	5.73	Hamilton R-factor (%)	2.85	2.43
0.446	5.6	5.59	5.60			
0.493	5.4	5.51	5.53			
0.553	5.6	5.45	5.48			
0.690	5.6	5.45	5.48			
1	6.2	—	6.20			

Source: Data from Kuvaeva et al. [15].

TABLE 5 Comparison between Experimental and Computed W-Values for the Exchange of Co^{2+} and H^+ on HDDNS in Heptane at 298 K

β	W exp	W_I	W_{III}	Residual analysis	No. of parameters	
					1	3
0	6.45	—	6.49	Residual-squares sum	7.64×10^{-2}	7.15×10^{-2}
0.117	6.30	6.36	6.38	Mean residual	7.00×10^{-2}	5.92×10^{-2}
0.164	6.54	6.34	6.36	Residual mean	-4.00×10^{-3}	-8.34×10^{-4}
0.301	6.30	6.34	6.34	Variance	7.62×10^{-3}	5.96×10^{-3}
0.345	6.30	6.36	6.36	Standard deviation	9.21×10^{-2}	8.91×10^{-2}
0.389	6.36	6.38	6.38	Hamilton R-factor (%)	1.345	1.170
0.458	6.57	6.44	6.43			
0.513	6.45	6.50	6.49			
0.630	6.66	6.68	6.66			
0.680	6.72	6.77	6.75			
0.721	6.78	6.85	6.83			
1	7.63	—	7.60			

Source: Data from Kuvaeva et al. [16].

C. Organic Ion Exchangers

Similar characterization of organic resins with the three-parameter model is illustrated by such treatment of data compiled for strong-acid, strong-base as well as weak-acid and chelating resins.

3. Strong-Acid Resins

  a. The Reaction $H^+ + LiR \rightleftharpoons HR + Li^+$. Bonner and Overton [17] studied the hydrogen-lithium exchange of 4%, 8%, and 16% DVB crosslinked Dowex 0°C, 25°C, 50°C, 75°C, and 98°C. The data used in this chapter were obtained by interpolation of Figs. 1-3 in their paper. Parameters obtained by nonlinear regression are listed in Table 6 together with some of the quantities from the residual analysis. The 45 parameters given in Table 6 are needed to describe the above system at the different temperatures and crosslinkings. This number is greatly reduced in the following way. For each crosslinking the parameters of the model are fitted by linear regression to the van't Hoff equation

TABLE 6   Parameters Obtained by Fitting the Model to Ion-Exchange Data Computed for the H^+-Li^+-Dowex 50 Resin Systems

% DVB	Temp. (°C)	log κ(0)	log κ(1)	log κ_m	log K	U	R(%)	s(Y)
4	0	0.173	0.167	0.047	0.129	2.10×10^{-5}	1.45	0.003
4	25	0.128	0.154	0.049	0.110	9.40×10^{-5}	2.72	0.004
4	50	0.113	0.138	0.023	0.091	1.15×10^{-4}	4.88	0.006
4	75	0.104	0.125	0.001	0.076	2.15×10^{-4}	7.02	0.007
4	98	0.083	0.087	-0.003	0.056	8.40×10^{-5}	6.10	0.005
8	0	0.178	0.151	0.016	0.115	1.40×10^{-5}	1.51	0.003
8	25	0.141	0.120	0.030	0.097	1.00×10^{-5}	1.22	0.002
8	50	0.114	0.084	0.030	0.076	6.00×10^{-6}	1.36	0.001
8	75	0.088	0.065	0.011	0.055	1.04×10^{-4}	6.54	0.005
8	98	0.065	0.040	0.001	0.035	2.90×10^{-5}	5.28	0.002
16	0	0.370	0.145	0.030	0.181	1.72×10^{-4}	3.07	0.008
16	25	0.301	0.114	0.033	0.149	5.90×10^{-5}	1.58	0.003
16	50	0.234	0.103	0.027	0.121	2.00×10^{-5}	1.53	0.003
16	75	0.182	0.069	0.048	0.100	1.91×10^{-4}	5.64	0.008
16	98	0.157	0.046	0.024	0.076	1.50×10^{-5}	2.04	0.002

Source: Data from Bonner and Overton [17].

$$\log \kappa = a + \frac{b}{T} \tag{57}$$

This operation leads to the 18 constants given in Table 7 and provides the parameters and fit shown in Table 8. When comparing Tables 6 and 8, it is seen that the fit, while still acceptable, is poorer in Table 8.

In Figs. 5 and 6 experimental and computed log κ values are compared using the parameters in Table 6.

2.   *Strong-Base Resins*

a.   *The Reaction* $1 + RNO_3 \rightleftharpoons RI + NO_3^-$.   Gärtner [18], in his studies, included examination of the iodide-nitrate exchange on Wofatit SB and a pyridine resin at 298 K and an ionic strength, I, of 0.100 M $K(I, NO_3)$ in the aqueous phase. The data given in Table 9 for these systems were obtained by interpolation of Figs. 1 and 2 in Gärtner's

TABLE 7   Parameters Obtained by Least-Squares Fitting to
Eq. (57)

% DVB	log $\kappa(0)$ a	log $\kappa(0)$ b	log $\kappa(1)$ a	log $\kappa(1)$ b	log κ_m a	log κ_m b
4	-0.148	85.5	-0.103	75.7	-0.166	60.3
8	-0.244	115.3	-0.267	114.5	-0.038	17.7
16	-0.463	227	-0.216	99.3	0.032	0

TABLE 8   Parameters Obtained by Using Eq. (57) and the Constants
in Table 7

% DVB	Temp. (°C)	log $\kappa(0)$	log $\kappa(1)$	log κ_m	log K	U	R(%)	s(Y)
4	0	0.165	0.174	0.055	0.131	1.83×10^{-4}	4,27	0.008
4	25	0.139	0.151	0.036	0.109	3.18×10^{-4}	5.00	0.007
4	50	0.117	0.131	0.021	0.090	1.94×10^{-4}	6.29	0.008
4	75	0.098	0.114	0.007	0.073	3.38×10^{-4}	8.81	0.009
4	98	0.082	0.101	-0.004	0.062	3.17×10^{-4}	11.85	0.009
8	0	0.178	0.152	0.027	0.119	1.15×10^{-4}	4.32	0.008
8	25	0.143	0.117	0.021	0.094	7.20×10^{-5}	3.27	0.004
8	50	0.113	0.087	0.017	0.072	1.02×10^{-4}	5.62	0.006
8	75	0.087	0.062	0.013	0.054	1.12×10^{-4}	6.79	0.005
8	98	0.067	0.041	0.010	0.039	1.58×10^{-4}	12.32	0.006
16	0	0.368	0.147	0.032	0.182	1.90×10^{-4}	3.23	0.008
16	25	0.298	0.117	0.032	0.149	8.60×10^{-5}	1.91	0.004
16	50	0.239	0.091	0.032	0.121	1.66×10^{-4}	4.40	0.007
16	75	0.189	0.069	0.032	0.097	3.05×10^{-4}	7.12	0.010
16	98	0.149	0.052	0.032	0.078	1.37×10^{-4}	6.18	0.007

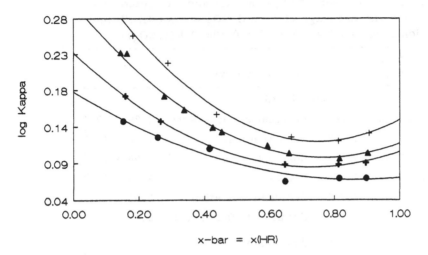

FIG. 5. Log κ plotted versus $\bar{x}_{HR}$ for the H^+-Li^+ system on Dowex 50 with 16% DVB at four temperatures. Ionic strength = 0.10 mol dm^{-3} (Li,H)Cl: +, 0°C; ▲, 25°C; +, 50°C; ●, 75°C. The curves have been computed with parameters in Table 6.

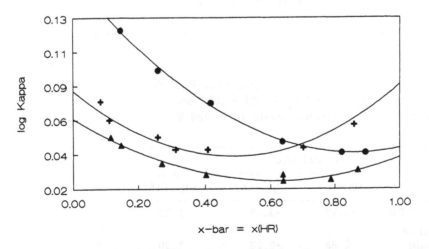

FIG. 6. Log κ plotted versus $\bar{x}_{HR}$ for the H^+-Li^+ system on Dowex 50 with three crosslinkings at 298 K. Ionic strength = 0.10 mol dm^{-3} (Li,H)Cl: +, 4% DVB; ▲, 8% DVB; ●, 16% DVB. The curves have been computed with parameters in Table 6.

TABLE 9 Comparison of Experimental and Computed ln κ
Values for the Exchange of I⁻ and NO₃⁻ on Wofatit SB and
a Pyridine Resin [T = 298 K, I = 0.100 M K(I,NO₃)]

$\bar{x}_{RI}$	ln κ exp	ln κ calc	Residual analysis	
		Wofatit BS		
0.150	1.18	1.186	Residual-squares sum	4.28×10^{-3}
0.362	1.52	1.492	Mean residual	2.36×10^{-2}
0.491	1.47	1.521	Residual mean	-8.00×10^{-4}
0.659	1.45	1.421	Variance	8.55×10^{-4}
0.863	1.06	1.064	Standard deviation	0.046
			Hamilton R-factor (%)	2.17
		Pyridine resin		
0.240	1.39	1.363	Residual-squares sum	1.93×10^{-3}
0.251	1.35	1.365	Mean residual	1.47×10^{-2}
0.365	1.35	1.370	Residual mean	-7.14×10^{-4}
0.480	1.34	1.335	Variance	2.74×10^{-4}
0.646	1.22	1.216	Standard deviation	0.022
0.804	1.04	1.027	Hamilton R-factor (%)	1.33
0.843	0.95	0.969		

Source: Data from Gärtner [18].

TABLE 10 Comparison of Integral Free Energies
of the Reaction I⁻ + RNO₃ ⇌ RI + NO₃⁻ on
Wofatit SB and a Pyridine Resin (T = 298 K,
I = 0.1 M)

Resin	ΔG° (kJ/equiv) Gärtner	Eq. (16)	ln K Eq. (18)
Wofatit SB	-3.18	-3.12	1.26
Pyridine resin	-2.93	-2.98	1.20

Source: Data from Gärtner [18].

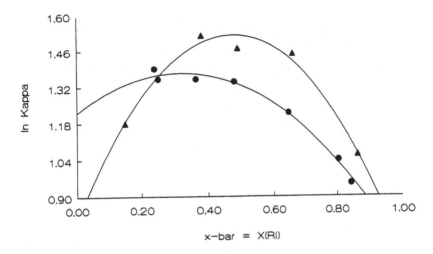

FIG. 7. ln κ plotted versus $\bar{x}_{RI}$ for the $I^--NO_3^-$ system on two strong-base resins. Temperature = 298 K. Ionic strength = 0.10 mol dm^{-3} K(I,NO_3): ▲, Wofatit SB; ●, pyridine resin. The curves have been computed from Eqs. (58) and (59). (Reproduced with permission from Z. Phys. Chem. (Leipzig).)

paper. By least-squares fitting of the model to these data the following equations were obtained ($\bar{x} = \bar{x}_{RI}$):

$$\ln \kappa = 0.68\bar{x} + 0.01(1 - \bar{x}) + 3.10\bar{x}(1 - \bar{x}) \quad \text{(Wofatit SB)} \quad (58)$$
$$\ln \kappa = 0.69\bar{x} + 1.22(1 - \bar{x}) + 1.48\bar{x}(1 - \bar{x}) \quad \text{(pyridine} \quad (59)$$
$$\text{resin)}$$

Experimental and computed ln κ values are compared in Table 9. The reasonable fit inferred from the residual analysis is also illustrated by Fig. 7. The two curves are very similar, and the integral free energies are close. Such closeness of integral free energies is seen in Table 10, where they are given together with the values of Gärtner, who integrated the curves in Fig. 7 using Eq. (18).

3. *Carboxylate Resins*

   a. *The Reaction* HR $\rightleftharpoons$ R^- + H^+(aq). Novitskaya [19] studied the dissociation of the Soviet carboxylate resin KB-4 in the presence of NaCl. The experiments were carried out at 296 K and at four ionic strengths: 0.02, 0.10, 0.50, and 2.50 mol dm^{-3} (Na)Cl.

TABLE 11   Parameters and Fit for pK_a of KB-4 [T = 296 $\mp$ K, I = 0.02 - 2.5 mol dm^{-3} (Na)Cl]

I	$pK_a(0)$	$pK_a(1)$	pK_{am}	U	R(%)	s'(pH)
0.02	6.10	7.50	7.22	3.68×10^{-2}	1.01	0.08
0.10	5.72	6.71	6.62	1.72×10^{-2}	0.77	0.06
0.50	4.98	6.03	6.18	2.48×10^{-2}	1.08	0.07
2.50	4.57	5.36	5.68	7.92×10^{-3}	0.64	0.04

The pH values interpolated at particular α values from Fig. 1 of the original paper (estimated accuracy 0.05 to 0.10 in pH) were used to compute pK_a with the Henderson-Hasselbalch equation

$$pK_a = pH + \log \left(\frac{1 - \alpha}{\alpha} \right) \tag{60}$$

These pK_a values were fitted at each ionic strength to the model giving the parameters in Table 11. The residual-squares sum and Hamilton R-factor are also given together with S'(pH), the standard deviation in pH.

The same data were recently treated in a slightly different way [20]. The equilibrium quotient, κ, of the reaction

$$H^+ + NaR \rightleftharpoons HR + Na^+ \tag{61}$$

is given by

$$\log \kappa = pH + \log \left(\frac{1 - \alpha}{\alpha} \right) + \log[Na^+] \tag{62}$$

Equations (60) and (62) differ only by $\log[Na^+]$. Since the ionic strength I $\simeq$ [Na$^+$] is kept constant, the fit to either quantity is the same.

For reaction (61) the three parameters of the model were correlatable with I, giving

$$\log \kappa(0) = 4.47 + 0.31 \sqrt{I} \tag{63a}$$

$$\log \kappa(1) = 5.75 \mp 0.04 \tag{63b}$$

$$\log \kappa_m = 5.47 + 8.43 \sqrt{I} \tag{63c}$$

No such correlation is obtainable for pK_a. Additional research is needed to show if the correlation with $\sqrt{I}$ is fortuitous. The experimental and computed pK_a and pH values are compared in Figs. 8 and 9. The agreement between the measured and computed values is fair in view of the uncertainty in the interpolation of these data from the figure.

4. Chelating Resins

a. *The Reaction* $HR \rightleftharpoons R^- + H^+(aq)$. Krasner and Marinsky [21] studied the protonation of the Dowex Al containing iminodiacetate groups. The experiments were carried out at 298 ∓ 3 K and at a (Na)Cl molality of 0.10. In their paper the intrinsic dissociation constant was estimated by relating reaction (20) to the protons in the resin phase. Their pK_a (intrinsic) value of 2.8 is in reasonable agreement with that of the monomer 2.5-3.0 [22]. In this chapter the three-parameter model is applied to the data. The pK_a for reaction (20), computed with Eq. (60) is, in this instance, based on the proton presence in solution and the model is fitted to such data by nonlinear regression. The parameters and fit are given in Table 12. In Fig. 10 experimental and computed pK_a values are compared. A fair fit is found. The value 4.09 given in Table 12 refers to reaction (20) and is not directly comparable with the resin protonation-based estimates of Krasner and Marinsky. Krasner and Marinsky also measured the water uptake. In Table 12 parameters and fit are given for a straight line as well as a second-degree polynomial. Both curves give an acceptable fit. The second-degree polynomial is used in Fig. 10.

b. *Water Uptake by Sephadex* Marinsky et al. [23] titrated linear carboxymethyldextran with standard base at 298 K and I = 0.10 mol dm^{-3} (Na)ClO$_4$. In connection with this work the water uptake by Sephadex gels is given for comparison. These data have been treated by the model giving the parameters and fit reported in Table 13. In Fig. 11 experimental and computed W values for Sephadex CM-50 and CM-25 are computed. In view of the relatively large

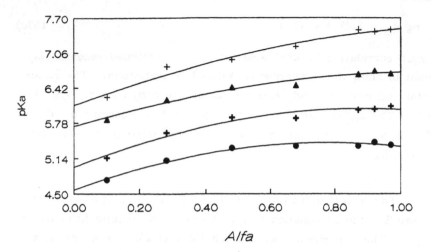

FIG. 8. pK_a plotted versus α for the carboxylate resin KB-4 at four ionic strengths.   Temperature = 296 ∓ 1 K: +, I = 0.02; ▲, I = 0.10; +, I = 0.50;● , I = 2.50.   The curves have been computed with parameters in Table 11.

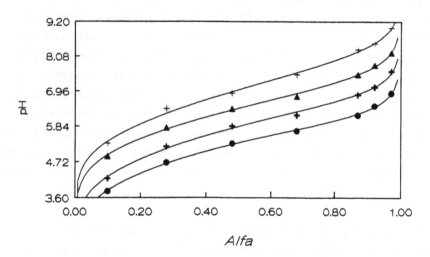

FIG. 9.   pH plotted versus α for KB-4.   Temperature = 296 ∓ 1 K: +, I = 0.02;▲ , I = 0.10; +, I = 0.50; ●; I = 2.50.   The curves have been computed with parameters in Table 11.   (Reproduced with permission from *Acta Chemica Scandinavica.*)

TABLE 12   Parameters and Fit for Dowex Al (T = 298 ± 3, I = 0.010 mol dm^{-1})

pK$_a$	pK$_a$(0)	pK$_a$(1)	pK$_m$	$\bar{\text{p}K}_a$ Eq. (16)	U	R (%)	s'(pH)
	4.27	4.47	3.54	4.09	7.60 × 10^{-3}	0.71	0.031
W	W(0)	W(1)	W$_m$	—	U	R(%)	s'(W)
2 p	7.24	36.16	21.70	—	3.41	2.68	0.65
3 p	5.62	34.54	24.50	—	1.01	1.46	0.36

Source: Data from Krasner and Marinsky [21].

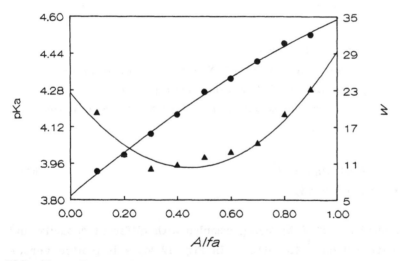

FIG. 10.   pK$_a$ and W plotted versus α for Dowex Al.   Temperature = 298 ∓ 3 K.   Ionic strength = 0.1 mol kg^{-1} (Na)Cl.   The curves have been computed with parameters in Table 12:   ▲, pK$_a$;   ●, W. (Reproduced with permission from *Journal of Physical Chemistry*.)

TABLE 13 Parameters and Fit for Water Uptake by Sephadex Gels
[T = 298 K, I = 0.10 mol dm^{-3} (Na)ClO$_4$]

Gel	W(0)	W(1)	W_m	U	R (%)	s(W)
CM-50	20.6	84.3	113.3	249.7	4.55	3.95
CM-25 2 p	17.9	60.4	39.2	23.41	4.01	1.83
CM-25 3 p	12.9	54.8	48.1	0.850	0.765	0.376

Source: Data from Marinsky et al. [23].

experimental uncertainty with the highly swelling CM-50, the fit is
acceptable. Table 13 shows that a straight line gives a fair fit to
the data for CM-25. The curve in Fig. 11 is that of the second-
degree polynomial.

5. Solvent-Impregnated Resins
    a. The Reaction NA$^+$ + HX $\rightleftharpoons$ NaX + H$^+$. Muraviev [24,25]
studied the properties of XAD-2 beads impregnated with HD. The
stability of the impregnated beads was of prime interest in this
study. The reaction

$$Na^+ + HX \rightleftharpoons NaX + H^+ \tag{64}$$
(HX = HD on bead)

was examined at 296 K by using samples with different capacity and
at I = 0.100 mol dm^{-3} (Na,H)Cl. In Fig. 12 log κ is plotted versus
$\bar{x}_{NaX}$ for four capacities. The curves in Fig. 12 have been computed from the model using the parameters given in Table 14.
    The selectivity is seen to change from preference of H$^+$ to a
preference of Na$^+$ with increasing capacity. Muraviev [25] explained
this selectivity reversal within the framework of the Gorshkov-
Eisenman theory [26,27]. The dependence of the parameters upon
capacity is illustrated in Fig. 13. The selectivity decreases with
s_0 and flattens out at high s_0-values in agreement with the model
discussed by Muraviev [25].

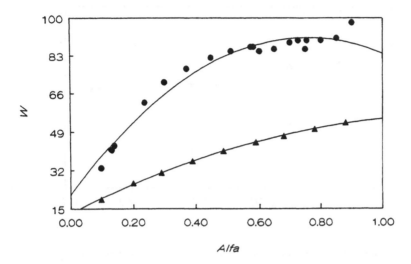

FIG. 11. W plotted versus α for Sephadex gels. Temperature =
298 K. Ionic strength = 0.10 mol dm^{-3} (Na)ClO$_4$: ●, CM-50;
▲, CM-25. The curves have been computed with parameters in
Table 13.

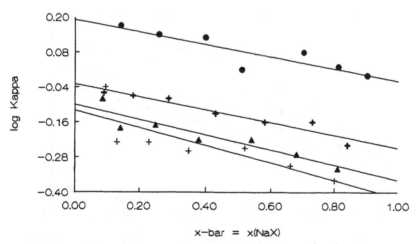

FIG. 12. Log κ plotted versus $\bar{x}_{NaX}$ for the Na$^+$-H$^+$ system on
XAD-2 impregnated with HD. Temperature = 293 $\mp$ 3 K. Ionic
strength = 0.10 mol dm^{-3} (Na,H)Cl: +, s_0 = 0.460; ▲, s_0 = 0.304;
+, s_0 = 0.159; ●, s_0 = 0.083. The curves have been computed
with parameters in Table 14. (Reproduced with permission from
Reactive Polymers.)

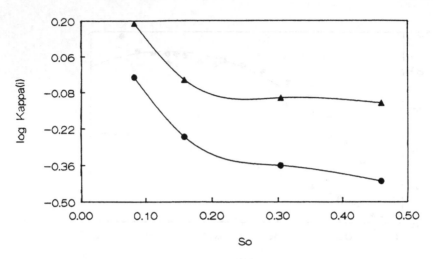

FIG. 13. Log $\kappa(0)$ and log $\kappa(1)$ plotted versus s_0 for impregnated
XAD-2: ▲, log (0); ●, log (1).

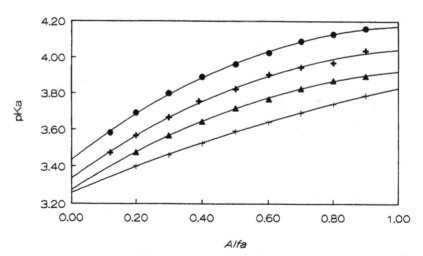

FIG. 14. pK_a plotted versus α for the Na^+-H^+ system on CmDx.
Temperature = 298 K.   Ionic strength 0.10 mol dm^{-3} (Na)Cl:
+, DS = 0.80; ▲, DS = 0.96; +, DS = 1.26; ●, DS = 1.70.   The
curves have been computed with parameters in Table 15.

TABLE 14 Parameters and Fit for XAD-2 Impregnated with HD
[$T \approx 293$ K, $I = 0.100$ mol dm^{-3} (Na,H)Cl]

S_0(meqv/g dry res)	log $\kappa(0)$	log $\kappa(1)$	log κ_m = log K	U	R (%)	s(log κ)
0.083 ± 0.001	0.19	-0.02	0.09	6.12×10^{-3}	28.94	0.03
0.159 ± 0.002	-0.03	-0.25	-0.14	1.91×10^{-3}	11.66	0.02
0.304 ± 0.002	-0.10	-0.36	-0.23	4.70×10^{-3}	11.76	0.03
0.460 ± 0.11	-0.12	-0.42	-0.27	2.04×10^{-2}	20.96	0.06

Source: Data from Muraviev and Högfeldt [24].

6. *Linear Polyelectrolytes*

a. *The Reaction* $R \rightleftharpoons R^- + H^+$. Marinsky et al. [28] studied
the protonation of linear carboxymethyldextran at 298 K and $I = 0.100$
mol dm^{-3} (Na)Cl. Data for samples with different degree of substitution (DS) were treated by the model. Parameters and fit are presented
in Table 15. In Fig. 14 the experimental data-based curves are compared with curves computed with the parameters in Table 15 to demonstrate the excellence of the fit. It can be concluded that the model
applies to uncrosslinked polymers as well as to gels.

According to the Gibbs-Donnan-based logic of Marinsky, $pK_a(0)$ is
an estimate of the intrinsic acidity constant, i.e., pK_a of the reaction

$$RH \rightleftharpoons R^- + H^+ \quad \text{(all in the polyelectrolyte "phase")} \quad (65)$$

TABLE 15 Parameters and Fit for CmDx with Different Degrees of
Substitution [$T = 298$ K, $I = 0.10$ mol dm^{-3} (Na)Cl]

DS	$pK_a(0)$	$pK_a(1)$	pK_{am}	$p\bar{K}_a$	U	R (%)	s(pK_a)
0.80	3.257	3.829	3.631	3.572	5.70×10^{-5}	0.074	0.083
0.96	3.273	3.922	3.833	3.676	8.20×10^{-5}	0.086	0.004
1.26	3.334	4.043	3.972	3.783	9.96×10^{-4}	0.277	0.013
1.70	3.433	4.172	4.131	3.912	2.70×10^{-4}	0.141	0.007
Average	3.324 ± 0.080						

Source: Data from Serjeant and Dempsey [22].

TABLE 16 Parameters and Fit for the Mn^{2+}, H^+, Stannic Arsenate System [T = 288 – 333 K, I = 0.01 mol dm^{-3} (Mn,H)Cl]

T °C	$\ln K_c(0)$ Nonlin. regress.	$\ln K_c(0)$ Eq. (67a)	$\ln K_c(1)$ Nonlin. regress.	$\ln K_c(1)$ Eq. (67b)	$\ln K_{cm}$ Nonlin. regress.	$\ln K_{cm}$ Eq. (67c)	U Nonlin. regress.	U Eq. (67)	R (%) Nonlin. regress.	R (%) Eq. (67)	$s(\ln K_c)$ Nonlin. regress.	$s(\ln K_c)$ Eq. (67)
15	1.018	1.024	2.291	2.241	0.969	0.876	9.38×10^{-3}	3.60×10^{-2}	2.06	4.04	0.037	0.072
25	0.939	0.930	1.941	1.967	0.716	0.825	7.84×10^{-3}	3.10×10^{-2}	2.36	4.70	0.033	0.067
40	0.799	0.799	1.518	1.587	0.720	0.755	2.35×10^{-3}	1.60×10^{-2}	1.56	4.06	0.018	0.048
60	0.642	0.644	1.186	1.137	0.716	0.671	1.08×10^{-2}	2.03×10^{-2}	4.05	5.56	0.039	0.054

Source: Data from Dabral and Högfeldt [29].

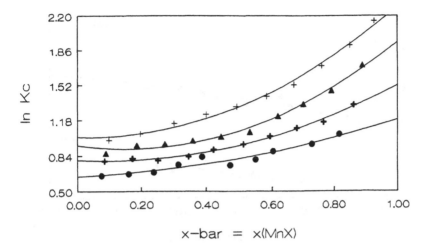

FIG. 15. Ln K_c plotted versus $\bar{x}_{MnX_2}$ for the Mn^{2+}-H^+ system on stannic arsenate at different temperatures. Ionic strength = 0.01 mol dm^{-3} (Mn,H)Cl: +, 15°C; ▲, 25°C; +, 40°C; ●, 60°C. The curves have been computed with parameters in Table 16. (Reproduced with permission from *Chemica Scripta*.)

In Table 15 the average $pK_a(0)$ = 3.324 agrees well with 3.32 used by Marinsky [23]. Observe that the present model treats the system as one phase as distinguished from the "two-phase" approach of Marinsky.

7. *Inorganic Ion Exchangers*

   a. *The Reaction* Mn^{2+} + 2HX $\rightleftharpoons$ MnX_2 + $2H^+$. Dabral [29] studied the exchange of several divalent metal ions for H^+ ion by equilibrating stannic arsenate at 15°C, 25°C, 40°C, and 60°C with 0.01 (Me)Cl. The Mn^{2+}-H^+ system is chosen for illustration. In Fig. 15, ln K_c of the reaction

$$Mn^{2+} + 2HX \rightleftharpoons MnX_2 + 2H^+ \tag{66}$$

is plotted versus $\bar{x}_{MnX_2}$. Equivalent fractions are used to express concentrations in both phases. The parameters were fitted to the van't Hoff equation, giving

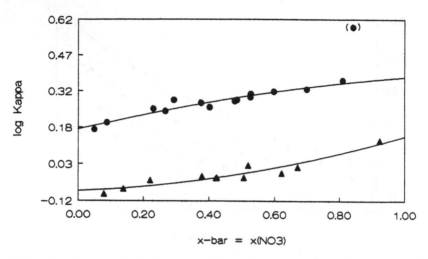

FIG. 16. Log κ plotted versus $\bar{x}_{NO_3}$ for the $Cl^--NO_3^-$ and $SCN^--NO_3^-$ systems on hydrous zirconia. Temperature = 298 K. Ionic strength = 0.10 mol dm^{-3} Na(A,NO$_3$): ▲, chloride; ●, thiocyanate. The curves have been computed with parameters in Table 17. The curve for thiocyanate has been computed with exclusion of the point within parentheses. (Reproduced with permission from *Journal of Inorganic and Nuclear Chemistry*.)

$$\ln K_c(0) = -1.797 + \frac{813}{T} \tag{67a}$$

$$\ln K_c(1) = -5.945 + \frac{2359}{T} \tag{67b}$$

$$\ln K_{cm} = -0.644 + \frac{438}{T} \tag{67c}$$

Their values and the fit obtained with Eqs. (67a-c) are recorded in Table 16. Further examples are given in Ref. 29.

b. *The Reaction* $A^- + XNO_3 \rightleftharpoons XA + NO_3^-$. Nancollas and Paterson [30] studied anion-exchange equilibria of hydrous zirconia at 298 K and I = 0.10 mol dm^{-3} Na(A,NO$_3$). In Fig. 16, log κ is plotted versus $\bar{x}_{NO_3}$ for A = Cl and SCN. Results are given in Table 17 for all experimental points and with one point excluded. The fit is greatly improved in the latter case, and the exclusion is considered acceptable on this basis.

TABLE 17   Parameters and Fit for Some Hydrous Zirconia Anion-Exchange Equilibria Data [T = 298 K, I = 0.1 mol dm^{-3} Na(A,NO$_3$)]

Anion	log $\kappa(0)$	log $\kappa(1)$	log κ_m	log K	U $\times 10^3$	R (%)	s(log κ)
Cl$^-$	-0.080	0.142	-0.051	0.004	3.25	30.9	0.028
SCN$^-$	0,201	0.575	0.213	0.330	29.5	14.2	0.046
SCN^{-a}	0.169	0.382	0.328	0.293	2.70	4.98	0.016

[a]One experimental point excluded from calculation.

Source: Data from Nancollas and Paterson [30].

The log K values computed with Eq. (16) are given in Table 17. Fair agreement is found with the values given by Nancollas and Paterson: Cl-NO$_3$:   0.017 $\mp$ 0.042, SCN-NO$_3$:   0.288 $\mp$ 0.045.

8.   *Mixed Solvents*

   a.   *The Reaction* Li$^+$ + HR $\rightleftharpoons$ LiR + H$^+$.   Novitskaya [31] studied the lithium-hydrogen exchange of the 6.5% DVB crosslinked Soviet sulfonate resin, KU-2, in H$_2$O-Me$_2$CO mixtures at 298 K and at I = 0.01 mol dm^{-3} (Li,H)Cl.   Three mixtures were studied where the mole fraction of acetone corresponded to 0.1, 0.3, and 0.5, respectively.   In Fig. 17, log κ is plotted versus $\bar{x}_{LiR}$ for the three mixtures.   The curves have been computed with the parameters given in Table 18.   From examination of this table it is evident that a good fit is obtained.   In order to relate the parameters to the solvent composition, they were fitted by linear regression to straight lines described in acetone mole fraction terms, X_{Me_2CO}, as shown:

$$\log \kappa(0) = 0.066 + 1.595 X_{Me_2CO} \tag{68a}$$

$$\log \kappa(1) = -0.101 + 1.385 X_{Me_2CO} \tag{68b}$$

$$\log \kappa_m = -0.060 + 1.220 X_{Me_2CO} \tag{68c}$$

The parameters computed from Eqs. (68a-c) are also given in Table 18 together with the fit.   A poorer but still acceptable fit is obtained. Equations (68a-c) can be used to estimate equilibrium quotients in other H$_2$O-Me$_2$CO mixtures in the range studied.   Extrapolations to higher acetone concentrations should proceed with caution, however.

TABLE 18 Parameters and Fit for the Exchange of Li^+ amd H^+ in Me_2CO-H_2O Mixtures Equilibrated with the 6.5% DVB Crosslinked KU-2 Resin [T = 298 K, I = 0.01 mol dm^{-3} (H,Li)Cl]

x_{Me_2CO}	log k(0) Nonlin. regress. (68a)	log k(1) Nonlin. regress.	log k(1) Eq. (68b)	log k_m Nonlin. regress.	log k_m Eq. (68c)	U Nonlin. regress.	U Eq. (68)	R (%) Nonlin. regress.	R (%) Eq. (68)	s(log k) Nonlin. regress.	s(log k) Eq. (68)
0.1	0.111	0.047	0.038	0.020	0.062	7.50×10^{-5}	9.56×10^{-4}	4.80	17.13	0.004	0.013
0.3	0.379	0.295	0.315	0.389	0.306	5.15×10^{-5}	3.34×10^{-3}	0.667	5.40	0.003	0.024
0.5	0.749	0.601	0.592	0.508	0.550	3.12×10^{-4}	1.20×10^{-3}	0.957	1.87	0.007	0.014

Source: Data from Novitskaya [31].

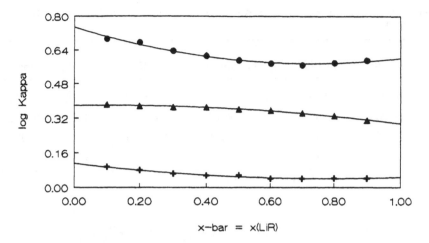

FIG. 17. Log κ plotted versus $\bar{x}_{LiR}$ for the Li^+-H^+ system on KU-2 with 6.5% DVB in water-acetone mixtures. Temperature = 298 K. Ionic strength = 0.01 mol dm^{-3} (Li,H)Cl: +, X_{Me_2CO} = 0.1; ▲, X_{Me_2CO} = 0.3; ●, X_{Me_2CO} = 0.5. The curves have been computed with parameters in Table 18.

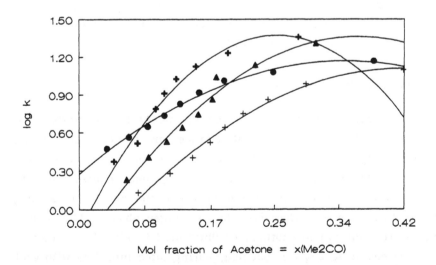

FIG. 18. Log k of reaction (69) plotted versus X_{Me_2CO} for various ionic forms. Temperature = 298 K: +, HR; ▲, LiR; +, NaR; ●, KR. The curves have been computed with parameters in Table 19.

TABLE 19 Parameters and Fit for Reaction (69) on KU-2, 6.5% DVB
(T = 298 K)

Ionic form	log k(0)	log k(1)	log k_m	log K	U	R (%)	s(log k)
HR	-0.423	-1,830	3.229	0.325	8.26×10^{-3}	4.32	0.037
LiR	-0.323	-3.959	4.355	0.024	2.33×10^{-2}	6.07	0.062
NaR	-0.176	-11.932	5.912	-2.065	2.02×10^{-2}	5.03	0.058
KR	0.273	-2.113	2.895	0.351	3.96×10^{-3}	2.45	0.026

Source: Data from Novitskaya [31].

   b. *The Reaction* $H_2O_{(soln)} + Me_2CO_{(res)} \rightleftharpoons H_2O_{(res)} +$
$Me_2CO_{(soln)}$. Novitskaya [31] also studied the extraction at 298 K
of water and acetone by 6.5% DVB crosslinked KU-2 in various ionic
forms. In Fig. 18, log k of the reaction

$$H_2O_{(soln)} + Me_2CO \rightleftharpoons H_2O_{(res)} + Me_2CO_{(soln)} \qquad (69)$$

is plotted versus X_{Me_2CO} for HR, LiR, NaR, and KR. Parameters
and fit are given in Table 19. Within the mole fraction range studied the fit is acceptable, but extrapolations to higher acetone concentrations still need to be entered with caution.

V. CONCLUDING REMARKS

The three-parameter model gives an acceptable fit for all kinds of ion-
exchange equilibria. The fact that it takes care of nonideal behavior
in a much simpler way than activity coefficients is illustrated above.
The applicability of the model implies that neighboring sites interact
with each other as suggested in the outline of the model. Ions carry
water of hydration with them into the exchanger phase. This water
changes from one ion to another, as illustrated in Fig. 4. It now
seems reasonable to assume that neighboring sites interfere with each
other through their water of hydration. When exchanging one ion for
another, the rearrangement of the water molecules around each ion is
dependent on the ions at the nearest sites.

The model is very convenient. It allows replacing experimental data with a few parameters with simple physical meaning.

The drawback is that the model has no predictive qualities. The parameters have to be obtained by curve-fitting procedures, a fate it shares with most solution equilibria, such as complex formation and protonation equilibria.

When a large body of parameters has been collected, attempts will be made to correlate them to various physical quantities.

REFERENCES

1. E. Högfeldt, *Acta Chem. Scand. A, 33*:557 (1979).

2. E. Högfeldt and V. S. Soldatov, *J. Inorg. Nucl. Chem.*, *41*:575 (1979).

3. E. Högfeldt, *Reactive Polymers, 11*:199 (1989).

4. E. A. Guggenheim, *Mixtures*, Clarendon Press, Oxford, 1952, Chap. 4.

5. E. Ekedahl, E. Högfeldt, and L. G. Sillén, *Acta Chem. Scand.*, *4*:556 (1950).

6. E. Högfeldt, E. Ekedahl, and L. G. Sillén, *Acta Chem. Scand.*, *4*:828 (1950).

7. E. Högfeldt, E. Ekedahl, and L. G. Sillén, *Acta Chem. Scand.*, *4*:829 (1950).

8. W. J. Argersinger, Jr., A. W. Davidson, and O. D. Bonner, *Trans. Kansas Acad. Sci.*, *53*:404 (1950).

9. E. Högfeldt, *Ark. Kemi*, *5*:147 (1952). (Reproduced in Ref. 10.)

10. E. Högfeldt, in *Chemistry in Soil Solutions* (A. M. Elprince, ed.), Van Nostrand Reinhold, New York, 1986, pp. 61-85.

11. G. L. Gaines and H. C. Thomas, *J. Chem. Phys.*, *21*:714 (1953).

12. L. Clavatta, *Ann. Chim. (Rome)*, *70*:550 (1980).

13. A. V. Milulich, Thesis, Inst. Gen. Inorg. Chem., Academy of Sciences, BSSR, Minsk, 1982.

14. V. S. Soldatov, A. V. Mikulich, and E. Högfeldt, *Chem. Scripta*, *25*:386 (1985).

15. Z. I. Kuvaeva, A. V. Popov, V. S. Soldatov, and E. Högfeldt, *Solv. Extr. Ion Exch.*, *4*:361 (1986).

16. Z. I. Kuvaeva, A. V. Milulich, A. V. Popov, and E. Högfeldt, *Hydrometallurgy*, *21*:73 (1988).

17. O. D. Bonner and J. R. Overton, *J. Phys. Chem.*, *65*:1599 (1961).

18. K. Gärtner, *Z. Phys. Chem. (Leipzig)*, *223*:132 (1963).

19. V. S. Soldatov and L. V. Novitskaya, *Zhur. Fiz. Khim.*, *39*:2720 (1965).

20. E. Högfeldt and L. V. Novitskaya, *Acta Chem. Scand. A*, *42*:298 (1988).

21. J. Krasner and J. A. Marinsky, *J. Phys. Chem.*, *67*:2557 (1963).

22. E. P. Serjeant and B. Dempsey, *Ionisation Constants of Organic Acids in Aqueous Solution*, Pergamon Press, 1979, p. 95.

23. J. A. Marinsky, T. Miyajima, E. Högfeldt, and M. Muhammed, *Reactive Polymers*, *11*:291 (1989).

24. D. N. Muraviev and E. Högfeldt, *Reactive Polymers*, *8*:97 (1988).

25. D. N. Muraviev, *Chem. Scripta*, *29*:9 (1989).

26. V. I. Gorshkov and G. M. Panchenkov, *Dokl. Akad. Nauk SSSR*, *114*:575 (1957).

27. G. Eisenman, *Biophys. J.*, *2*:259 (1962).

28. E. Högfeldt, T. Miyajima, J. A. Marinsky, and M. Muhammed, *Acta Chem. Scand.*, *43*:496 (1989).

29. S. Dabral and E. Högfeldt, *Chem. Scripta*, *29*:139 (1989).

30. G. H. Nancollas and R. Paterson, *J. Inorg. Nucl. Chem.*, *29*:565 (1967).

31. L. V. Novitskaya, private comm., Minsk, 1990.

3

Description of Ion-Exchange Equilibria by Means of the Surface Complexation Theory

WOLFGANG H. HÖLL, MATTHIAS FRANZREB, JÜRGEN HORST, and SIEGFRIED H. EBERLE Kernforschungszentrum Karlsruhe, Karlsruhe, Germany

I. INTRODUCTION

Numerous investigations of liquid systems containing dissolved species have shown that there is a great variety of interactions possible between solvent and solute or between dissolved species. Within a homogeneous phase these interactions are caused by, e.g., electrostatic and van der Waals forces, heteropolar and covalent binding, or by coordination forces, and they occur regardless of whether the species are neutral or not. Ion exchangers, both inorganic and organic, are bearing excess charge within their matrices due either to substitution of structural atoms of a particular valence, e.g., Si, with atoms of a different valence, e.g., Al, or to the presence of functional sites. As a consequence, physical and chemical interactions will also occur in heterogeneous systems between exchangers, solvent, and dissolved species. The interactions take place at the interface between the phases, i.e., at the surface of the exchanger.

Surface phenomena have been studied extensively for more than 100 years. The charge density of the charge-bearing surface

151

generates an electric potential normal to the surface. Consideration
of the resulting electrostatic interactions has led to the introduction
of several important models. Helmholtz postulates that the surface
charges are compensated by counterions which are located in an ordered (outer Helmholtz) layer with a distance from the surface depending on the hydration of the species [1]. Gouy and Chapman consider
the counterions as mass points which are statistically distributed over
a diffuse layer. Electric potential and charge density are linked by
the Poisson equation [2,3]. Since hydrated ions have a certain spa-
tial extension and are in thermal motion, the Stern model uses a combination of both approaches. A fraction of the counterions is located
in an ordered layer of closest approach, the rest of the counterions
are distributed over the ensuing diffuse layer [4]. More recent
models take into account additional chemical interactions, e.g., the
solvation energy of metal ions [5-7], the protonation of ligands [8],
or specific adsorption energies [9].

A second group of models considers the adsorption of counterions
to be mainly a result of chemical interaction between surface and
solute. The amphoteric behavior of hydrous oxides is treated as a
local equilibrium reaction. Protons or hydroxyl ions are considered
the source of specific interactions that lead to changes in the surface
charge, whereas nonspecific interactions are presumed to lead to the
formation of ion pairs at the surface that are called surface complexes
[10]. The additional contribution of electrostatic interactions has been
taken into account as well [11-16]. In order to overcome the problem
of unknown capacity some authors apply a constant-capacity model
[13,17,18]. These aspects have been discussed in detail by Davis
[14], Sigg [19], and Balistrieri [20].

Spectroscopic investigations of surfaces indicate that more than
a single layer has to be assumed to account for the adsorption of
counterions [21]. Influenced by these observations Yates [22] and
Davis et al. [14] assume a triple-layer model with a surface plane,
a β-plane and a diffuse layer. In their treatment of counterion adsorption ions can be adsorbed in both the outer (β-) plane and the
surface plane in this model [23].

Since hydrous oxides (or inorganic exchangers in general) have well-defined surfaces the above-mentioned models have been successfully applied to sorption phenomena by many authors. However, surface phenomena, as mentioned at the beginning, also occur with organic exchange resins whose functional sites can dissociate or adsorb counterions. The surface adsorption of counterions onto exchange resins has been studied by Cantwell et al., who applied the Gouy-Chapman model to weakly functionalized Amberlite XAD-2 resin [24]. It seems reasonable that the surface complexation approach should be well suited for consideration of individual properties of different functional sites. This possibility was first appreciated by Horst [25], who applied and extended Davis' triple-layer model to equilibria with commercially weak-acid resins.

In the present chapter the development and application of surface complexation theory for the description of the equilibria of representative arbitrary organic ion-exchange resins will be presented for comparison with pertinent experimental data.

II. THEORY

A. Model Assumptions

Organic ion-exchange resins consist of a matrix of irregular hydrocarbon chains to which the functional sites are attached. Uniform distribution of these sites by the matrix is assumed. In spite of its nonregular and nonplanar structure, the surface complexation approach envisions the exchange resin as a structure with a real or imaginary surface over which the functional groups are uniformly distributed. Throughout the theoretical considerations it is assumed that there are no swelling or shrinking effects.

Functional sites can undergo dissociation/association of protons. If protons dissociate, a negatively charged surface is generated and the resin acts as an cation exchanger. If the functional site is protonated, it obtains a positive charge and the resin is an anion exchanger. The surface charge has to be compensated by specific sorption of counterions within the hydrodynamic shear plane, and this leads to the formation of an electric double layer.

According to the Stern model it is assumed that part of the functional sites are neutralized by counterions located in an ordered layer parallel to the surface, whereas the remainder is distributed over a diffuse Gouy-Chapman layer. The surface and ordered layer of counterions can be considered as the two plates of an electric capacitor with a certain electric capacitance. Between counterions within one Stern layer and/or within adjacent Stern layers there are to be no repulsive phenomena. Any ordered layer has the same area as the surface. For the mathematical treatment activity coefficients in the resin phase are assumed to be 1.*

Dissociation or association of protons generate a surface charge density and an electric potential with a gradient normal to the surface. Surface charge and electric potential are noted by a negative sign for cation exchangers and a positive sign for anion exchangers. The electric potential drops from its value at the surface to a fixed slower value in the Stern layer and then decreases further across the diffuse layer until it reaches the free-solution value. It is assumed that the relative permittivity remains constant.

The distance between surface and counterions depends on the solvation of the ion and its size prior to its solvation. Since each ion is consequently different as a result, it has to be assumed that each kind of counterion must be located in an separate Stern layer with a characteristic distance from the surface. Functional sites and counterions form ion pairs, called *surface complexes*. According to Davis et al. all kinds of physical and chemical interactions are included in this classification [14]. Certain counterions can be adsorbed directly onto the surface while losing their hydration shell. This is the case for protons in contact with carboxylic acid groups of weak-acid resins; the reaction between hydroxyl ions and protonated amino groups can be interpreted in the same way. Undissociated carboxylic groups and unprotonated amino groups are also classified as surface complexes.

---

*See discussion in Sec. III.A.

To provide an example of surface complex formation the sorption of monovalent counterions A by the functional sites of a cation exchanger, R^-, is considered as a local equilibrium reaction

$$R^- + A^+ \rightleftharpoons R - A \tag{1}$$

which is then described using the mass-action law:

$$K^A = \frac{[R - A]}{[R^-][A^+]_{St}} \tag{2}$$

According to this relationship one must discriminate between the two different states of ions A in the Stern layer: $[R - A]$ represents the species in its ion pair form by binding to functional sites, while $[A^+]$ designates the "free" species, which are also located in the ordered layer but without forming ion pairs. The equilibrium constant K^A characterizes the ratio of both A species.

The K^A parameter is thermodynamically based. It is expressed in terms of concentrations or activities at the surface which, for the counterion, correspond to concentration in the appropriate Stern layer. The K^A characterizes the affinity of the resin for counterion A^+. As a consequence, individual properties of functional groups and counterions, e.g., specific interactions between weak-acid resins and protons or between weak-base resins and hydroxyl ions, are correlatable through the numerical value of K^A.

Concentrations or activities in the Stern layers close to the surface of the resin cannot be measured. Therefore they have to be expressed as a function of the corresponding quantities measured in the free solution next to the Stern layer and the electric potential in the plane of the Stern layer. Their interdependence is given by the Boltzmann relationship:

$$c(A)_{St} + c(A)_L \exp\left\{- \frac{z(A)F}{RT} \, \Psi_{St}\right\} \tag{3}$$

In the absence of specific interactions such as encountered between counterions (e.g., protons) and surface (e.g., weak-acid resin), the electric potential at the surface has a constant value which is proportional to the total capacity of the resin. It drops

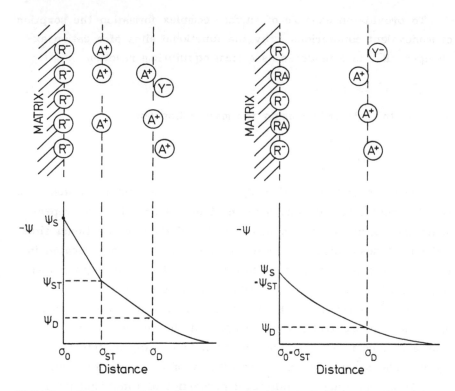

FIG. 1. Schematical surface concept without (left) and with (right) specific interaction of counterions with functional sites (Y^- designates a monovalent coion).

from this value to $\Psi_{St,A}$ in the Stern layer, to Ψ_D in the diffuse layer, and finally to zero in the free solution. With weak electrolyte resins and protons or hydroxyl ions the situation is different. In these cases H^- and OH^- ions are adsorbed in the surface plane. In terms of the surface complexation model the bond and free-ion layers are the same distance away from the surface. Adsorption of these species reduces the number of dissociated fixed sites and therefore decreases the surface potential Ψ_s. This parameter is a function of the loading with specifically adsorbed counterions and therefore depends on the pH of the liquid phase. Figure 1 shows schematically this model-based concept for a cation exchanger without (left) and with specific interaction (right) together with their respective developments of potential profiles.

B. Binary Exchange Equilibria

For discussion of binary equilibria we consider a system with counterions A and B of valencies $z(A)$ and $z(B)$. These valencies are assumed to be positive for cations and negative for anions. The functional sites of the resin are designated as R with a valency $z(R)$, which is negative for cation exchangers and positive for anion exchangers. The numerical value of $z(R)$ equals the smallest common multiple of the valencies of the two counterions.

We define stoichiometric factors W_A and W_B:

$$W_i = - \frac{z(R)}{z(i)} \qquad i = A, B \tag{4}$$

By this means the binary equilibrium can be written in the form

$$W_A A^{z(A)} + R - B_{W_B} \iff W_B B^{z(B)} + R - A_{W_A} \tag{5}$$

Using the stoichiometric factors results in one surface complex, $R - A_{W_A}$, being replaced by another surface complex, $R - B_{W_B}$*. Table 1 summarizes some examples of ion and resin valencies in binary exchange equilibria.

Let us consider the equilibrium state of a the resin loaded with two different counterions. We can then assume that counterions A, located in a first Stern layer are closer to the surface than ions B, located in a second layer, a larger distance away from the surface. This is an extension of Davis' triple-layer model to a multiple-layer approach. Part of the counterions A and B is found in the diffuse layer. Corresponding with the sequence of layers the potential drops from its value at the surface to the value in the ordered and diffuse layers and finally to zero in the free solution. Figure 2 shows the arrangement of layers and the development of corresponding potentials.

---

*In a previous paper Horst et al. [27] considered the weak-acid resin as a bivalent acid. The above representation is a more general one which reduces the exchange of ions of arbitrary unequal valencies to a "unit exchange."

TABLE 1   Counterion and Resin Valencies

| Valencies of the binary system | | | |
z(A)	z(B)	Examples	Valency of the resin z(R)
+1	+1	H^+, Na^+	-1
+1	+2	H^+, Ca^{2+}	-2
+2	+2	Ca^{2+}, Mg^{2+}	-2
+1	+3	H^+, Cr^{3+}	-3
+2	+3	Ca^{2+}, Cr^{3+}	-6
-1	-1	OH^-, Cl^-	+1
-1	-2	Cl^-, SO_4^2	+2

The formation of two different surface complexes can be characterized by two complexation constants:

$$K^A = \frac{[R - A_{W_A}]}{[R][A^{z(A)}]_{St,A}^{W_A}} \tag{6}$$

$$K^B = \frac{[R - B_{W_B}]}{[R][B^{z(B)}]_{St,B}^{W_B}} \tag{7}$$

Similar relationships with activities replacing of concentrations in the liquid phase have been developed by Koopal et al. [26] in a general discussion of the triple-layer model. Since the two constants express the affinity of the resin for the two counterions the ratio of both constants has to reflect the exchange equilibrium. Introducing concentrations $c(i)$ (without valency symbols) instead of symbols in brackets yields

$$K_B^A = \frac{c(R - A_{W_A})[c(B)_{St,B}]^{W_B}}{c(R - B_{W_B})[c(A)_{St,A}]^{W_A}} \tag{9}$$

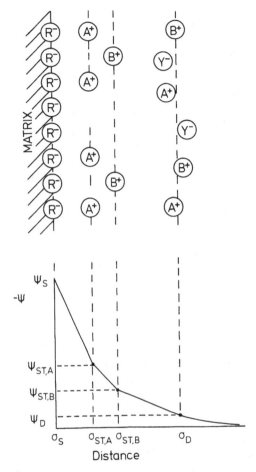

FIG. 2. Stern and diffuse layers and development of electrical potential for a binary system.

In this relationship $c(i)_{St,i}$ designates the concentrations of counterions in the respective Stern layers that their surface complexes form. Expressing these inaccessible concentrations as products of their concentrations in the free solution and the exponential of the electric potential at the respective Stern layer planes we obtain the relationships

$$c(R - A_{W_A}) = K^A c(R) c(A)^{W_A} \exp\left\{ + \frac{z(R)F}{RT} \Psi_{St,A} \right\} \qquad (10)$$

$$c(R - B_{W_B}) = K^B c(R)c(B)^{W_B} \exp\left\{+ \frac{z(R)F}{RT} \Psi_{St,B}\right\} \tag{11}$$

In Eqs. (10) and (11) the electric potentials $\Psi_{St,A}$ and $\Psi_{St,B}$ are also not measurable. In order to circumvent this problem, charge-balance relationships have to be considered.

The surface charge density σ_S is defined by the electric potentials Ψ_S and $\Psi_{St,A}$ and the electric capacitance $C(S,A)$ formed by the surface and the layer of counterions A:

$$\sigma_S = C(S,A)(\Psi_S - \Psi_{St,A}) \tag{12}$$

The charge density in the Stern layers of counterions A and B is obtained in an analogous way from

$$\sigma_{St,A} = C(S,A)(\Psi_S - \Psi_{St,A}) + C(A,B)(\Psi_{St,A} - \Psi_{St,B}) \tag{13}$$

The charge densities of the surface and the counterion layers have opposite signs and can be linked by the relationship

$$\sigma_S + \sigma_{St,A} + \sigma_{St,B} + \sigma_D = 0 \tag{14}$$

whose basis is the condition of electroneutrality that prevails in the resin phase. The charge density σ_S at the surface equals from the sum of functional groups and the counterions in the diffuse layer:

$$\sigma_S = \frac{FL_S}{A_0 S} [z(R)c(R - A_{W_A}) - z(A)c(A)_{St,A}$$
$$+ z(R)c(R - B_{W_B}) - z(B)c(B)_{St,B}] - \sigma_D \tag{15}$$

The charge densities $\sigma_{St,A}$ and $\sigma_{St,B}$ in the Stern layers are equal to the charge equivalents of counterions A and B:

$$\sigma_{St,A} = \frac{FL_S}{A_0 S} [z(R)c(R - A_{W_A}) - z(A)c(A)_{St,A}] \tag{16}$$

and

$$\sigma_{St,B} = - \frac{FL_S}{A_0 S} [z(R)c(R - B_{W_B}) - z(B)c(B)_{St,B}] \tag{17}$$

In Eqs. (15) to (17) the proportionality factor FL_S/A_0S converts molar concentrations to charge densities. The quantity L_S designates the surface volume element by which the Stern layer concentrations are defined. Since L_S is eliminated in the course of the mathematical development there is no need for its estimate.

If the electric potentials $\Psi_{St,A}$ and $\Psi_{St,B}$ are known, the concentrations of counterions in the Stern layers for the A and B counterions can be calculated with Eqs. (10) and (11). The concentrations of counterions distributed over the diffuse layer are obtained from the theory of Gouy and Chapman. Integration across the entire thickness yields

$$\sigma_D = \int_d^\infty \varepsilon \frac{d^2\psi}{dx^2}\, dx = -\varepsilon\left[\frac{d\psi}{dx}\right]_{x=d} \tag{18}$$

The gradient of the potential, $d\psi/dx$, is obtained from an integration of the Poisson-Boltzmann distribution:

$$\frac{d\psi}{dx} = \left\{\frac{2RT}{\varepsilon} \sum_i z(i)c(i)\left[\exp\left(-\frac{z(i)F}{RT}\psi\right) - 1\right]\right\}^{1/2} \tag{19}$$

Substitution of this expression for $d\psi/dx$ in Eq. (18) leads to the relationship

$$\sigma_D = \left\{2RT\varepsilon \sum_i z(i)c(i)\left[\exp\left(-\frac{z(i)F}{RT}\psi_D\right) - 1\right]\right\}^{1/2} \tag{20}$$

Integration of Eq. (20) yields the quantity of counterion across the diffuse layer, and this in turn yields the surface charge equivalents which are needed to neutralize the counterions i in the diffuse layer:

$$\Gamma_i = \int_d^\infty z(i)[c(i)_\psi - c(i)]\, dx \tag{21}$$

Conversion to molar concentrations is achieved with Eq. (22):

$$c(i)_D = \frac{SA_0\Gamma_i}{L_S z(i)} \tag{22}$$

Using the Boltzmann relationship and substituting (19) for $d\psi/dx$ leads to the integral equation

$$c(i)_D = \frac{A_0 S}{L_S} \int_\Psi^0 \frac{c(i)[\exp(- (F/RT)z(i)\Psi) - 1]}{\left\{ \frac{z(R)RT}{\varepsilon} \sum_j z(j)c(j)[\exp(- (F/RT)z(j)\Psi) - 1] \right\}^{1/2}} d\Psi$$

$$\text{(23)}$$

Normally this equation has to be solved numerically. However, for symmetric electrolytes an analytical solution exists with which the concentration of counterions in the diffuse layer can be approximated [25]:

$$c(i)_D = \frac{2A_0 S c(i)}{z(i)L_S} \sqrt{\frac{RT\varepsilon}{2F^2 I}} \left[\exp\left(- 0.5 \frac{z(i)F}{RT} \Psi_D \right) - 1 \right] \qquad (24)$$

The concentration ratio of two ions that results from such an approach is [25]

$$\frac{c(j)_D}{c(i)_D} = \frac{z(j)c(i)[\exp\{- 0.5(z(j)F/RT)\Psi_D\} - 1]}{z(i)c(j)[\exp\{- 0.5(z(i)F/RT)\Psi_D\} - 1]} \qquad (25)$$

Equation (25) which applies to counterions of equal valency, however, can be used as well to approximate ratios of ions which have different valencies. From (25) it follows that for ions of equal valency the concentration ratio in the diffuse layer is the same as in the free solution. If the valencies are different the counterion with larger $z(i)$ will be preferred. The second column of Table 2 lists the Eq. (25)-based ratio $c(Na^+)/c(Ca^{2+})$ in the diffuse layer of a cation exchanger as a function of the double-layer potential. Values presented in the first column were based on calculations provided by the exact solution of Eq. (23). It becomes obvious that at small potentials the difference between the two solutions is negligible. If we tolerate differences of 5% the less precise equation (25) can be applied up to potentials of -25 mV [25].

The equivalents of counterions that are needed in the diffuse layer to neutralize the surface charge not compensated by species in the Stern layers are given by

$$z(R)c(R) + z(A)c(A)_D + z(B)c(B)_D = 0 \qquad (25a)$$

or

TABLE 2   Concentration Ratio for Monovalent
and Divalent Counterions in the Diffuse Layer,
$c(NaCl) = c(CaCl_2) = 20$ mmol/l

Ψ_D, V	$c(Na)_D/c(Ca)_D$		Deviation, percent
	Eq. (23)	Eq. (24)	
-0.002	0.9803	0.9805	0.02
-0.005	0.9500	0.9514	0.15
-0.010	0.8975	0.9030	0.61
-0.020	0.7878	0.8078	2.54
-0.025	0.7316	0.7614	4.07
-0.050	0.4630	0.5486	18.49
-0.100	0.1326	0.2500	88.54

$$c(R) - \frac{1}{W_A} c(A)_D - \frac{1}{W_B} c(B)_D = 0 \qquad (25b)$$

For the concentration of counterions in the diffuse layer one
finally obtains the relationship

$$c(i)_D = \frac{c(i)}{z(i)F} \frac{\sigma_D A_0 S}{\sum\limits_j c(j)F(j;i)L_S} \qquad (26)$$

where

$$F(j;i) = \frac{\exp\{- 0.5z(i)F\Psi_D/RT\} - 1}{\exp\{- 0.5z(j)F\Psi_D/RT\} - 1} \qquad (27)$$

Equations (10)-(19) and Eq. (21) (the last two for both counterions) form a set of 12 equations that is usable for the resolution of
12 unknown quantities as long as additional unknowns are not introduced by chemical equilibria simultaneously encountered in the liquid
phase. If the exchange equilibrium is further complicated by chemical equilibria in the liquid phase which involve the exchangeable
species, the above set of equations has to be extended to include
additional relationships for chemical equilibria. Reaction-coupled exchange equilibria can easily be described in this manner. This system
of nonlinear equations has to be solved numerically. A detailed description of the computation has been presented in a previous publication [27].

C. Multicomponent Exchange Equilibria

According to the development outlined in Sec. II.B, N different
counterions are adsorbed in N different layers as shown in Fig. 3
for a three-component (N = 3) system. The description of such a
multicomponent system is straightforward. The sorption of each of
the counterion ions is again treated as a local equilibrium and characterized by each of the respective formation constants. There are
N Stern layers with charge densities, $\sigma_{St,n}$, for N different kinds
of counterions. Between the surface and each of the layers N electric capacitors with capacitances C(S,n) ($1 \leq n \leq N$) can be assigned.
Furthermore, N-1 capacitors are formed by two adjacent counterions,
I and J, layers with the capacitances C(I,J). The exercise that follows includes the case where two or more counterions are located
within the same layer.

For these systems the condition of electroneutrality is

$$\sigma_S - \sum_{k=1}^{N} \sigma_{St,k} - \sigma_D = 0 \qquad (1 \leq n \leq N) \tag{28}$$

Charge densities, capacitances, and potentials are linked in the following way:

$$\sigma_S = C(S,A)(\Psi_S - \Psi_{St,A}) \tag{29}$$

$$\sigma_{St,1} = C(S,B)(\Psi_{St,A} - \Psi_S) + C(S,B)(\Psi_{St,A} - \Psi_{St,B}) \tag{30}$$

$$\sigma_{St,i} = C(S,I)(\Psi_{St,I} - \Psi_{St,I-1}) + C(I,J)(\Psi_{St,I} - \Psi_{St,J}) \tag{31}$$

$$\sigma_{St,N} = C(S,N)(\Psi_N - \Psi_{N-1}) + C_D(\Psi_N - \Psi_D) \tag{32}$$

$$\sigma_D = C_D(\Psi_D - \Psi_N) \tag{33}$$

with I and J once again identifying two counterions in adjacent layers.
The charge balances for the various layers are

$$\sigma_S = \frac{FL_S}{A_0 S} \left[z(R)c(R) + z(R) \sum_{I=A}^{N} c(R - I^{W_I}) \right] \tag{34}$$

$$\sigma_{St,k} = \frac{FL_S}{A_0 S} [-z(R)c(R - K_{W_K}) + z(K)c(K)_{St,K}] \tag{35}$$

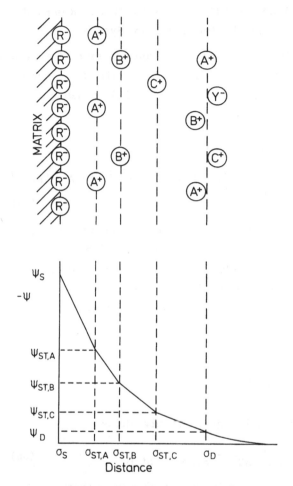

FIG. 3.   Stern and diffuse layers and development of electrical potential for a ternary system.

Expressions giving the concentration of each counterion in each of the Stern layers are found as shown earlier by using the definition of the respective surface complex and the Boltzmann relationship. The molar concentrations of counterions which are located in the diffuse layer can be approximated with Eq. (25). Once again a set of equations similar to the set in Sec. II.B become available for computation of the quantities being sought.

III. DERIVATION OF RESIN-SPECIFIC EQUILIBRIUM PARAMETERS

A. The Generalized Separation Factor in Binary Systems

According to Sec. II.A the surface complexation constant K' expresses the affinity of the resin for counterions i. The ratio of the surface complexation constants for the pair of ions A and B consequently reflects the exchange equilibrium:

$$K_B^A = \frac{K^A}{K^B} = \frac{c(R - A_{W_A})[c(B)]^{W_B}}{c(R - B_{W_B})[c(A)]^{W_A}} \exp\left\{ z(R) \frac{F}{RT} (-^\Psi_{St,A} + {}^\Psi_{St,B}) \right\}$$

(36)

The first factor in Eq. (36) contains equilibrium loadings and concentrations which can be measured. Thus, it is an equilibrium parameter very similar to the separation factor $\alpha(A,B)$ which is used for empirical characterization of binary equilibria [28]. It is therefore designated as a *generalized separation factor* and defined by

$$Q_B^A = \frac{c(R - A_{W_A})[c(B)]^{W_B}}{c(R - B_{W_B})[c(A)]^{W_A}}$$

(37)

Resolving Q_B^A and expressing it logarithmically yields

$$\log Q_B^A = \log K_B^A - \frac{z(R)}{\ln 10} \frac{F}{RT} (-^\Psi_{St,A} + {}^\Psi_{St,B})$$

(38)

The summation of counterions in the Stern layers (both surface complexes and noncomplexed species) and in the diffuse layer yields the resin loadings. If concentrations are expressed in mol/l and loadings in eq/g, respectively, we obtain

$$z(R)c(R - A_{W_A}) + z(A)c(A)_{St,A} + z(A)c(A)_d = q(A)\frac{S}{L_S}$$

(39)

and

$$z(R)c(R - B_{W_B}) + z(R)c(B)_{St,B} z(B)c(B)_d = \frac{1}{z(R)} q(B)\frac{S}{L_S}$$

(40)

As stated by Davis et al. [14], the charge density in the diffuse layer is small compared to that in the Stern layers. Furthermore,

except for a few cases, the concentrations of noncomplexed species
in the Stern layers are small compared to the concentration of surface
complexes. Thus, the free ions in both regions can be neglected. As
a consequence, the generalized separation factor can be directly expressed in terms of resin loadings with species A and B. The difference of the unknown potentials in Eq. (1) can be expressed by means
of Eqs. (13) and (15) as

$$\Psi_{St,A} - \Psi_{St,B} = -\frac{FL_S}{A_0 SC(A,B)}[z(B)c(B)_{St,B} - z(R)c(R - B_{W_B})] \tag{41}$$

Instead of expressing the difference of potentials in terms of the
surface charge density like Koopal et al. [26], the resin loadings can
be obtained in this much simpler way. Using Eqs. (39) and (40),
the expression in brackets in Eq. (41) becomes

$$[z(R)c(R) - z(R)c(R - B_{W_B})] = \bar{c}(B) = y(B)q_{max}\frac{S}{L_S} \tag{42}$$

with q_{max} corresponding to the total capacity of the resin and $y(B)$
to the equivalent fraction of species B in the resin phase. Eq. (36)
can finally be transformed to

$$\log Q_B^A = \log K_B^A - \frac{z(R)}{\ln 10}\frac{F^2 q_{max}}{A_0 C(A,B)RT}y(B) \tag{43}$$

As long as the premise that there are negligible amounts of counterion in the diffuse layer is valid, this equation of a straight line
expresses the logarithm of the generalized separation factor Q_B^A as a
function of the loadings with ion B, i.e., of the dimensionless loading with the ion of the outer of the two layers. The ratio of affinities is given by the intercept of the vertical axis, while the y(b) multiplication factor is the slope of the straight line (Fig. 4). Unlike the
conventional separation factor $\alpha(A,B)$, the above definition therefore
yields a quantity which takes into account the total concentration of
the liquid phase and the composition of the resin phase.

As is obvious from the theoretical developments, use of activity
coefficients in both phases has not been considered. The generalized
separation factor is defined by means of concentrations and resin

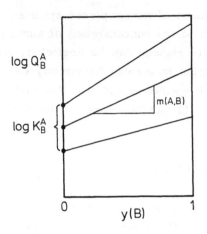

FIG. 4.   Geometric interpretation of Eq. (43).

loadings to facilitate the mathematical treatment.   Nevertheless, there
has to be an influence of ionic strength.   We can assume that this
influence is implicit in the numerical values of the equilibrium parameters log K_B^A and the multiplication factor of y(B) (slope) in Eq. (43).
So far no attempts have been made to derive relationships in which
activity coefficients are introduced similar to various approaches reported in the literature [26,29,30].

     For an exchange of ions of equal valency, Q_B^A is identical to
$\alpha(A,B)$ and does not depend on total normality or resin composition,
whereas for ions of unequal valency these dependencies are inherent.
In terms of the surface complexation model the increase of log Q_B^A
with y(B) is proportional to the distance between A and B Stern
layers.   With equal amounts of the counterions in the two layers the
capacitance of the resulting capacitor becomes smaller if the distance
between the layers increases.   As a consequence the slope of the
straight line has to increase, too.   In the opposite case, if both
ions are located within the same layer, i.e., the distance becomes
zero, then the capacitance tends toward ∞ and the slope becomes
zero.   As a consequence, it can be concluded that constant values
of Q_B^A or $\alpha(A,B)$ are obtained only if the counterions have equal
valencies and are sorbed at the same distance from the surface.

Numerical values of the ratio of affinities and the slope are easily obtained from experimental results by plotting log Q_B^A as a function of $y(B)$. Apart from the capacitance the expression for the slope contains quantities that cannot become negative. If the sequence of layers is assumed correctly, $C(A,B)$ is positive and therefore the slope has to be positive, too.

B. Generalized Separation Factors in Multicomponent Systems

For consideration of generalized separation factors in multicomponent systems we assume a sequence of layers as follows:

$S/A/.../I/J/.../N/D$

For such a system with N components N-1 subsequent binary equilibria I/J can be considered. Accordingly, N-1 subsequent ratios of affinities K^I/K^J can be defined.

$$K_J^I = \frac{K^I}{K^J} = \frac{[R - I_{W_I}][c(J)^{W_J}]_{St,J}}{[R - J_{W_J}][c(I)^{W_I}]_{St,I}} \quad (44)$$

Expressing the Stern layer concentrations by combining ion concentrations in the free solution with the electric potential yields

$$K_J^I = \frac{[R - I_{W_I}][c(J)^{W_J}]}{[R - J_{W_J}][c(I)^{W_I}]} \exp\{z(R)A_0(-\Psi_{St,I} + \Psi_{St,J})\} \quad (45)$$

The first term is the generalized separation factor Q_J^I which describes the binary equilibrium I/J in the multicomponent system. By using Eqs. (22)-(26) while once again neglecting the amount of counterions in the diffuse layer, the potential difference can be expressed by

$$\Psi_{St,I} - \Psi_{St,J} = -\frac{FL_S}{A_0 SC(I,J)}\left[-z(R)\sum_{k=J}^{N} c(R - K) + \sum_{k=j}^{N} z(K)c(K)_{St,K}\right] \quad (46)$$

with K referring to counterions J to N. The following expression
for the logarithm of the generalized separation factor is obtained by
employing the same approximations as for the binary case:

$$\log Q_J^I = \log K_J^I - \frac{z(R)}{\ln 10} \frac{F^2 q_{max}}{RTA_0 C(I,J)} \sum_{k=j}^{N} y(k) \tag{47}$$

In general, N-1 relationships can be derived to describe the equilibrium. Mathematically these equations again represent straight lines
when $\log Q_J^I$ is plotted versus $\sum_{i+1}^{N} y(k)$.

In the case of a ternary system with counterions A, B, and C,
one obtains the equations

$$\log Q_B^A = \log K_B^A - \frac{z(R)}{\ln 10} \frac{F^2 q_{max}}{A_0 C(A,B)RT} [y(B) + y(C)] \tag{48}$$

and

$$\log Q_C^B = \log K_C^B - \frac{z(R)}{\ln 10} \frac{F^2 q_{max}}{A_0 C(B,C)RT} y(C) \tag{49}$$

It is apparent that for each layer i summation of loadings has to
be made for the ions from i + 1 to N. Comparison of Eq. (48) with
Eq. (43) reveals that the parameters K_B^A and the slope of the two
straight lines are exactly the same as in the pure binary case. As
a consequence a plot of $\log Q_B^A$ versus $[y(B) + y(C)]$ using ternary
equilibrium data should lead to the same ratio of affinities and to the
same straight line as a similar evaluation of pure binary data.

A set of N-1 equilibrium relationships (47) contains 2(N-1)
equilibrium parameters which have to be derived from binary equilibria. However, there is no need to measure all possible binary
subsystems. If, e.g., the equilibrium parameters of the exchange
of B and C for A are known for the ternary system, then the
parameters for the binary subsystem B/C can easily be derived.
The unknown parameter K_C^B results from the definition of the complex formation constants:

$$K_C^B = \frac{K_C^A}{K_B^A} \tag{50}$$

Thus we obtain the simple relationship

$$\log K_C^B = \log K_C^A - \log K_B^A \tag{51}$$

For derivation of the unknown slope

$$m(I,J) = - \frac{z(R)}{\ln 10} \frac{F_2 q_{max}}{RTA_0 C(I,J)} \tag{52}$$

we use the well-known relationship for a series of electric capacitors,

$$\frac{1}{C(A,C)} = \frac{1}{C(A,B)} + \frac{1}{C(B,C)} \tag{53}$$

from which we obtain

$$m(B,C) = m(A,C) - m(A,B) \tag{54}$$

The parallel relationships for more than three components can be derived in a similar way. Generally the equilibrium parameters from N-1 arbitrary binary subsystems are required to predict an N-component system. These subsystems have to include all counterions of the system.

C. Parameters in Multicomponent Systems

As shown in Sec. B the equilibrium in a multicomponent system is considered as a sequence of coupled binary equilibria. In each of these binary subsystems the resin valency is the smallest common multiple of the two counterions. As a consequence the resin valency can vary within the series of binary equilibria. In order to avoid the problem of different sets of equilibrium parameters (e.g., for the equilibrium between one monovalent ion and a divalent ion in one subsystem and another monovalent ion in the adjacent subsystem), all parameters are transformed to a theoretical monovalent functional group. With $z(R)_p$ being the resin valency for an arbitrary binary subsystem the transformed parameters (designated by a hat symbol) are obtained from [31]

$$\log \hat{K}_{i+1}^i = \frac{\log K_{i+1}^I}{z(R)_i} \tag{55}$$

and

$$\hat{m}(i, i + 1) = \frac{m(i, i + 1)}{z(R)_i} \tag{56}$$

These transformed parameters can be treated as shown in the previous section. For example, for a system with SO_4^{2-}, NO_3^-, and Cl^- ions we obtain

$$\log \hat{K}_{NO_3}^{SO_4} = \log \hat{K}_{Cl}^{NO_3} + \log \hat{K}_{Cl}^{SO_4} \tag{57}$$

which corresponds to

$$\frac{\log K_{SO_4}^{NO_3}}{2} = \frac{\log K_{Cl}^{NO_3}}{1} + \frac{\log K_{SO_4}^{Cl}}{2} \tag{58}$$

In an analogous way the slopes can be correlated:

$$\hat{m}(NO_3, SO_4) = \hat{m}(NO_3, Cl) + \hat{m}(Cl, SO_4) \tag{59}$$

or

$$\frac{m(NO_3, SO_4)}{2} = \frac{m(NO_3, Cl)}{1} + \frac{m(Cl, SO_4)}{2} \tag{60}$$

D. Superposition of Several Exchange Equilibria

Binary equilibrium parameters can be applied for the prediction of multicomponent equilibria only if the counterions are of equal valency, e.g., when only monovalent ions are encountered with simple binary subsystems. Difficulties arise, however, if we consider the exchange of species with different valencies. In such cases a "mixed" exchange can occur. If, e.g., a strong-base resin in the mixed $Cl^-NO_3^-$ form is contacted with a sulfate-bearing solution, then one sulfate ion can replace either two nitrate species or two chloride species or one nitrate and one chloride ion. This mixed-exchange scenario has not been taken into account so far. For its consideration the dimensionless loadings of the two exchangeable monovalent species will have to be replaced by the corresponding loading probabilities [31].

The above example leads to a relatively simple mathematical description. In systems of higher complexity which arise from the presence of trivalent species more possibilities must be considered.

One trivalent ion can be replaced by only monovalent species or by one divalent and one monovalent ion. If we consider the replacement of two trivalent species there are further possibilities. Mathematically this leads to rather complicated interdependencies. Fortunately, only monovalent and divalent species are encountered in most of the exchange processes. As a consequence, we can simplify the operation by assuming that competition occurs only between ions of equal valency.

For derivation of the loading probabilities of the exchange, we first summarize the dimensionless loadings of ions of equal valency.

$$y_S(z) = \Sigma \ y(i^z) \tag{61}$$

with $z(i)$ being the valency of the ions. For the above example we obtain

$$y_S(1) = y(NO_3) + y(Cl) \tag{62}$$

$$y_S(2) = y(SO_4) \tag{63}$$

The loading probability, $y_{R,i}$, which replaces the pure dimensionless resin loading, y_i, becomes

$$y_{R,i} = \left[\frac{y(i^2)}{y_S(z)}\right]^{W_i} y_S(z) \tag{64}$$

If we have species of equal valencies all components have the identical stoichiometric factor $W_i = 1$. Thus, the probability $y_R(i)$ is identical with the dimensionless loading $y(i)$.

E. Graphical Representation of Equilibria

Equations (44) and (49) can be used to project theoretically based equilibria. For graphical representation of equilibria the resin composition is usually plotted as a function of the composition of the liquid phase. If the normality of the solution and the exchange capacity of the resin remain unchanged the dimensionless concentrations and resin loadings (equivalent fractions) are most often applied.

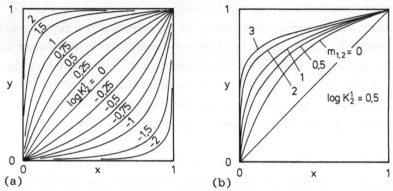

FIG. 5. Dimensionless isotherms for the exchange of counterions of equal valency at the same distance (left) and at different distances from the surface (right).

The usual square or triangular diagrams are obtained for binary and ternary systems, respectively. No graphical representation has been developed for systems with more than three components. Because graphical representation of ternary systems is seldom applied and because there is a rather large number of parameters encountered in ternary systems the discussion which follows is confined to binary equilibria.

If we assume an exchange of counterions of equal valency and their sorption within the same Stern layer the generalized separation factor is constant and does not depend on the liquid-phase normality. As a consequence we obtain symmetrical Langmuir-type isotherms. Their shape and curvature depend on the ratio of affinities for the counterions (see Fig. 5a). If both counterions have the same valency but are not located at the same distance from the surface, $\log Q_B^A$ will not be constant. In this case the isotherms become nonsymmetric and the deformation increases with increasing distance between the Stern layers (Fig. 5b).

If we consider an exchange of monovalent for bivalent counterions there is no constant generalized separation factor. Isotherms for the monovalent ion at a certain normality of the liquid phase and with the simplifying assumption that both species are located at the same distance from the surface are plotted in Fig. 6. If the

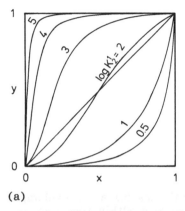 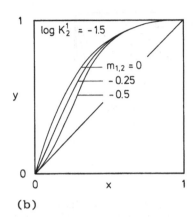

(a)   (b)

FIG. 6. Dimensionless isotherms for a 1/2 exchange with sorption in the same layer (left) and for a 2/1 exchange with the monovalent ion being adsorbed closer to the surface (right), total normality = 0.01 eq/l.

monovalent ion is strongly preferred (log $K_B^A \geq 3$) the isotherm will resemble the upper curves of the diagram on the left. Such isotherms are obtained for the uptake of H^+/OH^- by weak-electrolyte resins [32]. However, if there is a stronger affinity for the bivalent ion the isotherms become concave. This is consistent with the properties of experimental data compiled for the Na^+/Ca^{2+} exchange on strong-acid resins [33].

In systems with 1/2 exchange the monovalent species may be located closer to the surface than the divalent ones. If we consider the uptake of the divalent counterion another kind of nonsymmetric isotherm is encountered (Fig. 6b). The pertinent equilibria have been reported in the literature for the SO_4^{2-}/HCO_3^- exchange [34].

For an exchange of monovalent for bivalent ions we should also consider the influence of the normality of the solution. If we again assume that both ions are located within the same Stern layer [m(A,B) = 0], we obtain the isotherms as plotted in Fig. 7. It becomes apparent that the uptake of monovalent ions predicted to be low at small concentrations, does, however, increase with increasing normality. This is in agreement with projections based on consideration of electroselectivity with the Gibbs-Donnan approach [35].

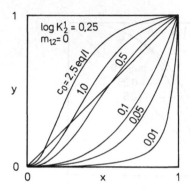

FIG. 7. Dependency of dimensionless isotherms for a 1/2 exchange on the total normality (sorption of both species within the same layer assumed).

The development of isotherms is in good qualitative agreement with data reported in the literature [33].

IV.   EVALUATION OF EQUILIBRIUM DATA

A.   General

Calculations of binary or multicomponent exchange equilibria for arbitrarily selected conditions require the following parameters:

1.   Surface complex formation constants for the various ionic species
2.   Surface parameter A_0
3.   Electric capacitances, e.g., $C(S,I)$ and C_D

Measurement of these quantities has to be accomplished experimentally. Detailed descriptions of the experimental approaches used for this purpose are presented in earlier publications [27,36]. The resins and their descriptions are listed in Table 3.

B.   Evaluation of Binary Equilibria

1.   *Cation Exchangers*

We see from Table 3 that the cation exchangers used in the experiments described next were the weak-acid resins Lewatit CNP 80, Amberlite IRC 50, and Lewatit OC 1058 and the strong-acid resins Lewatit S 100 and Dowex HCRS. Among the weak-acid resins

TABLE 3 Exchange Resins and Characteristic Data

Resin	Type	Functional site
LEWATIT CNP 80	weak-acid	carboxylic
AMBERLITE IRC 50	weak-acid	carboxylic
LEWATIT OC 1058	weak-acid	carboxylic
LEWATIT S 100	strong-acid	sulfonic
DOWEX HCR-S	strong-acid	sulfonic
DOWEX MWA-I	weak-base	tertiary amine, polystyrene
AMBERLITE IRA 93-SP	weak-base	tertiary amine, polystyrene
AMBERLITE IRA 68	weak-base	tertiary amine, acrylic
AMBERLITE IRA 410	strong-base	quarternary amine, polystyrene, type II
AMBERLITE IRA 458	strong-base	quarternary amine, acrylic type I
AMBERLITE IRA 958	strong-base	quarternary amine, polystyrene, type I
AMBERLITE IRA 996	strong-base	quarternary amine, acrylic type I
LEWATIT TP 207	chelating	iminodiacetate
AMBERLITE IRC 718	chelating	iminodiacetate

the dissociation of acrylic resins is stronger than that of methacrylic exchangers whereas the strong-acid resins are fully dissociated and exhibit similar properties.

Most series of experiments were carried out with the resin orig-
inally in a certain reference form. The protonated (free-acid) form
was selected for the experiments with weak-acid resins, whereas the
strong-acid resins were used in the Ca^{2+} form. Mass-balance estimates based on analyses of the liquid phase were used to evaluate
loadings. The concentrations and loadings were expressed in mol/L
and eq/kg, respectively, and pH was approximated by pH = -log c(H)
in the generalized separation factor, given by the relationships

$$\log Q_A^H = \log c(A) + \log \frac{q(H)}{q(A)} - z(R)pH \qquad (65a)$$

for weak-acid resins and

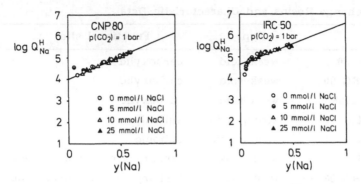

FIG. 8.   Generalized separation factor log Q_{Na}^{H} vs. y(Na) with the weak-acid resins Lewatit CNP 80 and Amberlite IRC 50.   (Partially reproduced from [27] with permission of Elsevier Science Publishers.)

$$\log Q_i^{Ca} = \log c(i) + \log \frac{q(Ca)}{q(i)} + z(R) \log c(Ca) \qquad (69b)$$

for strong-acid resins.   Both equations can be derived from Eq. (37).

a.   Weak-Acid Resins   Results from the experiments with Lewatit CNP 80 and Amberlite IRC 50 are presented in Fig. 8 for the exchange of sodium for protons and in Fig. 9 for the exchange of divalent counterions for protons.*

In both cases the logarithm of the generalized separation factor bears a linear relationship to y(A) once y(A) > 0.2.   Each linear dependency holds for a sizable concentration range.   We conclude that at y(A) > 0.2 and within the range of concentrations studied ionic strength has no influence.   The straight line has a positive slope indicating that the Stern layer for the metal ions is some distance from the surface, whereas protons undergo specific interactions with the functional sites and are located directly in the surface plane.   In the case of the exchange of Cu^{2+} for H^+ the occurrence of two straight lines leads to the assumption that copper species may be located in two different layers: the first one very close to the surface

---

*With respect to the more general treatment the resins are monovalent for monovalent counterions and bivalent for bivalent ones.   As a consequence the development of the generalized separation factor for the Na^+/H^+ equilibrium is different from that in Ref. 27.

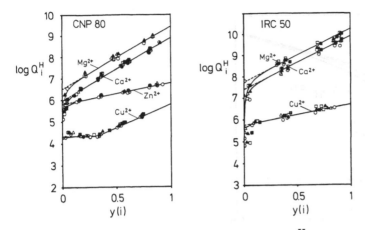

FIG. 9. Generalized separation factor log Q_i^H vs. y(i) from the exchange of protons for bivalent counterions with the weak-acid resins Lewatit CNP 80 and Amberlite IRC 50. (Partially reproduced from [27] with permission of Elsevier Science Publishers.)

(deduced from the almost horizontal development), and the second one with a larger distance away from the surface. Occupation of the second layer apparently occurs only after y(Cu) > 0.33. This phenomenon is in excellent agreement with measurements of copper complex formation in polymethacrylic gels [37]. Similar studies demonstrate that Ca^{2+} and Zn^{2+} ions form only one kind of complex. With methacrylic resins sorption in two different Stern layers is not observed.

Extrapolation of the straight portion to intercept the ordinate axis at y(i) = 0 leads to the ratio of affinities log K_i^H which is equal to log K^H - log K^i. The position of the lines relative to each other reveals this expected selectivity pattern with copper more strongly adsorbed compared to the other metal ions.

However, such an extrapolation does not permit an independent calculation of both log K^H and log K^A. In order to obtain both values use is made of the development of Q_i^H at small resin loadings. By varying K^H at constant K_i^H until the calculated development of Q_i^H follows the experimentally based values of Q_i^H in the region y(i) < 0.2, the unknown complexation constants being sought can

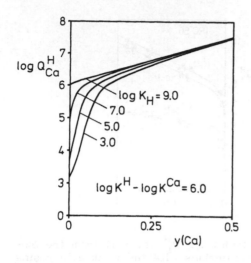

FIG. 10.   Theoretical development of log Q_{Ca}^{H} vs. y(Ca) for Lewatit CNP 80.   (Reproduced from [27] with permission of Elsevier Science Publishers.)

be deduced.   Figure 10 provides an example of such use of the cal-culated development of log Q_i^{H} versus y(i) at constant K_i^{H} for the Ca^{2+}/H^{+} exchange on Lewatit CNP 80.

The development of the series of curves indicates that with in-creasing log K^{H} more and more counterions are fixed in Stern layers, whereas at stronger dissociation values an increasingly large fraction of the counterion is located in the diffuse layer.   Greater deviation from linear development of log Q_i^{H} relative to y is due to this effect as well.   The formation constants for surface complexes with H^{+} and other counterions is calculable in this manner.   The numerical values of formation constants, so obtained, are listed in Table 4 [27].

The complex formation constants, K^{H}, listed in Table 4 are well in the range of dissociation constants for weak-acid resins or of polyacrylic or polymethacrylic gels [38-40].

It is apparent from Eq. (44) that the numerical values of the surface parameter A_0 and the capacitances of the electric double layers C(S,I) cannot be obtained independently.   Only their product, $A_0C(S,I)$, is accessible from the slope of the straight line.   For

TABLE 4 Formation Constants of Surface Complexes on Weak-Acid Resins (Reproduced from [27] with Permission of Elsevier Science Publishers)

Exchange equilibrium	log K^i			z(R)
	CNP 80	IRC 50	OC 1058	
$RCOO^- + H^+ \Leftrightarrow RCOOH$	3.40	4.13	4,30	-1
$R(COO^-)_2 + Cu^{2+} \Leftrightarrow R(COO)_2Cu(1)$	2.45	—	—	-2
$R(COO^-)_2 + Cu^{2+} \Leftrightarrow R(COO)_2Cu(2)$	1.80	2.53	—	-2
$R(COO^-)_2 + Zn^{3+} \Leftrightarrow R(COO)_2Zn$	1.00	—	—	-2
$R(COO^-)_2 + Ca^{2+} \Leftrightarrow R(COO)_2Ca$	0.95	0.85	0.80	-2
$R(COO^-)_2 + Mg^{2+} \Leftrightarrow R(COO)_2Mg$	0.31	0.52	0,60	-2
$RCOO^- + Na^+ \Leftrightarrow RCOONa$	-0.25	-0.25	-0.25	-1

Cu(1) and Cu(2) are designating the first and second Stern layers.

assessments relative to the hypothetical surface it can be assumed, e.g., that Na^+ ions are sorbed in their hydrated forms. Since the diameter of water molecules and Na^+ ions is 0.3 nm and 0.095 nm, respectively, it can be assumed that the distance, d, between sodium ions and the surface is about 0.4 nm. Application of the capacitor relationship

$$C(H,Na) = \frac{\varepsilon_o \varepsilon_r}{d} \tag{66}$$

while assuming that the relative permittivity of water also is applicable in the Stern layer permits calculation of the capacitance of the surface/sodium layer capacitor. By using this estimate of capacitance and the product, $A_0 C(S,I)$, defined by the slope of the straight line a specific surface which can be considered the effective surface is resolved. No estimates of the electric capacitance for the surface/diffuse layer capacitor were made. Instead the C_D value of 20 $\mu F/cm^2$ favored by Davis et al. was adopted. Table 5 summarizes resin capacities, capacitances of capacitors, and surface complexation parameters made available as described and used for further predictions of equilibria [27].

TABLE 5   Parameters of the Surface Complex Formation Model for Cation Exchangers (Partially Reproduced from [27] with Permission of Elsevier Science Publishers)

Resin	Form	q_{max} meq/g	Exchange	$C(i,j)$ $\mu F/cm^2$	$\log K_i^I$	$m(i,j)$	$z(R)$
CNP 80	H^+	6.25	H^+/Na^+	176	3.65	2.78	-1
	Mg^{2+}	4.20	H^+/Mg^{2+}	340	6.49	2.88	-2
	Ca^{2+}	4.98	H^+/Ca^{2+}	340	5.95	2.88	-2
	Zn^{2+}	5.01	H^+/Zn^{2+}	1100	5.80	0.89	-2
	Cu^{2+}/H^+	5.10	H^+/Cu^{2+}	∞	4.35	0	-2
	Cu^{2+}/H^+	5.10	H^+/Cu^{2+}	1000	5.00	2.13	-2
			$[y(Cu) > 0.33]$				
IRC 50	Na^+	2.57	H^+/Na^+	170	4.38	2.38	-1
	Mg^{2+}	3.20	H^+/Mg^{2+}	340	7.73	2.32	-2
	Ca^{2+}	3.38	H^+/Ca^{2+}	340	7.40	2.32	-2
	Cu^{2+}/H^+	3.69	H^+/Cu^{2+}	940	5.72	0.84	-2
OC 1058	Na^+	2.34	H^+/Na^+	195	4.50	2.15	-1
	Mg^{2+}	3.24	H^+/Mg^{2+}	340	8.00	2.47	-2
	Ca^{2+}	3.40	H^+/Ca^{2+}	340	7.80	2.47	-2
S 100	H^+	2.72	Ca^{2+}/H^+	n.c.	0.55	1.35	-2
	Na^+	2.72	Ca^{2+}/Na^+	n.c.	0.23	1.47	-2
	Mg^{2+}	2.72	Ca^{2+}/Mg^{2+}	n.c.	0.40	0	-2
HCR-S	H^+	2.65	Ca^{2+}/H^+	n.c.	0.60	1.49	-2
	Na^+	2.65	Ca^{2+}/Na^+	n.c.	0.20	1.77	-2
	Mg^{2+}	2.65	Ca^{2+}/Mg^{2+}	n.c.	0.42	0	-2

(n.c. = not calculated)

   b.   *Strong-Acid Resins*   The development of the generalized separation factor as a function of the loadings with magnesium, sodium, and hydrogen ions is plotted in Fig. 11a-c for strong-acid resins [41].

   In the case of the binary equilibrium Ca^{2+}/Mg^{2+} the development of the generalized separation factor as a function of $y(Mg)$ produces horizontal lines to indicate that both Ca^{2+} and Mg^{2+}

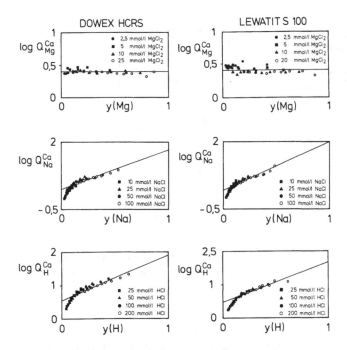

FIG. 11. Development of log Q_i^{Ca} vs. y(i) for the strong-acid resins Lewatit S 100 and Dowex HCR-S.

species are sorbed in identical layers as they were with weak-acid resins. Numerical values of log K_{Mg}^{Ca} are relatively small. Furthermore, there is no evidence of any curvature in the equilibrium line at small resin loadings.

For the exchange of calcium for sodium ions or protons there are no regions of constancy in the generalized separation factor. This result is to be expected from theoretical considerations. For both resins the development of log Q versus y shows a marked curvature for small loadings: in the systems with sodium up to y(Na) ≃ 0.15, and up to y(H) ≃ 0.20 in the systems with protons. It can therefore be concluded that there are more protons than Na$^+$ ions in the diffuse layer at low loadings. For larger resin loadings straight lines are found. Extrapolation of these lines to y = 0 lead to the ratios of affinities which demonstrate that protons have a smaller affinity for the resin than Na$^+$ ions. The preference for

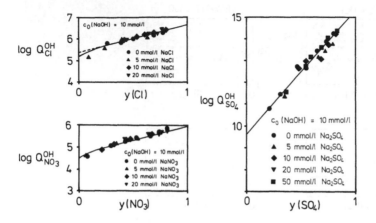

FIG. 12. Generalized separation factor log Q_i^{OH} vs. y(i) from the exchange of OH$^-$ ions for Cl$^-$, NO$_3^-$, and SO$_4^{2-}$ ions with the weak-base resin Amberlite IRA 93-SP.

Na$^+$ ions is well known from the usual selectivity series. However, the slopes for the systems with protons are smaller. The proton layer, consequently, is closer to the surface than the sodium-ion layer even though the proton ion is not preferred. As a consequence it follows that the preferred uptake of one species does not necessarily require its sorption at a smaller distance closer to the surface.

Calculations of individual complexation constants K^i and of capacitances $C(i,j)$ remain to be made.

2. Anion Exchangers

Anion-exchange equilibria were investigated using the weak-base resins Amberlite IRA 93 and Dowex MWA-1 and the strong-base resins Amberlite IRA 410, Amberlite IRA 958, and Amberlite IRA 996. Investigations of the exchange of OH$^-$ for Cl$^-$, NO$_3^-$, and SO$_4^{2-}$ on weak-base resins and of equilibria including NO$_3^-$, Cl$^-$, HCO$_3^-$, and SO$_4^{2-}$ on strong-base exchangers were presented. In addition, equilibria involving CuEDTA^{2-} and NiEDTA^{2-} complex species were studied. The experimental data were used to obtain generalized separation factors from the relationship

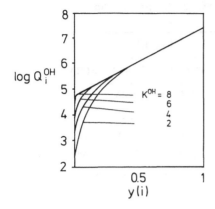

FIG. 13.   Theoretical development of log Q_i^{OH} vs. y(i) for monovalent counterions with weak-base resins.

$$\log Q_A^{OH} = \log c(A) + \log \frac{q(OH)}{q(A)} - z(R)pOH \qquad (67a)$$

for weak-base resins [with the simplification pOH = -log c(OH)] and from the relationship

$$\log Q_B^A = \log c(B) + \log \frac{q(A)}{q(B)} + z(R) \log c(A) \qquad (67b)$$

for strong-base resins.

   a.   *Weak-Base Resins*   Data compiled for the weak-base Amberlite IRA 93 resin at equilibrium are plotted in Figs. 12a-c [42].   As for the cation-exchange resins, plots of log Q as a function of dimensionless loading with anions yield straight lines when y(i) > 0.2; the deviation for smaller loadings is similar as well.   However, the deviation from the straight line is much smaller than with weak-acid resins. The formation constant for OH^- ions can be deduced just as for weak-acid resins if at constant K_i^{OH} the value of K^{OH} is varied until optimum agreement is obtained with the experimental development of log Q_i^{OH}.   Figure 13 shows the stepwise approach to the experimentally based development of log Q as a function of y for the exchange of monovalent species.   Comparison with experimental data yields the the relatively high value of 7.05 for both resins.   As a consequence, the degree of protonation of weak-base resins (which can be

TABLE 6   Formation Constants of Surface Complexes on
Weak-Base Resins

Exchange equilibrium	$\log K^i$		
	IRA 93	MWA-1	z(R)
$R-NR_3H^+ + OH^- \Leftrightarrow R-NR_3 \cdot H_2O$	7.05	7.05	+1
$R-NR_3H^+ + Cl^- \Leftrightarrow R-NR_3HCl$	1.70	1.20	+1
$R-NR_3H^+ + NO_3^- \Leftrightarrow R-NR_3HNO_3$	2.40	1.65	+1
$R-(NR_3H^+)_2 + SO_4^{2-} \Leftrightarrow R-(NR_3H)_2SO_4$	5.05	5.95	+2

interpreted as the degree of dissociation of OH^- ions) is much
smaller than that of weak-acid resins.   Complexation constants are
listed in Table 6, while equilibrium parameters are listed in Table 7.

   b.   *Strong-Base Resins*   Data compiled in studying the equilibria
of strong-base resins with common anions are presented in Figs.
14a-c [43].   Their evaluation is the same as for weak-base resins.
The log Q versus y plots obtained from these data yield straight
lines with no evidence of any deviation from linearity at small loadings.   For the exchange of nitrate for chloride the generalized separation factor is approximately constant; there is only a negligible
small positive slope.   For the exchange of NO_3^- for HCO_3^-, however,
the slope is slightly negative.   This may indicate that bicarbonate
species are located closer to the surface.   However, one must realize
that the concurrent presence of CO_3^{2-} species could contribute to
this dependency.   The straight lines of different slopes that are
observed for the exchange of both NO_3^- and Cl^- for SO_4^{2-} on Amberlite IRA 410 necessarily lead to the conclusion that sulfate species
are located in two different layers just like in the sorption of Cu^{2+}
by one of the weak-acid resins.   For the Type I resin Amberlite
IRA 996, no sorption of sulfate in two different layers is detectable.
   Data from the exchange of nitrate or chloride for sulfate ions
in both resins have been employed to evaluate log Q.   The straight
lines that develop from plots of log Q versus $y(SO_4)$ have a positive
slope.   As a consequence we have to assume that the sulfate layer

TABLE 7 Parameters of the Surface Complex Formation Model for Anion Exchangers

Resin	Form	q_{max} meq/g	Exchange	$C(i,j)_2$ μF/cm^2	log K_i^I	m(i,j)	z(R)
IRA 93	Cl^-	1.31	OH^-/Cl^-	160	5.35	1.32	+1
	NO_3^-	1.38	OH^-/NO_3^-	160	4.05	1.32	+1
	SO_4^{2-}	1.31	OH^-/SO_4^{2-}	104	9.65	5.65	+2
MWA-1	Cl^-	1.37	OH^-/Cl^-	1160	5.85	0.17	+1
	NO_3^-	1.44	Ohm/NO_3^-	1160	5.40	0.17	+1
	SO_4^{2-}	1.32	OH^-/SO_4^{2-}	108	15.85	3.95	+2
IRA 68	NO_3^-	1.67	OH^-/NO_3^-	n.c.	3.88	0.67	+1
IRA 410	Cl^-	2.15	Cl^-/HCO_3^-	n.c.	0.45	-0.26	+1
	NO_3^-	2.15	NO_3^-/HCO_3^-	n.c.	0.83	-0.01	+1
	NO_3^-	2.15	NO_3^-/Cl^-	n.c.	0.46	0.06	+1
	NO_3^-	2.15	NO_3^-/SO_4^{2-}	n.c.	0.47	2.62	+2
					-1.03	4.45	+2
				$[y(SO_4) > 0.6]$			
IRA 458	NO_3^-	1.79	$NO_3^-/C;^-$	n.c.	0.42	0	+1
	SO_4^{2-}	1.79	Cl^-/SO_4^{2-}	n.c.	-0.40	1.54	+2
IRA 996	NO_3^-	1.50	NO_3^-/Cl^-	n.c.	0.91	-0.08	+1
	SO_4^{2-}	1.50	NO_3^-/SO_4^{2-}	n.c.	2.60	2.98	+2
IRA 958	Cl^-	1.07	Cl^-/CuE^{2+}	n.c.	-1.66	1.46	+2
	Cl^-	1.07	Cl^-/NiE^{2+}	n.c.	-1.82	1.53	+2

($CuE^{2-} = CuEDTA^{2-}$; $NiE^{2-} = NiEDTA^{2-}$; n.c. = not calculated)

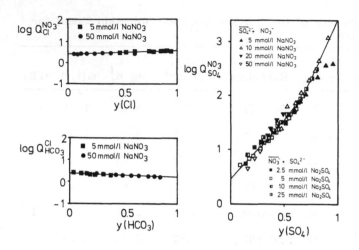

FIG. 14. Generalized separation factor log $Q_i^{NO_3}$ vs. y(i) from the exchange of NO_3^- ions for Cl^-, HCO_3^-, and SO_4^{2-} ions with the strong-base resin Amberlite IRA 410.

is further from the surface than the Cl^- ion. This again demonstrates that there is no strict interdependence between the selectivity order and the sequence of surface complex layers. Equilibria including OH^- ions have not been investigated. Therefore, no numerical values for the complexation or dissociation of hydroxyl ions can be given.

The exchange of $CuEDTA^{2-}$ and $NiEDTA^{2-}$ for Cl^- ions have been investigated for the resin Amberlite IRA 958 [44]. Representative results are plotted in Fig. 15. It is apparent that the surface complexation approach is applicable even for the sorption of such unusual and bulky anions. The development of positively sloped straight lines indicates that the complex species are preferred over chloride ions.

Parameters characterizing the above equilibria with strong-base resins are listed in Table 7.

3. Chelating Resins

Experiments with chelating resins were carried out using Lewatit TP 207 and Amberlite IRC 718. Both resins contain iminodiacetic-acid functional groups. The counterions employed were Na^+ and the

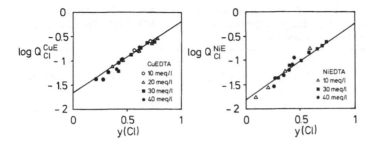

FIG. 15. Generalized separation factor $\log Q_{Cl}^{CuE}$ and $\log Q_{Cl}^{NiE}$ vs. y(Cl) for the strong-base resin Amberlite IRA 958.

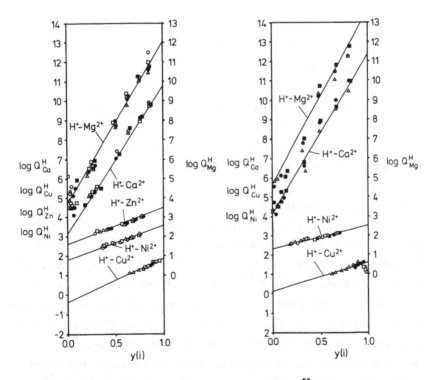

FIG. 16. Generalized separation factor $\log Q_i^H$ for the chelating resins Lewatit TP 207 and Amberlite IRC 718. (Reproduced from [36] with permission of Elsevier Science Publishers.)

TABLE 8   Parameters of the Surface Complex Formation Model for Chelating Resins (Partially Reproduced from [36] with Permission of Elsevier Science Publishers)

Resin	Form	q_{max} meq/g	Exchange	$C(i,j)_2$ µF/cm²	$\log K_i^I$	$m(i,j)$
TP 207	Mg^{2+}	2.638	H^+/Mg^{2+}	132	380	8.20
	Ca^{2+}	2.741	H^+/Ca^{2+}	141	315	5.55
	Cu^{2+}	2.320	H^+Cu^{2+}	478	-030	2.05
	Ni^{2+}	2.695	H^+/Ni^{2+}	n.c.	180	1.80
	Zn^{2+}	2.426	H^+/Zn^{2+}	n.c.	260	1.90
IRC 718	Mg^{2+}	1.961	H^+/Mg^{2+}	122	4.70	9.65
	Ca^{2+}	1.987	H^+/Ca^{2+}	132	4.12	8.23
	Cu^{2+}	1.923	H^+/Cu^{2+}	767	0.20	1.40
	Ni^{2+}	1.627	H^+/Ni^{2+}	n.c.	2.30	1.25

(n.c. = not calculated)

divalent Mg^{2+}, Ca^{2+}, Zn^{2+}, Ni^{2+}, and Cu^{2+}. The response of log Q_i^H to y(i) is plotted in Fig. 16 for the bivalent metal ions [36,45, 46]. The linearity of the response encountered once again holds for a broad range of concentrations. With both resins deviation from linearity is practically negligible. Thus, it can be concluded that the distribution of counterions in the diffuse layer is minimal. This is adequately well explained by the formation of strong coordination complexes between the divalent ions and the IDA sites. The experimental data extend only over the region of metal loading dominated by the heavy metals (Cu^{2+}, Ni^{2+}, Zn^{2+}) and the straight lines are justified only for this range. However, strong deviations or sorption in different layers is not likely. Thus the ratio of affinities can be deduced from the figures as they were for the other resins.

A comparison of the development of log Q versus y for representative systems again reflects the selectivity order with Cu^{2+} being strongly preferred. The ratio of affinities is smaller than in standard weak-acid resins indicating their preference for heavy-metal species. The large difference in the affinities of calcium and magnesium is due to the formation of weak complexes by Mg^{2+} ions.

The graphical representations of experimental data does not exhibit any curvature for small loadings. It is therefore impossible to
deduce numerical values of log K^H. Such absence of curvature must
be a consequence of strong sorption of divalent ions forcing the
counterions into fixed positions and of the small degree of dissociation of the protonated form which is considerably less than that of
ordinary weak-acid resins [47]. Equilibrium parameters describing
the exchange of protons for bivalent metal ions are listed in Table 8.

Assessment of data compiled at equilibrium for the exchange of
protons for Na^+ ions in the chelating resins reveals that the assumption of two different complexation constants is consistent with the
stepwise dissociation of IDA functional groups [36].

C. Evaluation of Multicomponent Equilibria

From the deviation of generalized separation factors in binary and
multicomponent systems it follows that the equilibrium A/B is characterized by the same parameters log K_B^A and m(A,B) [see Eqs. (43)
and (48)]. These parameters can, as a consequence, also be derived
from multicomponent equilibrium data if log Q_B^A is plotted versus the
sum of loadings y(B) + y(C) + $\cdots$ + y(N). Analysis of both binary
and multicomponent systems, therefore, has to lead to identical
parameters. Demonstration of this provides a very difficult test of
both the validity of the theoretical approach and the analytical accuracy.
The pertinent studies of binary and ternary anion-exchange equilibria
do show that there is an excellent agreement between both sets of
parameters to prove the validity of the theoretical approach [25,42,43].

V. COMPARISON OF EXPERIMENTAL AND PREDICTED EQUILIBRIA
A. Binary Equilibria

Equilibrium parameters for the different kinds of resins were obtained
from a number of binary experiments. One single set of parameters
was deduced for each binary exchange on each exchanger. These
sets, valid for any experimental condition, were used for the anticipation of equilibria on a theoretical basis. As becomes apparent in
each case, there is excellent agreement between predicted and

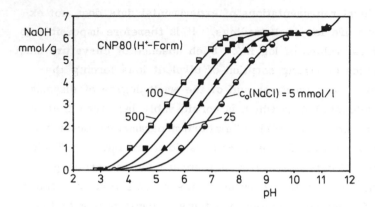

FIG. 17. Comparison of experimental and predicted titration curves for the weak-acid resin Lewaitit CNP 80.

experimental data. This result was not obtained with earlier surface complexation models, which sometimes fail even if one component is strongly preferred [20].

1. Weak-Acid Resins

Neutralization of weak-acid resins in the free-acid form with sodium hydroxide yields titration curves in which the (dimensionless) resin loading can be plotted as a function of the pH of the solution. This provides one possible representation of the exchange equilibrium. Figure 17 shows titration curves for Lewatit CNP 80 at different initial concentrations of NaCl.

Figure 18 provides a comparison of equilibria for an exchange of protons for Ca^{2+} ions at different initial conditions. The uptake of divalent species like that of monovalent counterions is strongly dominated by the proton concentration. As a consequence, these equilibria are also treated like the titration curves with the (dimensionless) resin loadings by the metal ions plotted versus the pH of the solution [27].

As mentioned in Sec. II the surface complexation approach allows chemical reactions or equilibria in the liquid phase to be taken into account using the same set of equilibrium parameters for the ion exchange part of the entire equilibrium. Figure 19 projects some

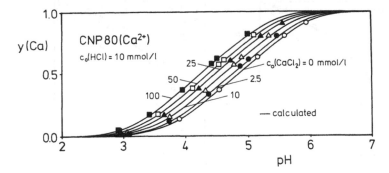

FIG. 18. Comparison of experimental and predicted equilibria H^+/Na^+ for the weak-acid resin Lewatit CNP 80. (Reproduced from [27] with permission of Elsevier Science Publishers.)

representative results for the exchange of Na^+ against H^+ at different initial NaCl concentrations in carbonic-acid-bearing systems [25]. Depending on the initial concentration of NaCl the isotherms are shifted to larger sodium concentrations. The results agree well with isotherms for the regeneration of weak-acid resins by means of carbonic acid [48].

2. Strong-Acid Resins

Equilibria with strong-acid resins are usually represented as square diagrams in which the dimensionless loading x(i) of the component i in the solution is plotted versus the dimensionless concentration in the

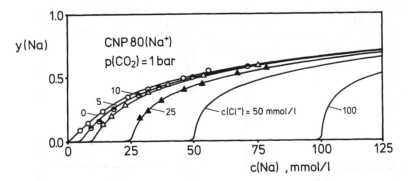

FIG. 19. Comparison of experimental and predicted equilibria H^+/Na^+ for the weak-acid resin Lewatit CNP 80 and systems with carbonic acid.

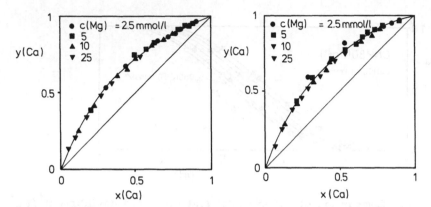

FIG. 20. Comparison of experimental and predicted equilibria Ca^{2+}/Mg^{2+} for the strong-acid resins Lewatit S 100 and Dowex HCR-S.

resin phase or the (dimensionless) resin loading y(i). Figure 20 shows results of the Ca^{2+}/Mg^{2+} equilibrium on Lewatit S 100 and Dowex HCRS. In accordance with the horizontal log Q versus y(Mg) dependencies symmetrical Langmuir isotherms which do not depend on the normality of the liquid phase were found [41]. For the exchange of Ca^{2+} for Na^+ or H^+, however, the isotherms are not symmetrical and they depend on total concentration. Representative results with Lewatit S 100 are presented in Fig. 21.

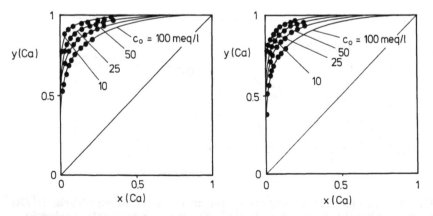

FIG. 21. Comparison of experimental and predicted equilibria of Ca^{2+} for Na^+ and H^+ ions for the strong-acid resin Lewatit S 100.

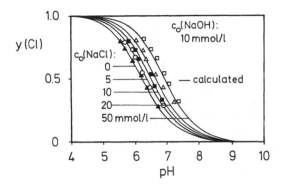

FIG. 22. Comparison of experimental and predicted equilibria of OH^-/NO_3^- for the weak-base resin Dowex MWA-1.

3. *Weak-Base Resins*

Weak-base resins can be titrated like the weak-acid resins by adding strong acid to the resin in the free-base form. Resin loading with, e.g., Cl^- ions is, as a result, obtained as a function of pH. The representative plots once again represent the protonation equilibrium of the functional groups which can formally be interpreted as the equilibrium behavior of the exchange of hydroxyl ions for anions of strong acids. Figure 22 contains one set of titration curves for the resin Dowex MWA-1 at different initial concentrations of NaCl [42].

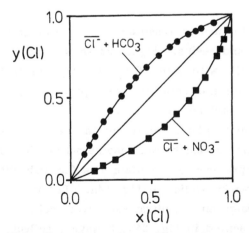

FIG. 23. Comparison of experimental and predicted equilibria of Cl^- for NO_3^- and HCO_3^- for the strong-base resin Amberlite IRA 410.

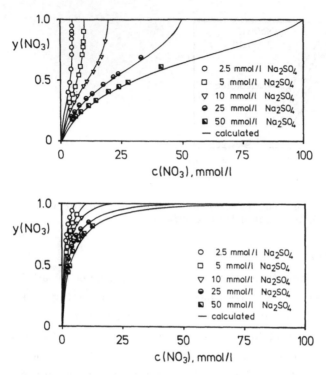

FIG. 24.   Comparison of experimental and predicted equilibria SO_4^{2-} / NO_3^- for the strong-base resins Amberlite IRA 410 and Amberlite IRA 996.

4.   Strong-Base Resins

Equilibria with strong-base resins are represented in square diagrams just as they were with strong-acid resins.   Figure 23 shows isotherms obtained from the exchange of Cl^- with both HCO_3^- and NO_3^- ions with Amberlite IRA 410.   In both cases the isotherms are approximately symmetrical and independent of total normality [43].

Selection of a semidimensionless plot to represent the exchange of nitrate for sulfate permits discrimination between isotherms at different liquid-phase normalities.   The preference of sulfate ions for Amberlite IRA 410 is clearly apparent from inspection of Fig. 24a.   The isotherms for the NO_3^-/SO_4^{2-} exchange on the nitrate-selective resin Amberlite IRA 996 that are presented in Fig. 24b, however, reflect the high affinity of this resin for nitrate ions [43].

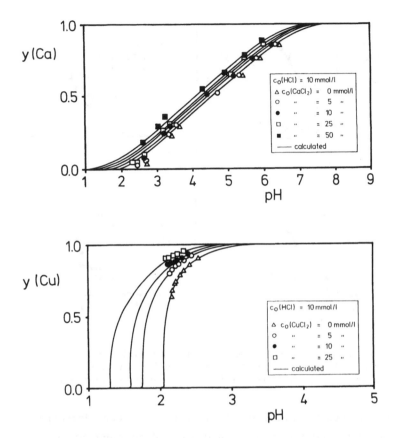

FIG. 25. Comparison of experimental and predicted equilibria H^+/Ca^{2+} and H^+/Cu^{2+} for the chelating resin Lewatit TP 207. (Reproduced from [36] with permission of Elsevier Science Publishers.)

5. Chelating Resins

Results obtained in studies of the exchange of proton for bivalent metal ions in chelating resins are presented in Fig. 25. Since these resins contain carboxylic-acid groups their performance is very similar to the standard weak-acid resin performance. The pH of the liquid phase also dominates these binary equilibria and their equilibria are consequently represented as titration curves as well. Due to the sizably different affinities for different metal ions these titration curves can be quite unusual in shape. In the case of Mg/H^+ equilibria the

198   HOLL ET AL.

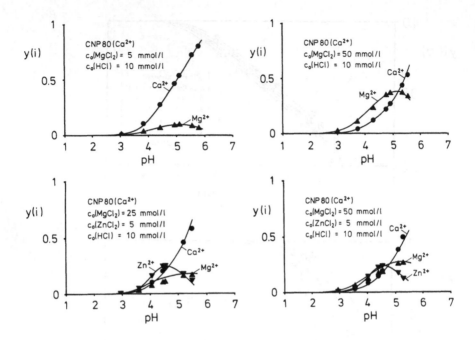

FIG. 26. Comparison of experimental and predicted ternary and
quarternary equilibria on the weak-acid resin Lewatit CNP 80.
(Reproduced from [27] with permission of Elsevier Science Publishers.)

development of experimental data gives evidence of two dissociation
steps of the resin and the predictive quality of only one set of
parameters is unsatisfactory [36,45].

B. Multicomponent Equilibria

A number of methods for the prediction of multiionic equilibria has
been proposed and discussed in the literature [50-63]. The proposed methods are restricted to ternary equilibria. Usually attempts
are made to apply binary equilibrium parameters in order to predict
the composition of the ternary system. Except for a few cases, however, the agreement with experimental data is not satisfactory. An
additional drawback of the approaches is the extra complication introduced to the computation. Much better results are obtained by applying the surface-complexation approach in which the binary equilibrium parameters are also applied to the prediction of multicomponent

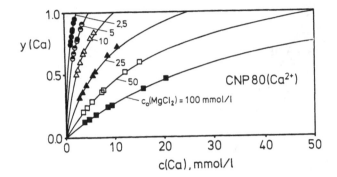

FIG. 27. Comparison of experimental and predicted Ca^{2+}/Mg^{2+} equilibria on the weak-acid resin Lewatit CNP 80. (Reproduced from [27] with permission of Elsevier Science Publishers.)

equilibria. Support for the above statement is available from investigations of ternary systems and from some quarternary systems with weak-acid resins. With few exceptions there is an excellent agreement between prediction and experiment.

1. Weak-Acid Resins

Weak-acid resins are highly selective for protons. As a consequence, the concentration of protons in systems with metal ions and weak-acid resins is usually much smaller than those of the metal ions. For this reason pH can be selected as the abscissa parameter in the representation of multicomponent equilibria encountered during neutralization studies. Results of such analysis of ternary and quarternary equilibria are plotted in Figs. 26a-d [27].

Due to the extremely high affinity of weak-acid resins for protons the equilibrium for divalent ion pair such as Ca^{2+}-Mg^{2+} is a ternary one with the third component (H^+) present at a negligible concentration level. Experimentally measured and predicted equilibria are compared in Fig. 27 in a semidimensionless representation. Normalization of the abscissa would yield a single symmetric isotherm in a square diagram [25].

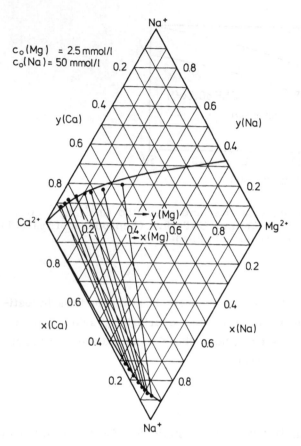

FIG. 28. Comparison of experimental and predicted $Ca^{2+}/Mg^{2+}/Na^+$ equilibria on the strong-acid resin Lewatit S 100.

2. Strong-Acid Resins

With strong-acid resin-salt systems containing cations like Ca^{2+}, Mg^{2+}, Na^+, and H^+ there is no special affinity of the resin for one species. The concentrations of all ions will, as a consequence, remain within the same relative level. As a consequence, the ternary equilibria can be represented in triangular diagrams. Figure 28 is used to show this for the Ca^{2+}, Mg^{2+}, and Na^+ ion, Lewatit S 100 resin system. Two "conjugated" triangles were employed for this representation of the composition of the liquid phase (lower triangle) and resin phase (upper triangle), respectively. It is obvious that

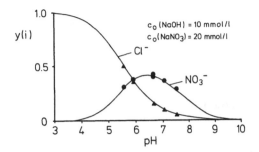

FIG. 29. Comparison of experimental and predicted $OH^-/Cl^-/SO_4^{2-}$ equilibrium on the weak-base resin Amberlite IRA 93-SP.

there are differences between the experimental and predicted projection. These have to be blamed, in part, on the neglect of nonnegligible amounts of counterions in the diffuse layer and upon inaccuracies of analyses. The composition of the liquid phase is much better predicted than that of the resin phase. For Dowex HCRS the differences are slightly larger [41].

3. *Weak-Base Resins*

The equilibria of weak-base resins depend strongly on the concentration of H^+ or OH^- ions in the liquid phase. As a consequence, pH can serve once again as an appropriate abscissa parameter. The experimental and predicted representation of y(i) versus pH for Amberlite IRA 93 and OH^-, Cl^-, and SO_4^{2-} counterions are compared in Fig. 29 [42].

4. *Strong-Base Resins*

Strong-base resins do not exhibit any special selectivity for most normal anions. Therefore, triangular diagrams can be used as they were for strong-acid resins, to plot solution and resin-phase compositions encountered in ternary equilibria. Figure 30 provides such a representation for a system with HCO_3^-, Cl^-, and NO_3^- ions and the Amberlite IRA 410 resin [43]. Systems that include sulfate ion ions have not yet been examined.

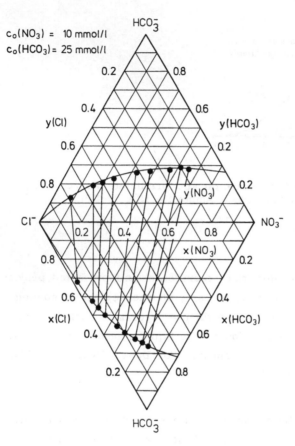

$c_o(NO_3) = 10 \text{ mmol/l}$
$c_o(HCO_3) = 25 \text{ mmol/l}$

FIG. 30.   Comparison of experimental and predicted NO_3^-/Cl^-/HCO_3^- equilibrium on the strong-base resin Amberlite IRA 410.

5.   Chelating Resins

Chelating resins, as pointed out earlier, are basically carboxylic-acid exchangers. As a consequence, the functional sites therefore strongly prefer protons. Thus, pH can be used as the abscissa parameter as it is with the standard carboxylic-acid resins. The experimental and predicted equilibrium properties for the ternary system H^+/Ca^{2+}/Cu^{2+}, Amberlite IRC 718 resin are compared in Fig. 31 [36].

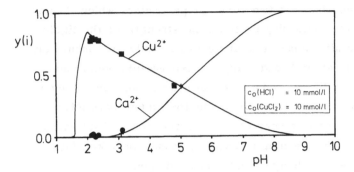

FIG. 31. Comparison of experimental and predicted $H^+/Ca^{2+}/Cu^{2+}$ equilibrium on the chelating resin Amberlite IRA 718. (Reproduced from [36] with permission of Elsevier Science Publishers.)

VI. CONCLUSIONS

It has been the object of this chapter to demonstrate the application of surface complexation theory for the interpretation of exchange equilibria of synthetic organic ion-exchange resins. Probably the most important abstraction adopted to facilitate this objective is that polymers with irregular hydrocarbon can be treated like planar surfaces of oxides. The uptake of counterions is considered as a local equilibrium reaction as it is for inorganic materials and the distribution of counterions near the surface is considered using Stern's theory. The triple-layer model of Davis et al. has been extended to a multilayer model. An important development to the deviation of a generalized separation factor which can be easily extracted from the experimental data. The fact that organic ion-exchange resins have an exchange capacity which is practically constant and much better defined than that of hydrous oxides is a sizable advantage in applying the model.

The development of the separation factor as a function of the resin phase composition provides a most useful characterization of the exchange equilibrium. A careful evaluation of data has demonstrated that most ions are located in ordered layers. As a consequence, binary equilibria can, except from a minor resin-phase composition

range, be characterized by two constant parameters which do not
depend on the total normality or the ionic strength of the liquid
phase.   This permits binary equilibria to be easily and accurately
predictable for a rather broad range of experimental conditions.
Multicomponent equilibria, considered as a superposition of several
binary equilibria are predicted by using the binary parameters
without any adaptation or correction.   Furthermore, simultaneous
chemical equilibria in the liquid phase can easily be taken into
account to facilitate exchange equilibria predictions.

In spite of its simplifications the surface complexation model provides an excellent description and prediction of counterion equilibria
with every kind of organic ion-exchange resin and facilitates the
accurate anticipation of their distribution in the resin and solution
phases.   This is of considerable importance for investigations of
multispecies exchange kinetics [31] and also for the prediction of
multispecies column performance [65].   Furthermore, on the basis
of a similar evaluation of limited data it has become evident that this
kind of approach can also be applied to equilibria with activated
alumina [66] and zeolites [67].   A similar assessment of the surface
complexation model to inorganic exchangers awaits additional
experimentation.

Quoting Balistrieri, surface complexation models are best guesses
of the physical, chemical, and electrical properties of the interfacial
region.   Hence, they may or may not have any relationship to reality.
They are, however, tools which can direct or shape our understanding of surface phenomena [20].   The results of this chapter demonstrate both the validity as well as the value of such a tool for the
description of exchange equilibria with organic ion-exchange resins.

ACKNOWLEDGMENT

The authors are much indebted to Mrs. Brigitte Kiehling for careful
drawing of the figures.

VII. APPENDIX

A_0	m^2/g	Surface parameter
$c(i)$	mol/l	Concentration of species "i^{z_i}"*
c_D	mol/l	Initial concentration of species "i"
$c(i)_0$	mol/l	Normality = $\Sigma\ z(i)c(i)$
$c(i)_D$	mol/l	Concentration of species "i" in the diffuse layer
C_D	$\mu F/cm^2$	Electric capacitance of a capacitor with the diffuse layer as one plate
$C(i,j)$	$\mu F/cm^2$	Electric capacity of the capacitor formed by the Stern layers of counterions "I" and "J"
d	m	Distance between two capacitor plates
F	C/eq	Faraday constant
$F(j;i)$	—	Distribution factor
I	mol/l	Ionic strength
K^i	1/mol	Surface complexation constant
K_j^i	—	Ratio of complexation constants
$\hat{K}_j^i$	—	Reduced ratio of complexation constants, defined by Eq. (55)
L	l	Volume of liquid phase
L_S	l	Volume of liquid phase close to the surface
$m(i,j)$	—	Slope of the straight line, defined by Eq. (43)
$\hat{m}(i,j)$	—	Reduced slope of the straight line, defined by Eq. (56)
$q(i)$	eq/kg	Loading of resin phase with species "i"
q_{max}	eq/kg	Exchange capacity
Q_j^i	—	Generalized separation factor, defined by Eq. (37)
R	J/mol K	Gas constant
S	kg	Resin quantity
T	K	Temperature
W_i	—	Stoichiometric factor, defined by Eq. (4)
x	m	Coordinate normal to surface

---

*Charges in concentration symbols have been deleted for simplicity.

$x(i)$ — Equivalent fraction in the liquid phase

$\qquad = z(i)c(i)/\Sigma_j\, z(j)c(j)$

$y(i)$ — Equivalent fraction in the resin phase

$\qquad = q(i)/q_{max}$

$y_{R,i}$ — Probability, defined by Eq. (64)

$y_s(z)$ — Sum of loadings with ions of valency "z"

z — Valency number

$\Gamma(i)$ $\quad$ eq/m^2 $\quad$ Surface charge equivalents

ε $\quad$ F/m $\quad$ Permittivity

ε_o $\quad$ F/M $\quad$ Electric field constant

ε_r $\quad$ — $\quad$ Relative permittivity

σ_D $\quad$ C/m^2 $\quad$ Charge density in the diffuse layer

$\sigma_{St,i}$ $\quad$ C/m^2 $\quad$ Charge density in Stern layer "i"

σ_s $\quad$ C/m^2 $\quad$ Charge density of the resin surface

Ψ_D $\quad$ V $\quad$ Electric potential in the diffuse layer

$\Psi_{St,i}$ $\quad$ V $\quad$ Electric potential in Stern layer "i"

Ψ_S $\quad$ V $\quad$ Electric potential at resin surface

REFERENCES

1. H. v. Helmholtz, *Wiss Abh. physik. techn. Reichsanstalt I*:925 (1879).

2. C. Gouy, *J. de Phys.*, 9:457 (1910).

3. D. I. Chapman, *Phil. Mag.*, 25:475 (1913).

4. O. Stern, *Z. Elektrochemie*, 30:508 (1924).

5. R. O. James and T. W. Healy, *J. Coll. Interf. Sc.*, 40:42 (1972).

6. R. O. James and T. W. Healy, *J. Coll. Interf. Sc.*, 40:53 (1972).

7. R. O. James and T. W. Healy, *J. Coll. Interf. Sc.*, 40:65 (1972).

8. J. W. Bowden, M. D. A. Bolland, A. M. Posner, and J. P. Quirk, *Nature Physical Science*, 245:81 (1973).

9. M. A. Anderson, J. F. Ferguson, and J. Gavis, *J. Coll. Interf. Sc.*, 54:391 (1976).

10. D. E. Yates, S. Levine, and T. W. Healy, *J. Chem. Soc. Faraday Trans. I*:1807 (1974).

11. W. Stumm, H. Hohl, and F. Dalang, *Croat. Chem. Acta*, *48*:491 (1972).

12. R. Kummert and W. Stumm, *J. Coll. Interf. Sci.*, *75*:373 (1980).

13. W. Stumm, R. Kummert, and L. Sigg, *Croat. Chem. Acta*, *53*:291 (1980).

14. J. A. Davis, R. O. James, and J. O. Leckie, *J. Coll. Interf. Sci.*, *63*:480 (1978).

15. J. A. Davis and J. O. Leckie, *J. Coll. Interf. Sci.*, *67*:90 (1978).

16. R. O. James and T. W. Healey, *J. Coll. Interf. Sci.*, *40*:65 (1972).

17. P. W. Schindler, B. Fürst, R. Dick, and P. Wolf, *J. Coll. Interf. Sci.*, *55*:281 (1976).

18. P. R. Anderson, M. M. Benjamin, in *Chemical Modelling of Aqueous Systems II* (D. C. Melchior and R. I. Bassett, eds.), American Chemical Society, Washington, DC, 1990, p. 272.

19. L. Sigg, *Die Wechselwirkung von Anionen und schwachen Säuren mit α-FeOOH in waßriger Lösung*, Thesis, Eidgenössische Technische Hochschule, Zürich, 1979.

20. L. S. Balistrieri and T. T. Chao, *Geochim. Cosmochim. Acta*, *54*:739 (1990) and references quoted.

21. K. F. Hayes, A. L. Roe, G. E. Brown, Jr., K. O. Hodgson, O. Leckie, and G. A. Parks, *Science*, *23*:783 (1987).

22. D. E. Yates, *The structure of the oxide/aqueous electrolyte surface*, PhD dissertation, University of Melbourne, 1973.

23. K. F. Hayes and J. O. Leckie, *J. Coll. Interf. Sci.*, *115*:563 (1987).

24. F. F. Cantwell, *Ion Exch. Solvent Extr.*, *9*:339 (1985).

25. J. Horst, *Beschreibung der Gleichgewichtslage des Ionenaustauschs an schwach sauren Harzen mit Hilfe eines Modells der Oberflächenkomplexbildung*, Thesis, University of Karlsruhe, 1988.

26. L. K. Koopal, W. H. van Riemsdijk, and M. G. Roffey, *J. Coll. Interf. Sci.*, *118*:117 (1987).

27. J. Horst and W. H. Höll, *Reactive Polymers*, *13*:209 (1990).

28. IUPAC, *Pure and Appl. Chem.*, *29*:619 (1972).

29. V. S. Soldatov and V. A. Bichkova, in *Ion Exchange Technology* (D. Naden and M. Streat, eds.), Ellis Horwood Ltd., Chichester, 1984, p. 179.

30. V. A. Bichkova and V. S. Soldatov, *Reactive Polymers*, *3*:207 (1985).

31. M. Franzreb, PhD Thesis, University of Karlsruhe, under preparation.

32. W. H. Höll, *Reactive Polymers*, 2 :93 (1984).

33. T. V. Arden, in *Ion Exchangers* (K. Dorfner, ed.), W. de Gruyter, Berlin, New York, 1991, Chap. 2.2.

34. W. H. Höll and B. Kiehling, *Water Research*, 15:1027 (1981).

35. F. Helfferich, *Ion Exchange*, McGraw Hill, New York, 1962.

36. W. H. Höll, J. Horst, and M. Wernet, *Reactive Polymers*, 14:251 (1991).

37. J. A. Marinsky, *J. Phys. Chem.*, 86:3318 (1982).

38. A. Chatterjee and J. A. Marinsky, *J. Phys. Chem.*, 67:41 (1961).

39. Y. Merle and J. A. Marinsky, *Talanta*, 67:41 (1963).

40. S. Alegret, M. T. Escalas, and J. A. Marinsky, *Talanta*, 67:683 (1984).

41. J. Günter, Diploma thesis, University of Karlsruhe, 1991.

42. M. Schilli, Diploma thesis, University of Karlsruhe, 1988.

43. M. Dumm, Diploma thesis, University of Karlsruhe, 1987.

44. G. Reize, Diploma thesis, University of Karlsruhe, 1991.

45. M. Wernet, Diploma thesis, University of Karlsruhe, 1989.

46. W. Widmer, Diploma thesis, University of Karlsruhe, 1990.

47. C. Berger-Wittmar, *Untersuchungen zur Regeneration schwach saurer Ionenaustauscher mit Kohlendioxid*, Thesis, University of Karlsruhe, 1976.

48. W. H. Höll, *Entwicklung und Grundlagen eines neuen Verfahrenskonzepts zur Teilentsalzung von Wasser mit Ionenaustauschern unter Verwendung von Kohlenstoffdioxid als Regenerierchemikalie*, Kernforschungszentrum Karlsruhe, Report No. 4022 (1985).

49. J. Dranoff and L. Lapidus, *Ind. Eng. Chem.*, 49:1297 (1957).

50. G. Jangg, *Österr. Chemiker Z.*, 59:331 (1958).

51. L. J. Pieroni and J. Dranoff, *AIChE J.*, 9:42 (1963).

52. M. Streat and W. J. Brignal, *Trans. Instn. Chem. Engrs.*, 48:T151 (1970).

53. R. P. Smith and E. T. Woodburn, *AIChE J.*, 24:577 (1977).

54. A. M. Elprince and K. L. Babcock, *Soil Science*, 120:332 (1975).

55. V. S. Soldatov and V. A. Bichkova, *Sep. Sci. Technol.*, 15:89 (1980).

56. J. Novosad and A. L. Myers, *Can. J. Chem. Engng.*, 60:500 (1982).

57. V. A. Bichkova and V. S. Soldatov, *Reactive Polymers*, 3:207 (1985).

58. M. Sengupta and T. B. Paul, *Reactive Polymers*, 3:217 (1985).

59. G. Vasquez, R. Mendez Pampin, and R. Blanco Caeiro, *Anales de Quimica*, 81:141 (1985).

60. A. L. Myers and S. Byington, in *Ion Exchange: Science and Technology* (A. Rodrigues, ed.), Martinus Nijhoff Publishers, Dordrecht / Boston / Lancaster, 1986.

61. V. A. Bichkova, V. S. Soldatov, and T. Y. Alefirova, *Z. Fiz. Khim.*, 60:1213 (1986).

62. V. S. Soldatov, V. A. Bichkova, V. P. Kol'nenkov, and T. Y. Alefirova, *Z. Fiz. Khim.*, 61:1855 (1987).

63. T. Zuyi and Z. Haimei, *Desalination*, 69:125 (1988).

64. D. C. Shallcross, C. C. Herrmann, and B. J. McCoy, *Chem. Eng. Sci.*, 43:279 (1988).

65. W. H. Höll, paper under preparation.

66. H. J. Fader, *Untersuchungen zur Kinetik der Sorption von Anionen aus wäßrigen Lösungen an Aktivtonerde*, Thesis, University of Karlsruhe (under preparation).

67. R. Harjula, Private communication.

4

Surface Complexation of Metals by Natural Colloids

GARRISON SPOSITO   University of California at Berkeley, Berkeley, California

I. INTRODUCTION

A. Metal-Natural Colloid Surface Reactions

Natural colloids are mixtures of aluminosilicates, hydrous oxides, and humus [1]. The surface reactivity of these particles is conditioned on the aqueous environment in which they reside and by the molecular structure of the colloid/aqueous solution interface [2-4]. Metal-ion reactions with natural colloid surfaces lead, in the simplest case, to an accumulation of matter at the interface without the development of three-dimensional structures. This process, termed adsorption [4], results in a change of surface properties that in turn affects the composition of the contiguous aqueous solution. Often, adsorption is the first step in a series of reactions that determine heterogeneous equilibria mediated by dissolution-precipitation, surface precipitation, and oxidation-reduction phenomena [5,6].

Molecular models have been developed throughout the present century in order to provide a mechanistic description of metal adsorption processes which has predictive power [7]. The oldest of these is the

211

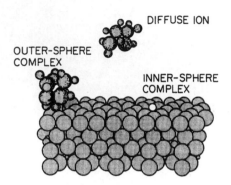

FIG. 1.   Three modes of metal ion adsorption by a smectite.

diffuse double-layer model, whose modern formulation continues to
be the object of intense study in statistical mechanics [8].   The
molecular species underlying this model is a solvated metal cation
that moves freely within a near-surface zone of an aqueous solution
phase, as illustrated in the upper portion of Fig. 1.   In this case,
the accumulation of metal ions that is the basis for adsorption derives from the restriction of the mobile adsorbate to a nanometer-
scale region controlled by the electric field of a charged surface.
This concept, despite its simplicity, has enjoyed success as a description of the adsorption of Li^+, Na^+, and, to some extent, Mg^{2+}
and Ca^{2+}, by natural colloids [9,10].   For metal ions whose chemistry is likely to involve more than electrostatic considerations
(e.g., Cu^{2+}), however, the diffuse double-layer model does not
suffice as a molecular theory of adsorption.   For these ions, the
consensus has developed that their adsorption reactions can be
interpreted accurately from the perspective of coordination chemistry [11-13].

B.   Surface Complexes

The concept of the functional group is familiar in the chemistry
of organic molecules, both monomeric and polymeric (e.g., ion
exchange resins).   Humus is a collection of such molecules found
as natural colloidal material in soils, sediments, and waters [14].

Of the variety of functional groups present in the organic compounds that polymerize ultimately to form humus (e.g., carboxyl and phenolic hydroxyl), some ultimately reside on the interface between the solid organic matter and a contacting aqueous solution phase. These molecular units that protrude from the solid surface are *surface functional groups*. Surface functional groups in general can be bound to either organic or inorganic solids, and they can have any molecular structural arrangement that is possible for them were they bound to small molecules instead of polymeric materials. Unlike the situation for small molecules in solution, however, functional groups on surfaces cannot be diluted, even in aqueous suspension. Unless the adsorbent to which they are bound decomposes, surface functional groups remain separated by more or less fixed distances, regardless of how dilute a suspension of the adsorbent may be. Thus the groups remain closely associated and can influence one another in most circumstances. Because of the variety in natural colloid composition, a broad spectrum of surface functional group reactivity is likely and a wide range of stereochemical and charge distribution characteristics are possible. It is conceivable that no particular type of surface functional group (e.g., hydroxyl) possesses unique quantitative chemical properties (e.g., a fixed proton dissociation equilibrium constant), but instead can be characterized only by ranges of values for these properties [14]. This "smearing-out" of their chemical behavior is another important feature that distinguishes surface functional groups from functional groups bound to small molecules.

Complexes that may form between surface functional groups and metal ions in an aqueous solution can be classified by analogy to the complexes that form among aqueous species [13]. For example, if no water molecule is interposed between the surface functional group and the ion or molecule it binds, the complex is *inner-sphere*. If at least one water molecule is interposed between the functional group and the bound ion or molecule, the complex is *outer-sphere*. As a general rule, outer-sphere surface complexes involve largely electrostatic bonding mechanisms and, therefore, are less stable than inner-sphere surface complexes, which usually involve either ionic or covalent bonding, or some combination of the two.

Examples of the two kinds of surface complex are shown schematically in Fig. 1. An inner-sphere surface complex involving a monovalent metal ion on a layer aluminosilicate appears on the right in the figure. This surface complex requires the coordination of the metal ion with six oxygen ions bordering one of the hexagonal cavities found in the surface of the mineral. Stereochemical enhancement of the stability of the complex occurs if the radius of the metal cation is close to that of the cavity (0.13 nm). An outer-sphere complex of a (solvated) metal ion is shown at the left in Fig. 1. This configuration of the metal ion differs from that in the diffuse ion swarm because an outer-sphere complex is immobilized on the adsorbent surface over the time scale for diffusive motions of solvated metal ions in aqueous solution (about 10 psec).

This chapter is intended as an introductory review of surface complexation as a mechanism of adsorption. The purpose of this review can be served by detailed consideration of the experimental characterization of a prototypical case, the adsorption of ions by members of the 2:1 layer aluminosilicate group, smectite. Excellent reviews accomplishing a similar objective for ion adsorption by hydrous oxides are available [12,15,16] and should be consulted to round out the perspective presented here.

II. SURFACE COMPLEXES OF CU(II) ON SMECTITES

A. Structural Chemistry of Smectites

Smectites are a group of 2:1 layer type aluminosilicates of widespread importance in geologic [17] and soil [18] weathering environments. From the point of view of surface chemistry, these minerals may be characterized as condensation sheet polymers of Si and metal hydroxides having the general unit-cell chemical formula [19]

$$C_{x/Z} [Si_{n1} Al_{8-n1}] (Al_{n1+n2-8} Fe(III)_{n3} Fe(II)_{n4} Mg_{n5} M_{n6}) O_{20} (OH)_4$$

where C represents a cation of valence Z on the cleavage surface, [ ] refers to cations in tetrahedral coordination, ( ) refers to metals in octahedral coordination, and x or n_i (i = 1 to 6) is a

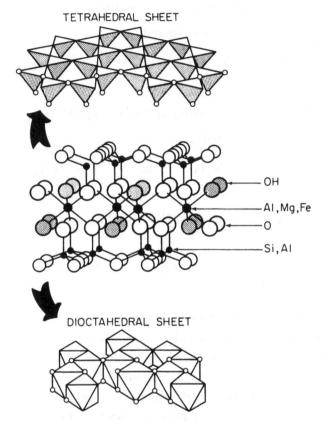

TETRAHEDRAL SHEET

OH

Al, Mg, Fe

O

Si, Al

DIOCTAHEDRAL SHEET

FIG. 2. Atomic structure of a smectite comprising two silica tetra-hedral sheets and an alumina (di-) octahedral sheet.

stoichiometric coefficient. The symbol M represents a metal other than Al, Fe, or Mg. The atomic structure of smectite is illustrated in Fig. 2. The cations in tetrahedral coordination with oxygen (Si or Al) form sheet structures that are bonded to a sheet of metals in octahedral coordination with oxygen or hydroxyl groups (Al, Fe, Mg, or M). In some instances, all three different sites of octahedral coordination evident in Fig. 2 are filled and the smectite is termed *trioctahedral*; if only two of the sites are filled (as in Fig. 2), the smectite is *dioctahedral*. For these two cases,

$$\sum_{i=1}^{6} n_i = \begin{cases} 14 & \text{trioctahedral smectite} \\ 12 & \text{dioctahedral smectite} \end{cases} \tag{1}$$

TABLE 1   Representative Chemical Formulas of Smectites

Smectite	Chemical formula
German beidellite	$Ca_{0.67}[Si_{7.12}Al_{0.88}](Al_{3.52}Fe(III)_{0.02}Mg_{0.46})$ $O_{20}(OH)_4$
Wyoming bentonite	$Na_{0.62}[Si_{7.8}Al_{0.2}](Al_{3.28}Fe(III)_{0.30}Fe(II)_{0.04}$ $Mg_{0.38})O_{20}(OH)_4$
Camp Berteau montmorillonite	$Na_{0.86}[Si_{7.86}Al_{0.14}](Al_{2.99}Fe(III)_{0.29}Mg_{0.72})$ $O_{20}(OH)_4$
Nontronite	$Na_{1.08}[Si_{7.14}Al_{1.06}](Fe(III)_{3.96}Mg_{0.02})$ $O_{20}(OH)_4$
Hectorite	$Na_{0.69}[Si_{7.98}Al_{0.02}](Mg_{5.33}Li_{0.67})O_{20}(OH)_4$

The occurrence of fractional values of ni in the chemical formula reflects the fact that smectites are (metastable) solid solutions [19,20].

The layer structure of smectite is bounded by planes of oxygen
ions that are termed *siloxane surfaces* [2].   Because of isomorphic
cation substitutions in the tetrahedral and octahedral sheets (e.g.,
Al^{3+} for Si^{4+} or Mg^{2+} for Al^{3+}), the net proton charge on the layer
structure is always negative and manifest on the two siloxane surfaces.   The cation C in the chemical formula balances the *layer
charge* x created in this way:

$$x = \begin{cases} 2(8 - n1) - n2 - n3 + (2-m)n6 & \text{trioctahedral} \\ 8 - n1 + n4 + n5 + (3-m)n6 & \text{dioctahedral} \end{cases} \quad (2)$$

where m is the valence of the unspecified metal M (e.g., 2 for Ni).
For smectites, $0.5 \leq x \leq 1.4$ by definition; values of $x > 1.4$ pertain
to other 2:1 layer-type aluminosilicates, such as vermiculite or illite
[2].

Examples of chemical formulas are listed in Table 1.   Among the
dioctahedral smectites containing Al in the octahedral sheet, the minerals are further classified as beidellite if $(8 - n1) > (n4 + n5)$ and

montmorillonite if (8 - n1) < (n4 + n5). Evidently beidellite has greater layer charge created in the tetrahedral sheet, whereas mont-morillonite has greater layer charge created in the octahedral sheet [17].

Primary smectite particles usually have the shape of thin, irregular disks or ribbons, with a length from 0.1 to 2.0 μm [17]. The edges of the particles do not make up a large fraction of the surface area, but they do bear hydroxyl groups and Lewis-acid sites resulting from unsatisfied metal valencies created by the fracture of crystallites [2].

B. Surface Functional Groups

The siloxane surface of a smectite is characterized by a distorted hexagonal symmetry among the constituent oxygen ions, and the reactive functional group associated with the siloxane surface is a roughly hexagonal cavity (Fig. 2) formed by the bases of six corner-sharing silica tetrahedra. This cavity has a diameter of about 0.26 nm and is bordered by six sets of "lone-pair" electron orbitals emanating from the surrounding ring of oxygen ions. The siloxane functional groups themselves also form an hexagonal array on the surface (as shown in Fig. 3).

The reactivity of the siloxane functional group depends on the nature of the electronic charge distribution in the layer silicate structure [2]. If there are no neighboring isomorphic cation substitutions to create local deficits of positive charge in the under-lying layer structure, the siloxane cavity will function as an electron donor that can bind only neutral, dipolar molecules, such as water molecules. If isomorphic substitution of Al^{3+} by Fe^{2+} or Mg^{2+} occurs in the octahedral sheet, the resulting excess negative charge on a nearby siloxane cavity allows it to form reasonably strong complexes with cations as well as dipolar molecules. If isomorphic substitution of Si^{4+} by Al^{3+} occurs in the tetrahedral sheet, the excess negative charge originates much nearer to the surface oxygen ions and much stronger complexes with cations and molecules become possible because of a greater localization of charge. Quantum chemical

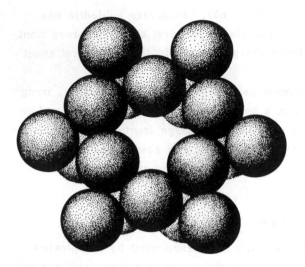

FIG. 3.   The surface functional group on a siloxane surface.

calculations, using qualitative perturbation theory and the extended
Hückel tight-binding method [21,22], have confirmed a greater de-
localization of surface charge than that expected from cation substitutions deep in the layer structure.   Experimental confirmation also
has come from infrared spectra of hydrogen bonds formed between
NH or OH groups and the surface oxygen ions, which show lower
stretching frequencies as the layer charge increases [2].

    At the edges of smectite layers exposed cations (e.g., Al^{3+})
can react with water to create solvated cations at the solid/aqueous
solution interface.   This combination of a cation and water molecule
at an interface is a *Lewis-acid site*, with the cation identified as the
Lewis acid.   Lewis-acid sites exist commonly, for example, on the
edge surfaces of gibbsite (γ-$Al(OH)_3$) and goethite (α-FeOOH), as
well as on the edge surfaces of clay minerals like kaolinite (Si_4Al_4
$O_{10}(OH)_8$).   These surface functional groups are very reactive,
since the positively charged water molecule is quite unstable and is
either deprotonated or exchanged readily for an organic or inorganic
anion in aqueous solution.

    The inorganic surface functional group of greatest abundance
and reactivity in clay-sized colloids is the hydroxyl group exposed

on the outer periphery of a mineral. This kind of OH group is found on metal oxides or hydroxides, and on the edges of smectite particles. Usually, more than one kind of surface OH group can be distinguished on the basis of stereochemistry, but little is known about those on smectite edges [2]. In any event, the importance of these groups and the Lewis-acid sites is diminished for metal-ion reaction on smectites by the relatively small fraction of the total surface area they occupy (usually <15%).

C. Spectroscopic Studies of Adsorbed Cu(II)

1. Smectites under Low Relative Humidity

Electron spin resonance (ESR) spectroscopy combined with x-ray diffraction (XRD) has proven to be a useful method for determining the speciation of Cu^{2+} on smectite surfaces. The basic principles of ESR spectroscopy are discussed by Vedrine [23] and Hall [24] with reference to clay minerals. Pinnavaia [25], Hall [26], and McBride [27] have reviewed the spectra of Cu^{2+} adsorbed by smectites and their interpretation in terms of surface complexes. Briefly, the ESR lineshape for Cu^{2+} in tetragonal (distorted octahedral) coordination with oxygen ligands exhibits a set of quadruplet "peaks" or a single "peak," corresponding to electron spin transitions with the symmetry axis of the complex parallel or perpendicular, respectively, to an applied magnetic field. If the complexes are oriented in essentially the same direction in a sample, changing the sample orientation relative to that of an applied magnetic field will shift the ESR lineshape from one form to the other. Otherwise, both forms will appear in the same spectrum.

Additional information about Cu(II) complexes can be obtained from electron spin-echo modulation (ESEM) spectroscopy [28]. This technique involves the application of three microwave pulses to a paramagnetic electron in order to generate an "echo" whose amplitude as a function of the time interval between the second and third pulses is modulated by dipolar interactions between the electron and neighboring magnetic nuclei (e.g., protons). The modulation pattern is sensitive to the number of nearest-neighbor nuclei

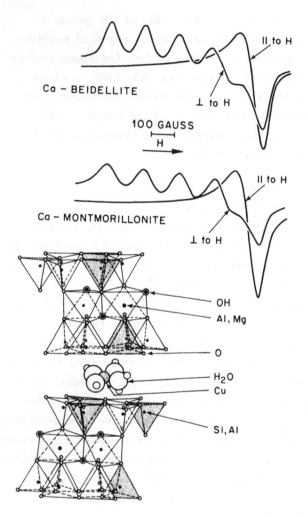

FIG. 4. ESR spectra of air-dry, Cu-doped Ca-smectites [27] and their molecular interpretation in terms of $Cu(H_2O)_4O_2$ surface complexes.

and their distance from the Cu^{2+} ion. By mathematical simulation of the modulation pattern one can estimate the coordination number of Cu^{2+} in terms of the nuclei. Brown and Kevan [29] have applied this technique along with ESR spectroscopy and XRD to determine the number of water molecules (actually D_2O) solvating Cu^{2+} adsorbed by montmorillonite.

Figure 4 shows ESR spectra of air-dry (<40% relative humidity) Ca-beidellite and Ca-montmorillonite samples in which Cu^{2+} has replaced a small fraction of the Ca^{2+} ions [27]. In both cases, the spectrum of the smectite sample comprising layers oriented perpendicularly to the direction of the applied magnetic field is essentially of the "four-peak" type, whereas that of the layers oriented parallel to the magnetic field is of the "one-peak" type. These results indicate that the symmetry axis of the Cu^{2+} complex is *perpendicular* to the smectite siloxane surface.

X-ray diffraction patterns of bivalent-cation-saturated smectites reveal that the layers are about 0.2-0.5 nm apart at low relative humidity [27,29], suggesting that only one layer of water molecules separates them. The ESEM modulation pattern [29] shows further that the number of nearest-neighbor water molecules is 4, with a Cu^{2+}-D^+ distance of 0.29 nm, which is consistent with the 0.198-nm Cu-O distance observed for solvated Cu^{2+} in aqueous solution [30]. These results and the ESR data are consistent with the binuclear, monodentate complex also illustrated in Fig. 3, which is inner-sphere relative to the siloxane functional group. The Cu^{2+} ion is coordinated to siloxane surface oxygen ions along the symmetry axis of the complex and to four water molecules in the equatorial plane parallel to the siloxane surface.

2. Smectites Fully Hydrated

Figure 5 shows ESR spectra of Cu^{2+}-doped Ca-beidellite [27] and Mg-montmorillonite [29] exposed to an atmosphere at 100% relative humidity. These spectra comprise only a single "peak" that has no significant dependence of intensity on the orientation of the smectite layers relative to the applied magnetic field. This result indicates that the Cu^{2+} complexes have no preferred orientation on the siloxane surfaces [27], at least not on the ESR time scale (motions occurring over about 10-100 psec). The same kind of spectrum is observed for the solvation complex, $Cu(H_2O)_6^{2+}$, in aqueous solution. Although an isotropic, single-peak ESR lineshape is also consistent with an immobile complex whose ligands undergo rapid exchange that distorts the configuration frequently enough to wash

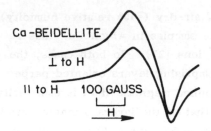

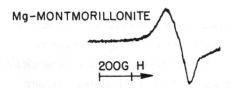

FIG. 5.   ESR spectra of hydrated, Ca-doped Ca- or Mg-smectite [27,29].

out structure in the spectrum, ancillary experiments with other paramagnetic cations suggest that the Cu^{2+} complex actually is mobile and tumbles slowly [25].   The rate of rotation of the complex is about two orders of magnitude less than in aqueous solution [31].

Simulation of the ESEM pattern for the Cu^{2+}-doped Mg-montmorillonite leads to a coordination number of six and a Cu^{2+}-D^{+} distance of 0.29 nm [29].   X-ray diffraction shows that the smectite layers are about 1.04 nm apart when the ESR lineshape becomes isotropic with a single peak [27,29].   This large spacing and the ESR/ESEM data are consistent with an interlayer $Cu(H_2O)_6^{2+}$ complex that tumbles sluggishly relative to a solvation complex in an aqueous solution.   This kind of complex is illustrated near the top of Fig. 1 as a diffuse-layer species.

Copper-doped Mg-hectorite whose layers are about 0.54 nm apart yields an ESR spectrum like those shown in Fig. 4, whereas when the layers are 1.04 nm apart, the spectrum is isotropic like those in Fig. 5 [25].   These results are consistent with an immobile, outer-sphere complex on the smectite surface.   Figure 6 illustrates the three Cu^{2+} surface complexes that appear successively

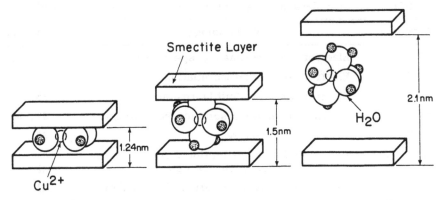

Smectite Layer

Cu^{2+}

INNER-SPHERE   OUTER-SPHERE   DIFFUSE SWARM

FIG. 6. Three modes of Cu^{2+} adsorption by smectites as revealed by ESR spectroscopy. (Layer spacing = 1 nm + interlayer spacing.)

as a smectite becomes increasingly hydrated. All three of the possibilities indicated in Fig. 1 can be realized depending on the degree of hydration of the clay mineral.

III. DETECTION AND QUANTITATION OF SURFACE COMPLEXES

The prototypical example of Cu(II) complexes on smectites illustrates most of the general characteristics of the experimental detection of surface species on natural colloids. Unambiguous results concerning the molecular structure and stability of these species can be obtained only with in situ, surface spectroscopic methods applicable to low concentrations and amenable to the need to investigate solid surfaces in the presence of water. Molecular models of surface speciation and invasive spectroscopic methods that require sample desiccation have heuristic value but cannot give definitive information [7].

Like ESR spectroscopy applied to the smectite/aqueous solution interface, in situ methods generally employ diagnostic probes that are sensitive to changes in near-neighbor molecular environments. When these environments in the vicinity of probe species are perturbed by external forces (which can be electromagnetic or

TABLE 2 In Situ Methodologies for Investigating Surface Complexes

Acronym	Methodology	References
ESR	Electron spin resonance spectroscopy	25, 27
ESEM	Electron spin-echo modulation	30
ENDOR	Electron nuclear double resonance	36
NMR	Nuclear magnetic resonance spectroscopy	37, 38
RAMAN	Raman spectroscopy	39
ATR-FTIR	Attenuated total reflection Fourier transform infrared spectroscopy	40
EXAFS	Extended x-ray absorption fine structure	41
XANES	X-ray absorption near edge structure	41
XRD	X-ray diffraction	42
NS	Neutron scattering (elastic and inelastic)	37
AFM	Atomic force microscopy	33

mechanical), the probes give characteristic information about local surface conditions. In the example of the previous section, the probe species was Cu^{2+}, in particular, its unpaired valence electron. Perturbation of this electron by electromagnetic fields leads to characterization, via the ESR spectral lineshape, of the local bonding environments of adsorbed Cu^{2+}, as depicted in Fig. 5. Additional insight is gained from the simultaneous application of electron spin-echo modulation techniques and x-ray diffraction methods. This kind of synergism reflects another important principle of surface species investigations: the use of different in situ spectroscopic methods on the same interfacial system to acquire a convergent set of molecular properties relevant to surface complex structure.

Table 2 lists a number of important and emerging in situ methodologies that have been applied to characterize surface complexes. The references indicated provide reviews of their applications to natural colloidal interfaces. Very recent among these methodologies is the atomic force microscope (AFM), which provides topographic images, at the nanometer scale, of the surfaces of solid materials. Since the invention of the AFM by Binnig et

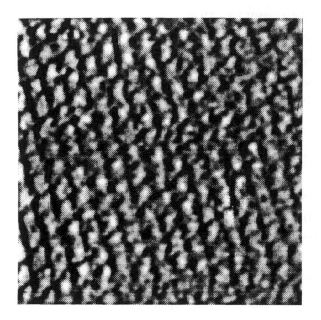

FIG. 7.   AFM image of a 6.6 × 6.6 nm patch on a montmorillonite
surface [35].   The hexagonal pattern of siloxane cavities (bright
"spots" on the surface) is evident.

al. [32], a variety of scanned-probe microscopes based loosely on
the principle of the AFM has been developed.   The AFM described
by Drake et al. [33] operates in the contact mode, wherein the
scanning tip gently touches the solid surface.   This type of AFM
has been used to investigate the atomic-scale structure of minerals
and biomolecule surfaces [34]. Sample preparation for AFM examination of mineral surfaces is relatively straightforward and surfaces
can be imaged under water or aqueous solutions.   Since the typical
scan time for a surface patch is on the order of seconds, direct
imaging of localized surface phenomena in real time is possible.
An AFM image of the surface of montmorillonite is shown in Fig. 7
[35].   The sophistication of these kinds of surface probe is expected to increase, until accurate molecular data on surface species
in heterogeneous colloidal systems become commonplace.

IV. MODELING SURFACE COMPLEXATION

A. Thermodynamic Approach

In order to visualize the surface complexation mechanism of adsorption on a macroscopic level one can picture Z moles of charged surface functional groups in reaction with one mole of metal cations of valence Z [43]:

$$A^{Z_A+}(aq) + Z_A X^-(s) = AX_{Z_A}(s) \tag{3}$$

where A^{Z_A+} is a metal cation and X^- represents a (charged) surface functional group, like the siloxane cavity on smectite surfaces. Equation (3) is analogous to the reaction of a solvated metal cation with a soluble ligand to form an aqueous complex. By analogy one can also define the equilibrium constant for this reaction as shown [44]:

$$K_1 = \frac{m_{AX} \gamma_{AX}}{(m_A \gamma_A)(m_X \gamma_X)^{Z_A}} \tag{4}$$

where m is molality and γ is an activity coefficient. The numerical evaluation of K_1 depends on the conventions adopted for determining the activity coefficients and the hypothetical molality m_X [43,44].

Fletcher and Sposito [44] have applied Eqs. (3) and (4) to the modeling of cation exchange on smectites. For this purpose, expressions identical to Eqs. (3) and (4) are provided for a second cation, B^{Z_B+}. When combined they yield the overall exchange reaction

$$Z_B A^{Z_A+}(aq) + Z_A BX_{Z_B}(s) = Z_A B^{Z_B+}(aq) + Z_B AX_{Z_A}(s) \tag{5}$$

The thermodynamic equilibrium constant for this reaction is [45]

$$K = \frac{(x_A f_A)^{Z_B}(m_B \gamma_B)^{Z_A}}{(x_B f_B)^{Z_A}(m_A \gamma_A)^{Z_B}} \tag{6}$$

where x_A and x_B are the mole fractions of ions on a solid exchanger, f_A and f_B are their respective activity coefficients, m_A and m_B are the molalities of cations in aqueous solution, and γ_A and γ_B are their respective single-ion activity coefficients. A conditional equilibrium constant for the overall exchange reaction given by Eq. (5) can be defined as [45]

$$K_v = \frac{x_A^{Z_B} (m_B \gamma_B)^{Z_A}}{x_B^{Z_A} (m_A \gamma_A)^{Z_B}} \tag{7}$$

which leads to the relationship

$$K = K_v \frac{f_A^{Z_B}}{f_B^{Z_A}} \tag{8}$$

Well-known experimental methods based on measurements of K_v are available for the determination of K and the "surface" activity coefficients [45]. In the model approach of Fletcher and Sposito [44], these methods are replaced by an algorithm for calculating K_1 in terms of the equilibrium constant K in Eq. (6). One begins by setting $f_A = f_B = \gamma_{AX} = \gamma_{BX} \equiv 1$, then rewriting Eq. (6) in terms of K_1 and the cognate equilibrium constant:

$$K_2 = \frac{m_{BX}}{(m_B \gamma_B)(m_X \gamma_X)^{Z_B}} \tag{9}$$

The result is [44]

$$K = \left(\frac{K_1^{Z_B}}{K_2^{Z_B}} \right) (m_{AX} + m_{BX})^{(Z_A - Z_B)} \tag{10}$$

For the case of heterovalent exchange ($Z_A \neq Z_B$) on an adsorbent of fixed total ionized functional groups, $m_{AX} + m_{BX}$ is dependent on the adsorbate composition, and the ratio

$$\frac{K_1^{Z_B}}{K_2^{Z_A}}$$

must also depend on both the adsorbate composition and the total quantity of adsorbent.

Both K_1 and K_2 are in principle unknown. However, if the value of one of them is given (K_2 for example), then the value of the other (K_1 for example) required to maintain consistency with a known value of K at any adsorbate composition (i.e., a value of $m_{AX} + m_{BX}$) is given by

$$K_1 = \left[K K_2^{Z_A} (m_{AX} + m_{BX})^{(Z_B - Z_A)} \right]^{1/Z_B} \tag{11}$$

In this way, the complexation reaction for B^{Z_B+} is used as a reference reaction with a constant value for K_2 to evaluate an internally consistent value of K_1. This latter value then can be used with K_2 to simulate cation exchange, given that one modifies K_1 continually in successive improvements of the predicted value of $m_{AX} + m_{BX}$ using Eq. (11). Such adjustment of K_1 parallels the activity coefficient correction made to a mass-action quotient for chemical speciation involving aqueous complex formation. The choice of reference reaction and equilibrium constant is arbitrary, but it is essential to use a constant which is large enough to ensure that the predicted value of the concentration of X^- is small enough to make a negligible contribution to mass balance. In this way the model of Fletcher and Sposito [44] describes cation exchange by implementing a convention based on the concept of surface complex formation. It is a quasi-thermodynamic approach designed to predict the macroscopic transfer of cations between two phases. It is not a description of molecular configurations in adsorbates at smectite surfaces. Detailed application of their model to binary and ternary cation-exchange data for montmorillonite are described by Fletcher and Sposito [44].

B. Statistical Mechanics Approach

In statistical mechanics, the modeling of surface complexes can be considered as a special case of a multispecies *lattice model*: surface complexes are molecular species immobilized on an array of sites that represent surface functional groups. If there are two different surface species on the sites (e.g., a protonated and an unprotonated surface hydroxyl group, or a free siloxane cavity and one that has complexed K^+), and if each site has z nearest neighbors, then any distribution of the two species (call them A and B) over the sites must satisfy the conditions [46]:

$$zN_A = 2N_{AA} + N_{AB} \tag{12}$$

$$zN_B = 2N_{BB} + N_{AB} \tag{13}$$

where N_A is the total number of A species, N_{AA} is the number of nearest-neighbor A pairs (with analogous definitions for N_B and N_{BB}), and N_{AB} is the number of nearest-neighbor AB pairs. It is assumed that each species binds to just one site. If the array of sites is not regular, z can be interpreted as the average number of nearest neighbors of a site.

The chemical properties of the distribution of the two species on the sites can be calculated with standard expressions in statistical mechanics once the canonical partition function is known [46]. For a two-species lattice model, this function has the form [46]

$$Q = q_A^{N_A} q_B^{N_B} \sum_{N_{AB}} g(N_A, M, N_{AB}) \exp\left(-\frac{W}{k_B T}\right) \tag{14}$$

where q is the canonical partition function for a single species on a single site, k_B is the Boltzmann constant, T is absolute temperature, and

$$W \equiv N_{AA}\varepsilon_{AA} + N_{BB}\varepsilon_{BB} + N_{AB}\varepsilon_{AB} \tag{15}$$

is the total energy of interaction between species on nearest neighbor sites. In Eq. (14), the function $g(N_A, M, N_{AB})$ is the number of ways that N_A species can be distributed on M total sites such that N_{AB} nearest-neighbor AB pairs occur. This function is subject to the constraint [46]

$$\sum_{N_{AB}} g(N_A, M, N_{AB}) = \frac{M!}{N_A! N_B!} \qquad (16)$$

where ! refers to the factorial function and the right side is the total number of ways that N_A indistinguishable species of type A can be distributed among $M = N_A + N_B$ sites. The sum in Eqs. (14) and (16) is over all possible numbers of nearest-neighbor AB pairs that can be placed on the sites.

The interaction energy W is assumed to depend only on the respective pair interaction energies ε, which can be positive, negative, or zero. Thus Eq. (14) can be interpreted as an expression of the relative likelihood of a given choice of N_A species-A molecules and N_B species-B molecules being found on an array of M sites, with emphasis on nearest-neighbor interactions only. If there were no interactions, $W = 0$ and Eq. (14) would reduce to the relative probability that there is independent occupancy of N_A sites by species A and N_B sites by species B. (Each q is interpreted in statistical mechanics as the relative probability of a single site containing a single species [46].) The function $g(N_A, M, N_{AB})$ appears when the site occupancy is influenced by nearest-neighbor interactions, with $\exp(-W/k_B T)$ being the appropriate Boltzmann factor for weighting each combinatorial term. Of principal concern in the Boltzmann factor is the relative importance of AA and BB interactions versus AB interactions. On defining [46]

$$\varepsilon \equiv \varepsilon_{AA} + \varepsilon_{BB} - 2\varepsilon_{AB} \qquad (17)$$

and making use of Eqs. (12) and (13), one can transform Eq. (14) to

$$Q = \left[q_A \exp\left(- \frac{z\,\varepsilon_{AA}}{2k_B T} \right) \right]^{N_A} \left[q_B \exp\left(- \frac{z\varepsilon_{BB}}{2k_B T} \right) \right]^{N_B}$$

$$\times \sum_{N_{AB}} g(N_A, M, N_{AB}) \exp\left(\frac{N_{AB}\varepsilon}{2k_B T} \right) \qquad (18)$$

Equation (18) exposes the significance of ε as an energy difference that determines the statistical mechanical weighting of each possible N_{AB} that contributes to Q.

The evaluation of the g-factor in Eq. (18) is a formidable task to perform in general. Thus, model simplifying assumptions are always made in order to facilitate the calculation of Q. One of the simplest is the *Bragg-Williams approximation*, which can be shown to be equivalent to the well-known van der Waals model of liquids [46]. It consists of replacing N_{AB} in Eq. (18) by its average value as computed for a random distribution of the two species over the M sites, namely,

$$N_{AB}^{BW} \equiv \frac{zN_A N_B}{M} \qquad (19)$$

In this approximation, all N_{AB} values are set equal to the product of the total number of nearest neighbors of A species (zN_A) and the fraction of total sites occupied by B species (N_B/M). This result can be strictly true only if $\varepsilon = 0$, so that no advantage in probability accrues to any particular nearest-neighbor association. Alternatively, it is possible to demonstrate that Eq. (19) is equivalent to the mathematical operation of letting z become infinite while ε goes to zero in such a way that the product $z\varepsilon$ remains finite [47,48]. This "van der Waals limit" is interpreted physically as the result of placing each surface species in an average potential energy field determined collectively by all of its neighbors on other surface sites [2].

The combination of Eqs. (16), (18), and (19) produces the Bragg-Williams partition function:

$$Q_{BW} = \left[q_A \exp\left(-\frac{z\varepsilon_{AA}}{2k_B T} \right) \right]^{N_A} \left[q_B \exp\left(-\frac{z\varepsilon_{BB}}{2k_B T} \right) \right]^{N_B}$$

$$\times \frac{M!}{N_A! N_B!} \left[\exp\left(\frac{z\varepsilon}{2k_B T} \right) \right]^{N_A N_B / M} \qquad (20)$$

All of the present generation of molecular models of surface complexation can be derived from Eq. (20). These derivations, as well as generalizations of Q_{BW} for an arbitrary number of sites, have been given by Sposito [2,48].

Chemical equilibrium conditions for surface species described by
Eq. (17) can be derived readily by application of the standard relation [46,48]:

$$\mu_A - \mu_B = -k_B T \left(\frac{\partial \ln Q_{BW}}{\partial N_A} \right) \tag{21}$$

where μ is a chemical potential and the derivative of Q_{BW} is calculated under the mass-balance constraint $dN_A = -dN_B$ (e.g., the reaction in Eq. (5) with $Z_A = Z_B$). The result is

$$\mu_A - \mu_B = -k_B T \ln \left[\frac{q_B \exp(-z\varepsilon_{BB}/2k_B T)}{q_A \exp(-z\varepsilon_{AA}/2k_B T)} \right]$$

$$+ k_B T \ln \left(\frac{x_A}{x_B} \right) + \frac{(x_A - x_B)z\varepsilon}{2}$$

$$= k_B T \ln \left(\frac{q_B x_A}{q_A x_B} \right) + x_A z\varepsilon + (\varepsilon_{AB} - \varepsilon_{BB}) \tag{22}$$

where $x = N/M$ is a mole fraction and $x_A = 1 - x_B$ has, along
with Eq. (17), been used to derive the second step. Chemical equilibrium between the surface species and the cognate aqueous species
exists if [45,48]

$$\mu_A - \mu_B = \mu[A_{aq}] - \mu[B_{aq}] = \mu_A^\circ - \mu_B^\circ + k_B T \ln \left(\frac{a_A}{a_B} \right) \tag{23}$$

where $\mu[\ ]$ is the chemical potential of an aqueous species whose
standard-state chemical potential is μ° and whose activity is a. Given
the conditional-equilibrium constant in Eq. (7) for the case $Z_A = Z_B$
in Eq. (5), namely

$$K_v \equiv \frac{x_A a_B}{x_B a_A} \tag{24}$$

it follows from Eqs. (22) and (23) that

$$K_v = \frac{K_A}{K_B} \exp \left(-\frac{z\varepsilon x_A}{k_B T} \right) \tag{25}$$

in the Bragg-Williams approximation, where

$$K_i \equiv q_i \exp\left(\frac{\mu_i^o}{k_B T}\right) \qquad (i = A, B) \qquad (26)$$

for either adsorbate species. (The difference between ε_{AB} and ε_{BB} has been ignored, consistent with the van der Waals limit [48].) Sposito [48] has derived the generalization of Eq. (25) when K_v contains arbitrary powers of the mole fractions and activities introduced by a reaction stoichiometry between A and B that differs from 1:1.

C. Applications to Smectites

The successful use of Eq. (25) and its generalization to heterovalent exchange to describe ion adsorption by hydrous oxides has been reviewed extensively in recent years [2-4,13,15,48]. Keeping to the lietmotif of the present chapter, one may illustrate the Bragg-Williams approximation by considering applications to cation adsorption by smectite. Primordial surface complexation models of this latter phenomenon were essentially simple elaborations of the Stern model [45]. Indeed, the Bragg-Williams approximation does not specify the molecular details of the surface complexes formed, but only requires that such complexes exist in some kind of array. The essential feature of the model, insofar as its application to cation exchange data is concerned, is the linear relation between $\ln K_v$ and the mole fraction x_A that is implied in Eq. (25) [2,15]. As emphasized by Johnston and Sposito [7], this relation always must be augmented by spectroscopic information concerning the molecular nature of surface complexes (inner-sphere or outer-sphere, charged or neutral, etc.) *before* the model algorithm can be implemented unambiguously.

James and Parks [49] have employed Eq. (25) to model Na^+-H^+ exchange on smectites and thereby calculate proton titration curves. In their model, Na^+ is assumed to adsorb via outer-sphere complexes and the diffuse ion swarm, while H^+ adsorbs only via inner-sphere complexes. The ratio K_A/K_B is termed an "intrinsic equilibrium constant" and the product $Z\varepsilon$ is equated to $e^2 M/C_1$, where e is the protonic charge and C_1 is a capacitance density for the region between the centers of a solvated Na^+ and an ionized surface functional group that binds it [2,48]. These identifications constitute a special case of the *triple-layer model* developed by Davis et al. [50]. James and Parks [49] apply Eq. (25) to the siloxane surface and to the hydroxylated edge surface of smectites to simulate proton titration curves. They also apply the model to K^+-H^+ and Na^+-K^+ exchange reactions. Nir [51] and Nir et al. [52] have used the same general approach to describe Na^+ exchange on smectites with a variety of metal cations, whereas Hirsch et al. [53] have done this for Cd^{2+}-Na^+ exchange reactions. The success of these model applications, of course, does *not* substantiate the assumption of outer-sphere surface complexation as an adsorption mechanism, which is made a priori.

ACKNOWLEDGMENTS

The preparation of this review was supported in part by NSF grant EAR-8915291. Gratitude is expressed to Ms. Joan Van Horn for excellent typing of the manuscript and to Mr. Frank Murillo for drawing the figures.

REFERENCES

1. D. J. Greenland and M. H. B. Hayes, eds., *The Chemistry of Soil Constituents*, Wiley, New York, 1978.

2. G. Sposito, *The Surface Chemistry of Soils*, Oxford University Press, New York, 1984.

3. J. A. Davis and K. F. Hayes, eds., *Geochemical Processes at Mineral Surfaces*, Amer. Chem. Soc., Washington, DC, 1986.

4. W. Stumm, ed., *Aquatic Surface Chemistry*, Wiley, New York, 1987.

5. W. Stumm, G. Furrer, and B. Kunz, *Croat. Chem. Acta*, 56:593 (1983).

6. W. Stumm, B. Wehrli, and E. Wieland, *Croat. Chem. Acta*, 60:429 (1987).

7. C. T. Johnston and G. Sposito, in *Future Developments in Soil Science Research* (L. L. Boersma, ed.), Soil Sci. Soc. Am., Madison, 1987.

8. S. L. Carnie and G. M. Torrie, *Adv. Chem. Phys.*, 56:141 (1984).

9. G. H. Bolt, ed., *Soil Chemistry. B. Physico-chemical Models*, Elsevier, Amsterdam, 1982.

10. G. Sposito, in *Rheology of Clay-Water Systems* (N. Güven, ed.), Clay Minerals Society, New York, 1992, pp. 139-170.

11. W. Stumm, *Geoderma*, *38*:19 (1986).

12. P. W. Schindler and G. Sposito, in *Interactions at the Soil Colloid-Soil Solution Interface* (G. H. Bolt, M. De Boodt, M. B. McBride, and M. H. B. Hayes, eds.), Kluwer, Dordrecht, 1991, Chap. 4.

13. G. Sposito, *Chimia*, *43*:169 (1989).

14. J. Buffle, *Complexation Reactions in Aquatic Systems*, Wiley, New York, 1988, Chap. 3.

15. W. Stumm and P. W. Schindler, in *Aquatic Surface Chemistry* (W. Stumm, ed.), Wiley, New York, 1987, Chap. 4.

16. W. Stumm, B. Sulzberger, and J. Sinniger, *Croat. Chem. Acta*, *63*: (in press).

17. N. Güven, in *Hydrous Phyllosilicates* (S. W. Bailey, ed.), Mineralogical Soc. Am., Washington, DC, 1988, Chap. 13.

18. G. Borchardt, in *Minerals in Soil Environments* (J. B. Dixon and S. B. Weed, eds.), Soil Sci. Soc. Am., Madison, 1989, Chap. 14.

19. G. Sposito, *Clays Clay Min.*, *34*:198 (1986).

20. H. M. May, D. G. Kinniburgh, P. A. Helmke, and M. L. Jackson, *Geochim. Cosmochim. Acta*, *50*:1667 (1986).

21. W. F. Bleam and R. Hoffmann, *Phys. Chem. Min.*, *15*:398 (1988).

22. W. F. Bleam and R. Hoffmann, *Inorg. Chem.*, *27*:3180 (1988).

23. J. C. Vedrine, in *Advanced Chemical Methods for Soil and Clay Minerals Research* (J. Stucki and W. Banwart, eds.), Reidel, Dordrecht, 1980, Chap. 7.

24. P. L. Hall, *Clay Min.*, *15*:321 (1980).

25. T. J. Pinnavaia, in *Advanced Chemical Methods for Soil and Clay Minerals Research* (J. Stucki and W. Banwart, eds.), Reidel, Dordrecht, 1980, Chap. 7.

26. P. L. Hall, *Clay Min.*, *15*:337 (1980).

27. M. B. McBride, in *Geochemical Processes at Mineral Surfaces* (J. A. Davis and K. F. Hayes, eds.), Amer. Chem. Soc., Washington, DC, 1986, Chap. 17.

28. L. Kevan and R. N. Schwartz, eds., *Time Domain Electron Spin Resonance*, Wiley, New York, 1979.

29. D. R. Brown and L. Kevan, *J. Am. Chem. Soc.*, *110*:2743 (1988).

30. Y. Marcus, *J. Solution Chem.*, *12*:271 (1983).

31. G. Sposito and R. Prost, *Chem. Rev.*, *82*:553 (1982).

32. G. Binnig, C. F. Quate, and C. Gerber, *Phys. Rev. Lett.*, *56*:930 (1986).

33. B. Drake, C. B. Prater, A. L. Weisenhorn, S. A. C. Gould,
     T. R. Albrecht, C. F. Quate, D. S. Cannell, H. G. Hansma,
     and P. K. Hansma, *Science*, *243*:1586 (1989).

34. S. A. C. Gould, B. Drake, C. B. Prater, A. L. Weisenhorn,
     S. Manne, H. G. Hansma, P. K. Hansma, J. Masse, M. Longmire,
     V. Elings, B. Dixon Northern, B. Mukergee, C. M. Peterson, W.
     Stoeckenius, R. T. Albrecht, and C. F. Quate, *J. Vac. Sci.
     Technol. A*, *8*:369 (1990).

35. H. Hartman, G. Sposito, A. Yang, S. Manne, S. A. C. Gould,
     and P. K. Hansma, *Clays Clay Min.*, *38*:337 (1990).

36. M. Rudin and H. Motschi, *J. Colloid Interface Sci.*, *98*:385
     (1984).

37. J. J. Fripiat, ed., *Advanced Techniques for Clay Mineral
     Analysis*, Elsevier, Amsterdam, 1982.

38. C. A. Weiss, Jr., R. J. Kirkpatrick, and S. P. Altaner,
     Geochim. Cosmochim. Acta, *54*:1655 (1990).

39. C. T. Johnston, G. Sposito, and R. R. Birge, *Clays Clay Min.*,
     33:483 (1985).

40. M. I. Tejedor-Tejedor and M. A. Anderson, *Langmuir*, *2*:203
     (1986).

41. G. E. Brown and G. A. Parks, *Rev. Geophys.* *27*:519 (1989).

42. G. W. Brindley and G. Brown, eds., *Crystal Structures of Clay
     Minerals and Their X-ray Identification*, Mineral Soc., London,
     1980.

43. G. Sposito and S. V. Mattigod, *Soil Sci. Soc. Am. J.*, *41*:323
     (1977).

44. P. Fletcher and G. Sposito, *Clay Min.*, *24*:375 (1989).

45. G. Sposito, *The Thermodynamics of Soil Solutions*, Clarendon
     Press, Oxford, 1981, Chap. 6.

46. T. L. Hill, *An Introduction to Statistical Thermodynamics*,
     Addison-Wesley, Reading, MA, 1960.

47. R. A. Fowler and E. A. Guggenheim, *Statistical Thermodynamics*,
     Cambridge University Press, London, 1949.

48. G. Sposito, *J. Colloid Interface Sci.*, *91*:329 (1983).

49. R. O. James and G. A. Parks, *Surface Colloid Sci.*, *12*:119 (1982).

50. J. A. Davis, R. O. James, and J. O. Leckie, *J. Colloid Interface
     Sci.*, *63*:480 (1978).

51. S. Nir, *Soil Sci. Am. J.*, *50*:52 (1986).

52. S. Nir, D. Hirsch, J. Navrot, and A. Banin, *Soil Sci. Soc. Am.
     J.*, *50*:40 (1986).

53. D. Hirsch, S. Nir, and A. Banin, *Soil Sci. Soc. Am. J.*, *53*:716
     (1989).

5

A Gibbs-Donnan-Based Analysis of Ion-Exchange and Related Phenomena

JACOB A. MARINSKY   State University of New York at Buffalo, Buffalo, New York

I. INTRODUCTION

When a charged polymer is dissolved or dispersed as a gel (colloid) in aqueous media the concentration of counterions associated with it will always be a sensitive function of the ionic strength of the solution.   That this is indeed the case may be immediately discerned by comparing the pH response of weak-acid polyelectrolytes, linear and crosslinked, to neutralization with standard base at different salt concentration levels.   When the value of pH or pK_a, computed with the Henderson-Hasselbalch equation at the different salt concentration levels,

$$pK_a = pH - \log \frac{\alpha}{1 - \alpha} \qquad (1)$$

is plotted versus α, the degree of dissociation of the acid, the displacement of the curves at a particular α value provides a direct assessment of the effect of the change in counterion concentration on the change in pH.

Examples of such sensitivity of charged polymer equilibria to counterion concentration levels are presented in Figs. 1-4.   These

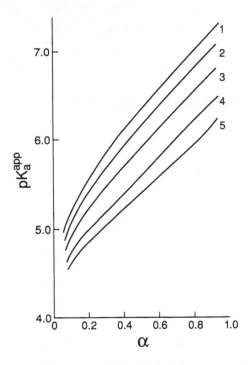

FIG. 1.   The effect of salt concentration levels on the dissociation
properties of poly(acrylic) acid; NaCl concentrations: 1, 0.00500,
2, 0.0100; 3, 0.0200; 4, 0.0500; 5, 0.100 M.

figures are based on potentiometric data compiled during neutralization of poly(acrylic) acid [1], carboxymethyldextran [2], and its
crosslinked analog, Sephadex CM-25 [3], and alginic acid [4] at
different salt concentration levels.   The sensitive response, during
neutralization, of pK_a/pH to change in salt concentration levels is
immediately apparent from the separation of the curves associated
with such change.

This sensitivity to salt concentration levels has interfered with
the quantitative interpretation of the physical-chemical behavior of
systems as diverse as polyelectrolytes, hydroxylated oxides, micelles,
ionomers, natural organic acids, clays, glasses, and ion-exchange
gels.   In each of these different systems interpretation of the observed behavior, when sought, has been influenced by considerations

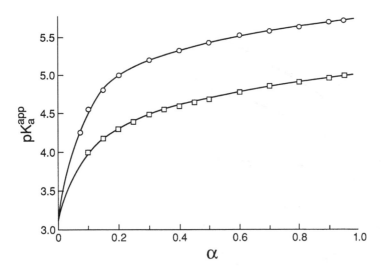

FIG. 2.   The effect of salt presence on the acid dissociation
properties of the polysaccharide, carboxymethyldextran; o,
no added salt; □, 0.020 M NaCl.

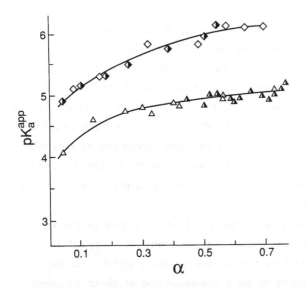

FIG. 3.   The effect of salt presence on the acid dissociation
properties of Sephadex CM-25, a crosslinked analog of carboxymethyldextran; $NaClO_4$ concentrations: ◇ and ◆, 0.0010;
△ and ▲, 0.010 M.

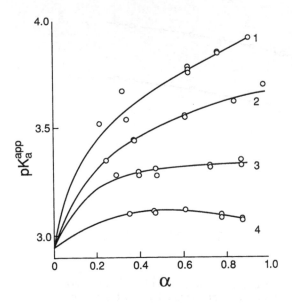

FIG. 4. The effect of salt presence on the dissociation properties
of alginic acid; NaCl concentrations: 1, 0.010; 2, 0.030; 3, 0.10;
4, 0.30 M.

presumed to be unique to the system under investigation. For example, with hydroxylated oxides, the attempts to rationalize the sensitivity of equilibria to neutral salt concentration levels have attributed
the observed response to complexation of the excess salt counterion
by the dissociated fraction of the hydroxylated oxide surface coupled
with counterion distribution estimates deduced from a Stern double-
layer model [5-7]. With polyelectrolytes, interpretation of this phenomenon has been sought in the Poisson-Boltzmann theory of cylin-
drical polyions [8,9] and alternatively in the counterion condensation
theory of Manning [10,11].

     The fact that the dissociation equilibria of charged polymers
respond similarly to counterion concentration changes, no matter
whether they are dissolved or suspended as gel particles in the
aqueous medium, is believed to be a consequence of their physical
resemblance. This assumption leads one to presume that the linear
polyelectrolyte analog of a charged gel in solution develops a

counterion-concentrating region next to it that closely resembles the
one associated with the gel itself. Since the separate phase properties of the gel, salt solution systems have been used to rationalize
the observed sensitivity of their equilibria to salt concentration
levels on a thermodynamically sound basis [3,12,13], one can envision similar advantage being taken of this description of the linear
polyelectrolyte in solution.

Our examination of this phenomenon on this basis has led to the
development of a unified model [14-16] which explains the extremely
sensitive response of counterion equilibria to simple neutral salt in
the presence of any charged polymeric molecule present as a gel or
presumably, uniformly dispersed in solution. Its use of Gibbs-
Donnan based concepts [17,18] seems to provide the most suitable
explanation of this phenomenon in the most straightforward manner.

To demonstrate this a review of the fundamental aspects that
provide the basis of the approach that facilitates the accurate
assessment of such behavior is presented first. A number of examples are then selected to demonstrate the validity of the claim made
above for the model developed by this analysis of these systems.

II. THE GIBBS-DONNAN MODEL [17-19]

The chemical potential, μ, of any electroneutral component, i, is
defined by Gibbs as

$$\mu_i = \left(\frac{\partial G}{\partial n_i} \right)_{P,T,n_j,n_k,\ldots} \tag{2}$$

where G = free energy, n_i = number of moles of component i, P =
pressure, T = absolute temperature, and subscripts j, k, ... refer
to the other electroneutral components of a particular system. The
pressure dependence of component i's chemical potential can be shown
to reduce to V_1, its partial molar volume, $(\partial V/\partial n_i)_{P,T,n_j,n_k,\ldots}$,
where V corresponds to volume. By assuming that partial molar
volumes are essentially independent of composition and pressure and
that the chemical potential in isothermal systems can be split into

two additive terms, one depending only on composition and the other only on pressure, the chemical potential, μ_i, in a solution of molality m and under a pressure P is

$$\mu_i(P,m) = \mu_i(P^\circ,m) + (P - P^\circ)V_i \qquad (3)$$

where P° is the standard pressure of 1 atm. The activity, a_i, of component i is defined by

$$\mu_i(P,m) = \mu_i^\circ(P) + RT \ln a_i \qquad (4)$$

where μ_i° = chemical potential of component i in the standard state. By combining Eqs. (1) and (2)

$$\mu_i(P,m) = \mu_i^\circ(P^\circ) + RT \ln a_i + (P - P^\circ)V_i \qquad (5)$$

With such representation of the chemical potential of electroneutral components the equilibrium distribution of diffusable components such as NX, MX, and H_2O between a crosslinked, polyelectrolyte gel in the N^+-ion form and a simple salt, MX, can be represented by equating the chemical potential of each diffusable component in the two phases as follows:

$$\bar{\mu}_{NX}(P,m) = \bar{\mu}_{NX}^\circ(P^\circ) + RT \ln \bar{a}_{NX} + (\bar{P} - \bar{P}^\circ)V_{NX} = \mu_{NX}(P,m)$$

$$= \mu_{NX}^\circ(P^\circ) + RT \ln a_{NX} + (P - P^\circ)V_{NX} \qquad (6a)$$

$$\bar{\mu}_{MX}(P,m) = \bar{\mu}_{MX}^\circ(P^\circ) + RT \ln \bar{a}_{MX} + (\bar{P} - \bar{P}^\circ)V_{MX} = \mu_{MX}(P,m)$$

$$= \mu_{MX}^\circ(P^\circ) + RT \ln a_{MX} + (P - P^\circ)V_{MX} \qquad (6b)$$

$$\bar{\mu}_{H_2O}(P,m) = \bar{\mu}_{H_2O}(P^\circ) + RT \ln \bar{a}_{H_2O} + (\bar{P} - \bar{P}^\circ)V_{H_2O} = \mu_{H_2O}(P,m)$$

$$= \mu_{H_2O}^\circ(P^\circ) + RT \ln a_{H_2O} + (P - P^\circ)V_{H_2O} \qquad (6c)$$

In these equations the bar is used to identify each component's association with the gel phase. By choosing the standard state to be the same in both phases ($\bar{\mu}^\circ = \mu^\circ$ and $\bar{P}^\circ = P^\circ$) and by equating ($\bar{P} - P$) to the "swelling pressure", π, the following convenient relationships are obtained:

$$\ln a_{NX} = \ln \bar{a}_{NX} + \frac{\pi}{RT} V_{NX} \qquad (7a)$$

$$\ln a_{MX} = \ln \bar{a}_{MX} + \frac{\pi}{RT} V_{MX} \qquad (7b)$$

$$\ln a_{H_2O} = \ln \bar{a}_{H_2O} + \frac{\pi}{RT} V_{H_2O} \qquad (7c)$$

Until this point we have treated components as if they are accessible in an isolated state in order to satisfy Gibbsian thermodynamics. However, when the three components of the two-phase system under scrutiny reach their equilibrium distribution the total concentration of M^+ and N^+ greatly exceeds the concentration of X^- in the gel phase. The extra electric charge and the resultant electrical force on the ions affect their equilibrium distribution and have to be taken into account. In correcting for this aspect thermodynamic rigor is necessarily lost in the process which consists of redefining equilibrium as that condition in which the electrochemical potential, E, of the separate ionic species are equal in both phases

$$E_{M^+} = \bar{E}_{M^+} \qquad (8a)$$

$$E_{N^+} = \bar{E}_{N^+} \qquad (8b)$$

$$E_{X^-} = \bar{E}_{X^-} \qquad (8c)$$

The electrochemical potential differs from the chemical potential by a term which depends on the electric potential, σ, of the phase

$$E_i = \mu_i + Z_i F \sigma \qquad (9)$$

Here Z_i represents the electrochemical valence of the i-th ion and F is the Faraday constant. By combining Eqs. (8), (9), and (5) the Donnan potential, $\bar{\sigma} - \sigma$, is resolved with Eq. (10). Equation (10) is applicable to all mobile-ion species present.

$$\bar{\sigma} - \sigma = \frac{1}{Z_i F} \left(RT \ln \frac{a_i}{\bar{a}_i} - \pi V_i \right) \qquad (10)$$

Representation of the equilibrium distribution of N^+, M^+, and X^- with this equation leads finally to Eq. (11) presented next.

$$\log \frac{(a_{M^+})(\bar{a}_{N^+})}{(\bar{a}_{M^+})(a_{N^+})} = \frac{\pi}{2.3\ RT}\ (V_M - V_N) \tag{11}$$

The activity ratio defined above is not directly measurable and Eq. (11) has been rearranged, as shown below, to provide a term that is

$$\log \frac{\bar{m}_N m_M}{m_N \bar{m}_M} = \log \frac{N_K}{M_{Ex}} = \frac{\pi}{2.3\ RT}\ (V_M - V_N) + \log \frac{\bar{\gamma}_{M^+}}{\bar{\gamma}_{N^+}} - 2\ \log \frac{\gamma_{\pm MX}}{\gamma_{\pm NX}} \tag{12}$$

In this equation K is the experimentally determined selectivity coefficient, m is the molality of each ion, $\bar{\gamma}$ is the activity coefficient of the ion in the gel phase, and $\gamma_\pm$ is the mean molal activity coefficient of the electrolyte in the external solution phase.

III. DOCUMENTATION OF THE GIBBS-DONNAN-BASED APPROACH

Use of the Gibbs-Donnan model has led to a capability for accurate anticipation of the chemical reaction paths of linear, weak-acid polyelectrolytes and their crosslinked gel analogs, of linear, fully dissociated polyelectrolytes and their crosslinked gel analogs, of polysaccharides, of proteins, of glasses, of clays, of hydrous oxide colloids and, indeed, of any charged polymeric material present in aqueous media as a gel, a colloid or as an apparently homogeneously dispersed solute. In the first example offered to validate this claim it is shown that the dissociation response of weak-acid, crosslinked polyelectrolyte gels to change in counterion concentration levels can be satisfactorily interpreted by employing the Gibbs-Donnan-based equations introduced in the preceding section [15]. Their applicability for the interpretation of metal-ion complexation by the weak-acid group repeated in the crosslinked gels and their linear polyelectrolyte analogs [20,21] is also included in the development of this documentation of the Gibbs-Donnan approach. Additional examples that are provided in this section to illustrate the quality of this approach are those involved first with the preliminary physical characterization of the counterion-concentrating domain of the linear polyelectrolyte. After demonstrating their facility for providing insights with respect to polyelectrolyte rigidity

or flexibility accurate estimates of solvent uptake by the solvation
sheath of the polyelectrolyte during its neutralization with base are
shown to be facilitated by refinements of the initial conformational
scans [15,16]. Further validation is then provided by the inter-
pretive quality exhibited by its anticipation of the ion exchange
properties of the Linde A zeolite [22], by its precise diagnosis of
the role of a fully dissociated, linear polyelectrolyte presence on the
distribution of pairs of ions between crosslinked ion-exchange gels,
also fully dissociated, and simple neutral electrolyte [23] and by its
accurate characterization of the counterion condensing domain of fully
dissociated, linear polyelectrolytes and the rest of the solution containing them [24].

A. Dependence of the Dissociation Response of Weak-Acid,
    Crosslinked Polyelectrolyte Gels to Changes in Counterion
    Concentration Levels [15]

Equation (11) can be used, as shown, to relate accurately the equilibrium distribution of two simple electrolytes, HX and MX, between
solution and crosslinked polyelectrolyte gel phases.

$$\log \frac{(a_{M^+})(\bar{a}_{H^+})}{(\bar{a}_{M^+})(a_{H^+})} = \frac{\pi(V_M - V_H)}{2.3\ RT} \tag{11}$$

With NaX the ionic strength-defining constituent of the aqueous phase
$V_M - V_H = 0.0012$ L [25] and the pressure-volume term is small enough
to neglect,* even when π reaches 300 atm, a value as high as one en-
counters with even the most highly crosslinked ion exchangers [19].
One can equate $\log (\bar{a}_{H^+})/(a_{H^+})$ to $\log (\bar{a}_{M^+})/(a_{M^+})$ and by substituting p for logarithm and { } for a, this equality can be conveniently
expressed as

$$p\{H^+\} - p\{\bar{H}^+\} = p\{Na^+\} - p\{\bar{N}a^+\} \tag{13}$$

Employment of the solution $p\{H^+\}$ in the well-known Henderson-

---

*$\dfrac{\pi(V_M - V_H)}{2.3\ RT} = \dfrac{300\ \text{atm}\ (0.0012\ \text{L})}{(2.3)(0.0821)\ \text{L atm deg}^{-1}\ (298\ \text{deg})} = 0.0064$

Hasselbalch equation [Eq. (1)] for evaluation of pK_a, where K_a is the dissociation constant of HA, the weak-acid group repeated in the gel, in the course of its neutralization with standard base, is inappropriate since it refers to the reaction $\overline{HA} \rightleftarrows H^+ + \overline{A}^-$, where barred species belong to the gel phase. It is the dissociation reaction in the gel phase, $\overline{HA} \rightleftarrows \overline{H}^+ + \overline{A}^-$ and the estimate of its $p\overline{K}_a$ value with Eq. (14),

$$p\{\overline{H}^+\} - \log \frac{\alpha}{1-\alpha} = p\overline{K}_a \tag{14}$$

that is of interest. Since $p\overline{H}$ cannot be measured experimentally for this purpose $p\overline{K}_a$ is evaluated in the following way: From Eqs. (1) and (13)

$$p\{\overline{H}^+\} - \log \frac{\alpha}{1-\alpha} + (p\{Na^+\} - p\{\overline{Na}^+\})_\alpha = pK_a \tag{15}$$

and

$$p\overline{K}_a = pK_a - (p\{Na^+\} - p\{\overline{Na}^+\})_\alpha \tag{16}$$

In Eq. (15), pK_a and $p\{Na^+\}$ are directly measurable while $p\{\overline{Na}^+\}$ has to be estimated by resolving both the concentration and activity coefficient of Na^+ ion in the gel phase. To this end the number of millimoles of A^- that form during titration of the gel with standard base is obtained first from the mass-balance condition

$$\overline{n}_{A^-} = C_b V_b + C_h(V_0 + V_b) \tag{17}$$

where

$\overline{n}_A$ = number of millimoles of A^-

C_h = concentration of H^+ ion in solution (moles/dm^3)

C_b = molarity of titrating base

V_0 = initial volume of aqueous phase (cm^3)

V_b = volume of base added (cm^3)

When the salt concentration in the aqueous phase is small, invasion of the gel phase can be neglected. Then, electroneutrality in the gel phase requires that

$$\overline{n}_{H^+} + \overline{n}_{Na^+} = \overline{n}_{A^-} \tag{18}$$

For weak-acid gels $\bar{n}_{H^+} \ll \bar{n}_{Na^+}$ and Eq. (18) reduces to

$$\bar{n}_{Na^+} = \bar{n}_{A^-} \tag{19}$$

By measuring W_g, the water content (gH_2O) associated with the gel at each α the molality of Na^+ in the resin phase, $\bar{m}_{Na^+}$, is obtained from

$$\bar{m}_{Na^+} = \frac{\bar{n}_{A^-}}{W_g} \tag{20}$$

The activity coefficient, $\bar{\gamma}_{Na^+}$ of Na^+ in the gel phase is estimated as shown:

$$\bar{\gamma}_{Na^+} = \frac{(\gamma_{\pm(NaCl)})^2}{\gamma_{\pm(KCl)}} \tag{21}$$

The $\gamma_\pm$ values for NaCl and KCl at $m_{Na^+} = \bar{m}_{Na^+}$ are obtained from mean-molal-activity-coefficient data in the literature [26]. In this approach to the estimate of $\bar{\gamma}_{Na^+}$ it is assumed that $\bar{\gamma}_{K^+} \approx \bar{\gamma}_{Cl^-} = \gamma_{\pm(KCl)}$ at $m_{Na^+} = \bar{m}_{Na^+}$. The transport numbers of K^+ and Cl^- are nearly the same in KCl over a sizable concentration range [27], implying that the hydrated ionic radii are practically the same. This lends some support to the extra thermodynamic assumption inherent in Eq. (21). The additional assumption that $\bar{\gamma}_{Na^+} = \gamma_{Na^+}$ when $\bar{m}_{Na^+} = m_{Na^+}$ that is also needed does not appear to be unreasonable.

In a number of instances a constant or nearly constant $\bar{pK}_a$ value was resolved with Eq. (16) when pH and W_g data were compiled in neutralization studies of crosslinked, weak-acid ion-exchange resins [13,15]. In these studies, summarized in Tables 1 and 2, the $\bar{pK}_a$ value of 3.40 ± 0.04 resolved for the crosslinked carboxymethyldextran, Sephadex CM-25, is based on numbers that exhibit no particular bias as α varies. The $\bar{pK}_a$ value of 5.03 ± 0.081 obtained for Amberlite IRC-50, the crosslinked polymethacrylic-acid ion-exchange resin, however, is based on numbers that tend to increase slowly in value as α increases.

The $\bar{pK}_a$ value of 3.4 resolved for the Sephadex CM-25 is in good agreement with the pK_a value of 3.3 measured for methoxyacetic

TABLE 1   Estimation of $p\bar{K}_a$ for Sephadex CM-25 (T = 298 K)

α	pK_a	$W'_g\left(\dfrac{KgH_2O}{mole\ HA}\right)$	$\bar{\gamma}_{Na^+}$	$p\{\bar{Na}^+\}$	$p\bar{K}_a$
		0.010 M NaClO$_4$; p{Na$^+$} = 2.046			
0.102	4.587	0.53	0.750	0.841	3.382
0.202	4.751	0.70	0.735	0.673	3.378
0.299	4.858	0.81	0.723	0.573	3.385
0.403	4.935	0.92	0.719	0.502	3.391
0.498	5.022	1.00	0.715	0.448	3.424
0.599	5.103	1.07	0.712	0.399	3.456
0.701	5.189	1.10	0.711	0.344	3.487
				Average	3.415 ± 0.04
		0.10 M NaClO$_4$; p{Na$^+$} = 1.102			
0.10	3.935	0.29	0.727	0.601	3.434
0.20	4.005	0.42	0.716	0.467	3.370
0.29	4.084	0.52	0.713	0.401	3.383
0.39	4.14	0.60	0.711	0.335	3.373
0.49	4.20	0.67	0.710	0.285	3.383
0.59	4.242	0.72	0.710	0.235	3.375
0.68	4.295	0.80	0.710	0.219	3.412
0.78	4.35	0.86	0.713	0.190	3.438
				Average	3.396 ± 0.028
			Overall average		3.405 ± 0.036

Source: Marinsky et al. [15].

acid in 0.10 M salt [28a].   Methoxyacetic acid resembles closely the repeating acid functionality of carboxymethyldextran and the pK_a values of the two are expected to be similar.

It is interesting to note that the average $p\bar{K}_a$ value of 5.03 ± 0.08 obtained for the Amberlite IRC-50 also compares favorably with estimates of the thermodynamic pK_a value of 4.8 to 5.0 made by using data obtained from the literature for isobutyric acid [28b], the simple weak acid resembled by the functionality repeated in the resin.

TABLE 2   Estimation of $p\bar{K}_a$ for IRC-50 (T = 298 ± 3 K)

α	pK_a	$p\{Na^+\}$	$W'_g \left(\dfrac{K_g\,H_2O}{mole\,HA}\right)$	$\bar{\gamma}_{Na^+}$	$p\{\bar{Na}^+\}$	$p\bar{K}_a$
			0.030 M NaPSS (MW ~70,000)			
0.101	6.850	2.027	0.11	0.714	0.167	4.990
0.200	7.098	2.044	0.12	0.749	0.093	4.961
0.300	7.116	2.065	0.16	0.771	-0.170	4.881
0.349	7.255	2.060	0.17	0.786	-0.210	4.985
0.401	7.341	2.070	0.18	0.801	-0.249	5.022
0.451	7.405	2.065	0.18	0.832	-0.319	5.021
0.500	7.513	2.072	0.18	0.880	-0.390	5.051
0.550	7.457	2.070	0.20	0.859	-0.367	5.032
0.600	7.544	2.072	0.20	0.886	-0.416	5.056
0.650	7.559	2.062	0.215	0.898	-0.433	5.064
0.700	7.655	2.060	0.216	0.932	-0.480	5.115
0.751	7.656	2.053	0.216	0.969	-0.527	5.076
0.849	7.84	2.065	0.24	0.981	-0.541	5.234
0.899	7.892	2.072	0.23	1.049	-0.615	5.205
					Average	5.050 ± 0.092
			0.030 M NaPSS (MW ~500,000)			
0.0289	6.463	1.964	0.062	0.717	0.476	4.975
0.0513	6.700	2.002	0.063	0.710	0.238	4.936
0.0756	6.811	2.010	0.084	0.712	0.192	4.993
0.100	6.886	2.005	0.093	0.717	0.114	4.995
0.200	7.088	2.014	0.137	0.736	-0.030	5.044
0.400	7.272	2.026	0.177	0.806	-0.260	4.986
0.800	7.647	2.012	0.208	1.036	-0.601	5.034
					Average	4.995 ± 0.036
					Overall average	5.031 ± 0.081

Source: Slota and Marinsky [13].

The slightly positive trend in $p\bar{K}_a$ with α, noticeable with the Amberlite IRC-50 resin but absent in the Sephadex CM-25 gel, is probably due to the fact that the possibility of error in estimate of $\bar{\gamma}_{Na}$ for evaluation of the $p\{\bar{Na}^+\}$ value to be used in the computation of $p\bar{K}_a$ with Eq. (16) is considerably larger in the IRC-50. The assumption that $\gamma_{K+} = \gamma_{Cl-}$ is undoubtedly less applicable in the more concentrated IRC-50 gel phase.

The possibility that $p\bar{K}_a$ values resolved with Eq. (16) actually correspond to the intrinsic pK of the repeating acid group in these kinds of systems seemed a reasonable conclusion on the basis of the above. Contradictory results obtained with the flexible Sephadex CM-50 gel [12] and in earlier studies with the Amberlite IRC-50 [29] were attributed to the macroporosity of these resins [30,31]. The macropores, presumed to function as an extension of the solution phase, were believed to result in overestimates of W_g, which, in turn, led to an anomolous increase in $p\bar{K}_a$ with α.

The projected characteristics of Eq. (16), i.e., (1) the constancy of $p\bar{K}_a$ as α is varied and (2) its approach in value to the thermodynamic dissociation constant of the weak acid that the repeating weak-acid functionality of the crosslinked gel most closely resembles have been rationalized in the following way: The constancy of $p\bar{K}_a$, which has to arise from the fact that the $\alpha/(1 - \alpha)$ term needs no modification, must be a consequence of the equivalent accessibility of A$^-$ and HA to the reaction being monitored. Their accessibility derives from the three degrees of freedom (vibrational, rotational, and translational) available to them in the system. With crosslinked polyelectrolyte gels all functional groups, dissociated and undissociated, that are repeated throughout the gel skeletal structure have only the vibrational mode essentially unhindered, their rotational and translational modes being effectively restricted by their three-dimensional attachment to each other. Since the functional groups are repeated a great number of times in the gel the average distribution of such vibratory movement of the functional sites is the same at a given temperature. This duplication of the vibratory motion spectrum which defines the average accessible concentration or the activity coefficient of functional sites ($\bar{A}^-$ or $\overline{HA}$) ensures their cancellation in the $\bar{A}^-$ to $\overline{HA}$ ratio so that $\alpha/(1 - \alpha)$ remains a term that is unaffected by nonideality

factors as neutralization of the weak-acid groups repeated in the
gel progresses.

Rationalization of the second observation is similarly based on
statistical arguments. The functional units are repeated in these
gel structures enough times to be statistically equivalent. On this
basis one should expect resolution of $pK_a(int)$ values that approach
those of the weak acid that closely resembles the weak-acid functionality repeated in the gel.

B. Use of the Gibbs-Donnan Model for Interpretation of Metal-Ion
   Complexation in Crosslinked, Weak-Acid Polyelectrolyte Gels

Additional experimental support for the above interpretation of the
physical-chemical characteristics of crosslinked polyelectrolyte gel,
salt mixtures is available from metal-ion-complexation studies carried
out with the IRC-50 gel [20] and the Sephadex CM-50 and CM-25
gels [32]. In these complexation studies the binding of divalent
metal ions by the weak-acid moiety repeated in the resin was studied polarographically [20] (IRC-50, $NiSO_4$, Na_2SO_4) or through the
use of radioactively tagged metal ions [32] [CM-25 and CM-50.
$Zn(ClO_4)_2$, $Zn^{65}(ClO_4)_2$, $NaClO_4$]. Batch equilibration of the mixtures were performed in the examination of an extended neutralization range. The computation of pK_a^{app} and $pK_{(MA^+)}^{app}$ with Eq. (1)
and Eq. (22)

$$pK_{(MA^+)}^{app} = p(M^+) - \log \frac{\bar{n}_{A^-}}{\bar{n}_{MA^+}} \tag{22}$$

used estimates of $\bar{n}_{MA^+}$, $\bar{n}_{A^-}$, and $\bar{n}_{HA}$ based on the following mass-
balance considerations:

$$\bar{n}_{MA^+} = (C_{M_0^{+2}})(V_0) - (C_{M_e^{+2}})(V_0 + V_b) \tag{23}$$

$$\bar{n}_{A^-} = C_b V_b + C_h(V_0 + V_b) - \bar{n}_{MA^+} \tag{24}$$

$$\bar{n}_{HA} = (\bar{n}_{HA_0}) - \bar{n}_{A^-} - \bar{n}_{MA^+} \tag{25}$$

Here $\bar{n}_{HA_0}$ corresponds to the number of millimoles of HA associated
with the resin phase prior to reaction with base and metal ion, $\bar{n}_{MA^+}$,

$\bar{n}_{A^-}$, and $\bar{n}_{HA}$ represent the number of millimoles of $\overline{MA}^+$, $\bar{A}^-$, and $\overline{HA}$ associated with the resin phase after reaction with base and metal ion, $C_{M_0^{+2}}$ and $C_{M_e^{+2}}$ refer to the concentrations of divalent metal ion in the solution prior to and after complexation, C_b is the molarity of the titrating base, C_h is the molarity of the H^+ ion in solution at equilibrium, V_0 is the initial volume of the aqueous phase (cm^3), and V_b is the volume of base added (cm^3).

Since statistical arguments used to rationalize the cancellation of nonideality in the $\bar{n}_{A^-}/\bar{n}_{HA}$ term of Eq. (1) are equally applicable to the $\bar{n}_{A^-}/\bar{n}_{MA^+}$ term of Eq. (22) it is reasonable to expect Eq. (26) to be as meaningful as Eq. (16).

$$p\bar{K}_{MA^+}^{int} = pK_{MA^+}^{app} - 2(p\{Na^+\} - p\{\overline{Na}^+\})_\alpha \qquad (26)$$

The Donnan term is squared in Eq. (26) to account for the double charge of the free metal ion.

If only the unidentate complex forms in detectable quantity and the nonideality of all resin species (A^-, HA, and MA^+) is identical, or nearly so, as we have presumed, plots of pK_a versus pK_{MA^+} must resolve straight lines with a slope equal to or very close to a value of 2.0. That this is indeed the case for data compiled with the IRC-50, $NiSO_4$, Na_2SO_4 [20], and the Sephadex CM-25 and CM-50, $Zn(ClO_4)_2$, $Zn^{65}(ClO_4)_2$, $NaClO_4$ [32] systems may be seen from inspection of Figs. 5 and 6. The straight lines with slopes of 1.7, 2.0, and 2.0 that are resolved indicate once again that the nonideality of each of the crosslinked polyelectrolyte gel species (A^-, HA, and MA^+) is indeed the same or nearly so. The validity of Eqs. (16) and (26) appears to be well documented by the results.

The assignment of intrinsic $p\bar{K}_{MA^+}$ values to the nickel carboxylate complex formed with the isobutyric-acid-like unit repeated in Amberlite IRC-50, a crosslinked poly(methacrylic acid) gel, and to the zinc carboxylate complex formed with the methoxyacetic-acid-like unit repeated in the Sephadex CM-25 and CM-50 crosslinked carboxymethyldextran gels, is accessible with Figs. 5 and 6 as well. This is accomplished by extending the straight line plots of these

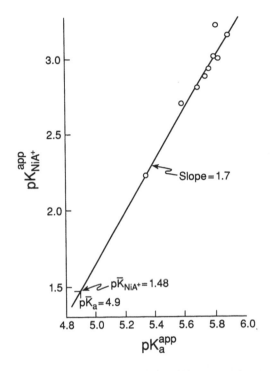

FIG. 5. Plot of $pK_{NiA^+}^{app}$ versus pK_a^{app}; system: Amberlite IRC-50, 0.050 M Na_2SO_4, 0.00486 M $NiSO_4$.

figures to the point where the value of the x coordinate corresponds to the intrinsic $p\bar{K}_a$ of the repeating weak acid. The value of the y coordinate at this point corresponds to the intrinsic $p\bar{K}_{MA^+}$ of the unidentate complex under investigation. The $p\bar{K}_{MA^+}$ value of 1.48 ± 0.1, resolved by using Fig. 5 in this way, is based on a $p\bar{K}_a$ assignment of 4.90 that is reached by using the $p\{H^+\}$, $\bar{n}_{A^-}$, and $\bar{n}_{HA}$ data, compiled in the course of the Ni^{+2}-ion complexation study, in Eq. (1). These data, reduced to the plot of pK_a versus α of Fig. 7, yield a curve whose extrapolation to intercept the ordinate axis at $\alpha = 0$ leads to this $p\bar{K}_a$ assignment. The value of 1.74 ± 0.1 resolved for the intrinsic $p\bar{K}_{ZnA^+}$ value in Fig. 6 is based on use of the intrinsic $p\bar{K}_a$ value of 3.4 resolved earlier (Table 1) for the weak-acid ensemble repeated in the Sephadex CM-25 gel.

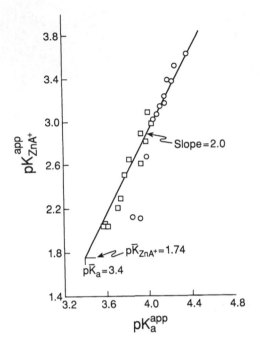

FIG. 6.   Plot of $pK_{ZnA^+}^{app}$ versus pK_a^{app}; system: Sephadex gels, $^{65}Zn(ClO_4)_2$, 0.10 M $NaClO_4$ - o CM-25; □ CM-50.

No pK_{MA^+} values are reported in the literature for the unidentate complexes that Ni^{+2} ion and isobutyric acid and Zn^{+2} ion and
methoxyacetic acid may form.   As a consequence, direct confirmation
of the model-based prediction that the $p\bar{K}_{MA^+}$ values resolved in
Figs. 5 and 6 should be identical with or very close to the values
obtained with the simple weak-acid molecules that closely resemble
the repeating monomer units of the crosslinked gels is not possible.
In spite of the absence of such data, however, the fact that the
$p\bar{K}_{MA^+}$ values resolved are in reasonable accord with the literature
values of 1.46 and 1.57 reported for the respective unidentate complexes formed with acetic acid by Ni^{+2} [33a] and Zn^{+2} [33b] can be
claimed to provide confirmatory evidence for the model.   Such a
claim is believed to have merit because it has been demonstrated
that different arrangements of the C, H, and O atoms associated

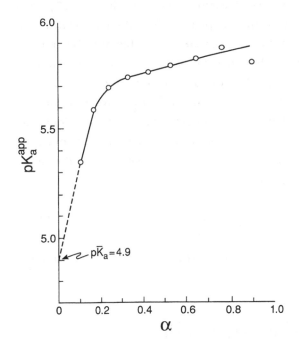

FIG. 7. Plot of pK_a versus α; system: Amberlite IRC-50, 0.050 M Na_2SO_4; 0.00486 M $NiSO_4$.

with a carboxylic acid moiety, while having a strong effect on the acid strength of the carboxylic acid constituent of the simple molecule, have little effect on the magnitude of the formation constant of the unidentate carboxylate complexes formed with Ni^{+2} and Zn^{+2} [33]. This is not the case with the unidentate carboxylate complexes of Cu^{+2} ion, their pK_{CuA^+} values increasing linearly with the pK_a of the weak-acid molecule being complexed [34]. The insensitivity of pK_{NiA^+} and pK_{ZnA^+} values to sizable change in pK_a values is undoubtedly a consequence of the small contribution of covalency to the M—O—CO bond.

Insight with respect to the magnitude of a, the gel species ($\overline{A^-}$, $\overline{HA}$, $\overline{MA^+}$, $\overline{MA_2}$) accessibility factor, has been gained from complexation studies carried out with the $CuSO_4$, Na_2SO_4, Amberlite IRC-50 system [35,36]. In this research formation of a bidentate species, undetectable in the Ni^{+2} and Zn^{+2} complexation studies [20,32], was observed. While estimates of the respective formation constants for the two species resulted in the resolution of a value for the uniden-tate species that was comparable in magnitude with the literature value for the unidentate copper(acetate)$^+$ complex, the formation constant resolved for the bidentate species, $\overline{CuA}_2$, was at least a factor of 50 smaller [36] than the value published for the bidentate copper(acetate)$_2$ complex [33c]. This result has to be attributable to the fact that while the availability factor, a, cancels in the esti-mate of $\bar{\beta}_{CuA^+}$ it remains unfactored

$$\bar{\beta}_{CuA^+} = \frac{(\overline{CuA}^+)(a)}{\{\overline{Cu}^{+2}\}(\bar{A}^-)(a)} = \frac{\bar{A}^+}{\{\overline{Cu}^{+2}\}(\bar{A}^-)} \tag{27}$$

in the denominator of the mass-action expression for the formation of $\overline{CuA}_2$.

$$\bar{\beta}_{CuA_2} = \frac{(\overline{CuA}_2)(a)}{\{\overline{Cu}^{+2}\}(\bar{A}^-)^2(a)^2} = \frac{\overline{CuA}_2}{\{\overline{Cu}^{+2}\}(\bar{A}^-)^2(a)} \tag{28}$$

The fact that $\bar{\beta}_{CuA_2}$ is approximately a factor of 50 smaller than expected leads to the assignment of a value of 0.02 to a, the resin species accessibility factor. The absence of bidentate species in the Ni^{+2} and Zn^{+2} complexation studies is understandable on this basis, their respective β_{MA_2} values being much too low to compensate for the low accessibility of $\bar{A}^-$.

C. Additional Evidence for the Physical Resemblance of Linear
    Polyelectrolytes to Their Crosslinked-Gel Analogs Through
    Use of Gibbs-Donnan-Based Concepts

The fact that the dissociation equilibria of linear, weak-acid polyelec-trolytes respond in exactly the same way to solution counterion con-centration changes as their crosslinked gel analogs (Figs. 1 to 4) provides strong evidence for the physical resemblance of the two [15,16]. On this basis a counterion concentrating solvated region next to the charged polyelectrolyte surface is considered to be the counterpart of the gel phase and the Gibbs-Donnan model is

presumed to be as applicable for the analysis of their equilibria.
That this is indeed the case has been unambiguously demonstrated
by the results obtained when the metal ion complexation properties
of the linear, weak-acid polyelectrolyte are examined [21] in the
same way as their gel analogs have been.

To facilitate this study the distribution, D_o, of trace-level concentrations of divalent metal ion (^{60}Co, [65]Zn, and [40]Ca) between a
well-defined quantity of 8% crosslinked Dowex 50 resin in the Na^+
ion form, expressed as activity per gram of dry resin, and a fixed
molality of sodium perchlorate solution, expressed as activity per gram
of solvent, was compared with the distribution, D, expressed in exactly the same way, for trace-level concentrations of the divalent
metal ion between the resin and solution, unchanged except for the
presence of a small, fixed quantity of poly(methacrylic) or poly-
(acrylic) acid at various degrees of dissociation. Values of
$\vec{n}_{MA^+}/m_{M^{+2}}$ were made available with these measurements in the
following way [37]:

$$\frac{D_o - D}{D} = \frac{\vec{n}_{MA^+}}{(m_{M^{+2}})} \tag{29}$$

and were combined with estimates of $\vec{n}_{A^-}$ in Eq. (22) to evaluate
pK_{MA^+}. The values of $\vec{n}_{A^-}$ were obtained by using the equilibrium
pH, measured concurrently for the polyelectrolyte, salt solution in
contact with the resin for the measurement of D, to facilitate their
assessment by interpolation of a pH versus $\vec{n}_A$ plot that was obtained
earlier for the equivalent solution during its titration with standard
base. In these computations quantities associated with the polyion
domain are identified by an arrow placed above them.

When these $pK_{MA^+}^{app}$ ($-\log (\vec{n}_{MA^+})/(m_{M^{+2}})(\vec{n}_{A^-})$) values are plotted
versus the pK_a^{app} values compiled concurrently (pH $- \log \vec{n}_A/\vec{n}_{HA}$)
the slope of the resultant straight line is once again found to approach a value of 2.0 (Figs. 8 and 9). The values of $p\vec{K}_{MA^+}^{int}$ resolved by extrapolating this line to the point corresponding to the
$p\vec{K}_a^{int}$ value of the weak-acid monomer repeated in the polyelectrolytes,

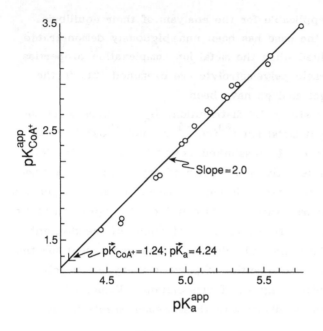

FIG. 8.   Plot of $pK_{CoA^+}^{app}$ versus pK_a^{app}; system: Dowex-50, 0.030 M poly(acrylic acid), ^{60}Co, 0.10 M NaClO$_4$.

are comparable to those obtained similarly for the gel analogs, which in turn, are in reasonable agreement with the $pK_{MA^+}^{int}$ values extractable from the literature [33b,d] for the corresponding M acetates$^+$.

Exactly the same result is obtained when the pK_a and pK_{MA^+} values obtained concurrently in this study are plotted versus α and extrapolated to intercept the ordinate axis at $\alpha = 0$ as shown in Figs. 10 and 11.   The resemblance of the pairs of curves in the two separate plots indicates, as well, that the availability factor, a, for the species associated with the polyelectrolyte domain cancel in Eq. (22) just as they did for the gel-phase species.

On the basis of these results one can conclude that the Gibbs-Donnan model, so successfully employed for interpretation of the various equilibria encountered with crosslinked, weak-acid polyelectrolyte gels, is as employable for interpretation of the various equilibria encountered with their linear, polyelectrolyte analogs.

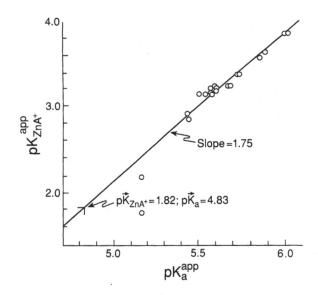

FIG. 9. Plot of $pK_{ZnA^+}^{app}$ versus pK_a^{app}; system: Dowex-50, 0.020 M poly(methacrylic acid), ^{65}Zn, 0.10 M $NaClO_4$.

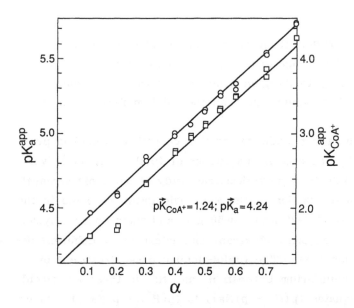

FIG. 10. Separate plots of $pK_{CoA^+}^{app}$ and pK_a^{app} versus α; system: 0.030 M poly(acrylic acid), ^{60}Co, 0.10 M $NaClO_4$; o pK_a^{app}; □ $pK_{CoA^+}^{app}$.

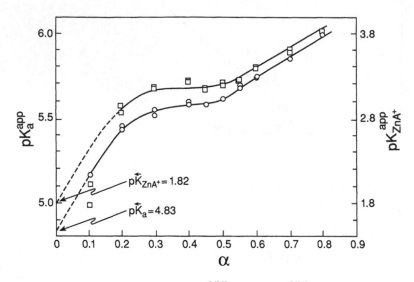

FIG. 11.   Separate plots of $pK_{ZnA^+}^{app}$ and pK_a^{app} versus α; system: 0.020 M poly(methacrylic acid), ^{65}Zn, 0.10 M $NaClO_4$; o, pK_a^{app}; □, $pK_{ZnA^+}^{app}$.

In order to rationalize these results one must envision the charged polyion to be circumscribed by a solvent sheath. The capture of counterions by this region produces a more highly concentrated solution of free mobile ions than the solution from which they have escaped.

Additional supporting evidence for this operational model is provided in Fig. 12 by the comparable shapes of $(p\{H^+\} - p\{Na^+\})$ versus α plots accessible from potentiometric study at one ionic strength (0.10 M) of the neutralization of the weak-acid group repeated in the crosslinked Sephadex CM-25 and CM-50 gels and their linear polyelectrolyte analogs, carboxymethyldextran, characterized by different degrees of substitution [15]. The Gibbs-Donnan-based treatment of the ion-exchange equilibrium between H^+ and Na^+ in these weak-acid gels [Eq. (13)] equates $(p\{H\} - p\{Na\})$ to $(p\{\vec{H}^+\} - p\{\vec{Na}^+\})$. Since $p\{\vec{H}^+\}$ is a unique function of α in this treatment the horizontal displacement of the curves in Fig. 12 has to be a consequence of

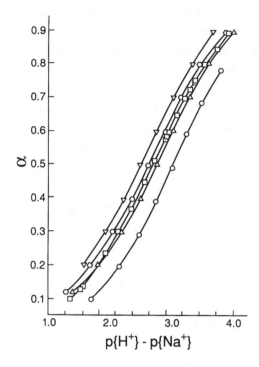

FIG. 12.   Plots of $\{pH^+\}$ - $\{pNa^+\}$ versus α for Sephadex CM-25 and CM-50 and several linear analogs, substituted carboxymethyldextran (C_mD_x), with various charge densities defined by the degree of substitution (DS).   The NaCl concentration is 0.10 M.   o, CM-25; DS = 1.05,   , CM-50; DS = 1.05, $\triangle$, C_mD_x; DS = 1.70, O, C_mD_x; DS = 1.26, ∇, C_mD_x; DS = 0.96.

differences in the value of $p\{\vec{Na}\}$.   Such differences reflect the differences in solvent content of the gel phase of the Sephadex CM-25 and CM-50 and the solvent sheath presumed by the operational model to envelop the substituted carboxymethyldextrans.

D.   Characterization of the Counterion-Concentrating Solvent
      Sheath of Linear Polyelectrolytes in Solution

Because the counterions contained by the solvent sheath are only free to move in a lateral direction the colligative properties of the polyelectrolyte, salt system are necessarily controlled by those counterions that remain in the external solution.   The osmotic coefficient, ϕ_p,

measured for the salt-free polyelectrolyte thus provides a measure
of the fraction of counterions that escape from the solvent sheath
[16]. With this picture the only difference between the polyelectrolyte, salt system, and its crosslinked gel analog, salt system is that
whereas the separate phases of the gel, salt system are electroneutral the two regions of the polyelectrolyte, salt system are not. In
the linear, weak-acid, polyelectrolyte, salt system the molality of the
counterion in the polyion domain is equal to $[\alpha(1 - \phi_p)/W_p]m_p + \vec{m}_s$
while in the external solution its molality is defined by $m_s + \alpha m_p \phi_p$.
In these expressions m_p refers to the molality of the polyelectrolyte
on a monomer basis, m_s and $\vec{m}_s$ corresponds to the molality of the
simple neutral salt in the two regions of the polyelectrolyte system
at equilibrium, and W_p represents the grams of solvent associated
with the solvated polyion region. For the weak-acid gel, salt system
the counterion molality of the gel phase is $\alpha\bar{m}_r + \bar{m}_s$ while in the solution its molality is equal to m_s.

The inability to see and mark the boundary of the two separate
regions presumed to prevail in linear polyelectrolyte, salt system has
necessitated the development of indirect paths to their physical description. Modification of the qualitative approaches to this problem
that are described first leads to the more quantitative scheme that is
presented in detail at the end of this section.

1. Conformational Insights

Gibbs-Donnan-based logic can be usefully employed to gain insight
with respect to the solvent uptake characteristics of linear, weak-acid
polyelectrolytes as a function of α and counterion, C^+, concentration
levels. For this purpose let us first consider the $p\{\vec{H}^+\}^* - \log \alpha/(1 - \alpha)$
term in Eq. (15). This term equates to the intrinsic $p\vec{K}_a$ of the weak-

---

*The arrow above a symbol in the equations used to describe poly-
electrolyte behavior identifies the particular parameter with the solvent sheath it has been assigned. The arrow is substituted for the
bar used to identify association with the gel phase in the earlier
treatment of the crosslinked polyelectrolyte, simple salt-solution
equilibria that the linear polyelectrolytes are presumed to mimic.

acid ensemble repeated in a linear, weak-acid polyelectrolyte. Because of its uniqueness this term cancels when $(p\{C^+\})_\alpha$ and $(pK_a^{app})_\alpha$ values obtained at a particular α at different salt concentration levels are divided. Use of such data in Eq. (16) to monitor the response of $(pK_a^{app})_\alpha$ to change in $(p\{C^+\})_\alpha$ affected by changes in salt concentration level determine the response of $p\{\vec{C}^+\}_\alpha$ as shown:

$$\Delta(p\{C^+\})_\alpha - \Delta(pK_a^{app})_\alpha = \Delta(p\{\vec{C}^+\})_\alpha \qquad (16a)$$

A comparison of the relative magnitudes of $\Delta(\{pC\})_\alpha$ and $\Delta(\{pK_a^{app}\})_\alpha$ provides insight with respect to the change in solvent uptake by the polyelectrolyte sheath that is affected at a particular α by changing the concentration level of simple salt. For example, if the skeletal network of the polyelectrolyte molecule is flexible one would expect the solvent content of the domain to be strongly affected by the increase or decrease in water activity arising from change in salt concentration levels. Rigidity of the polyelectrolyte structural network would, on the other hand, be expected to resist osmotically induced changes in solvent uptake. In the first case $\Delta(pK_a^{app})_\alpha$ will be significantly smaller than $\Delta(p\{C\})_\alpha$, whereas it will approach $\Delta(p\{C\})_\alpha$ when the molecule is inflexible.

A graphical approach which facilitates the elucidation of such conformational information for linear, weak-acid polyelectrolytes consists of the compilation of pK_a^{app} versus α plots provided by potentiometric measurements carried out at different salt concentration levels in the course of their neutralization with standard base. The vertical displacement of the curves at a particular α permit direct comparison of $\Delta(pK_a^{app})_\alpha$ with the experimentally controlled $\Delta(p\{C\})_\alpha$ values.

Plots of α versus $p\{H^+\}_\alpha - p\{C^+\}_\alpha$, compiled at different salt concentration levels, provide a second graphical approach to the assessment of the effect of ionic strength on polyelectrolyte conformation. The reason that $p\{H\}_\alpha$ can be substituted for $(pK_a^{app})_\alpha$ in Eq. (16), as shown, is that $\Delta(p\{H\})_\alpha = \Delta(pK_a^{app})_\alpha$.

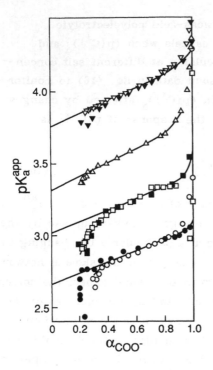

FIG. 13. The effect of salt concentration levels on the dissociation
properties of the carboxylic acid component of chondroitin sulfate A;
NaCl concentrations: ∇, ▼, 0.010 M NaCl; △, 0.030 M NaCl; ■, □,
0.10 M NaCl; ●, ○, 0.50 M NaCl.

$$\Delta(p\{C\})_\alpha - \Delta(p\{H\})_\alpha = \Delta(p\{\vec{C}\})_\alpha \qquad (16b)$$

In this mixture the horizontal shift in the curves at fixed α values
relates directly to the change in $(p\{\vec{C}\})_\alpha$ affected by the change in
salt concentration level.

Figure 4, presented earlier, and Fig. 13, presented next [4],
provide examples of the utility of the first graphical approach for
characterization of the conformational properties of two polysaccharides whose acid-dissociation responses are quite different. In Fig. 4
which examines the sensitivity of the acid-dissociation properties of
alginic acid* (AA) to sodium-ion concentration levels the curves

---

*Sodium form extract from Macrocystic Pyrifera, Sigma Chemical
Company, St. Louis, Missouri.

diverge with increasing α never reaching a point where the vertical displacement, $\Delta(pK_a^{app})_\alpha$ of the curves, even at the highest α value (0.9) and the two lowest salt concentration levels (0.010 and 0.030 M), approach the limiting value of 0.452[*] that would indicate that the solvent content of the solvent sheath of the AA molecule had not changed at that α value. The fact that the vertical displacement of these two curves at $\alpha = 0.9$ is only 0.24 indicates that sizable reduction (a factor of ~2) in the water content of the AA solvent sheath has been affected by the increase in salt concentration levels from 0.010 to 0.030 M.

The fact that $\Delta(pK_a^{app})_\alpha$ increases in magnitude with α for a particular $\Delta(p\{Na^+\})_\alpha$ shows that the change in solvent content of the counterion-concentrating domain of the polyelectrolyte at each α point is reduced by resistance to conformational change that increasing charge density promotes in a molecule even as flexible as alginic acid [4,38-40].

In this analysis of the response of the alginic-acid-dissociation equilibria to counterion concentration levels the change in $(pK_a^{app})_\alpha$ relative to the change in $p\{Na^+\}$ has been attributed solely to changes in $p\{\vec{N}a\}_\alpha$ affected by solvent transfer. However, this is an oversimplification, especially at the higher salt concentration levels (0.10 to 0.30 M), because of the increasingly important role that salt transfer to the solvated region of the polyion can be expected to play. The effect of such extra transfer of Na^+ ion at the higher ionic strength is to exaggerate solvent loss at the lowest α values and the highest salt concentration levels.

---

[*]$\Delta(p\{Na^+\}) = p(M_s)(\gamma_{\pm s}) - p(M_s')(\gamma_{\pm s}')$

$$= p(0.010)(0.905) - p(0.030)(0.848)$$

$$= 2.043 - 1.591 = 0.452$$

The response of the potentiometric properties* of chondroitin sulfate $A^\dagger$ (CSA) to counterion concentration levels during its neutralization with base [4] is examined similarly in Fig. 13. Unlike the AA, NaCl systems the curves tend to parallel each other over most of the α range studied. At the two lowest salt concentration levels (0.010 and 0.030 M) employed in these systems the vertical displacement of 0.40 to 0.435 pK units approaches the value of 0.452 expected when solvent content of the polyelectrolyte solvation sheath remains essentially unchanged at fixed α values. This result shows that the conformation of CSA in solution must remain essentially unaffected by the threefold change in ionic strength.

Solvent loss from the counterion-concentrating region of the polyelectrolyte domain is increasingly noticeable when the Na^+ ion concentration level is raised from 0.030 to 0.10 M and from 0.10 to 0.50 M. The respective $\Delta(pK_a^{app})_\alpha$ values of 0.35 to 0.38 and 0.35 to 0.36 observed diverge from the 0.486 and 0.658 values that correspond to solvent and Na^+ ion content constancy of the polyelectrolyte domain. Some of this divergence, especially in the highest salt concentration range, is, as we shall see, contributed by neglect of salt invasion.

The conformational properties discernible for the alginic-acid, sodium chloride system from scrutiny of the pK_a^{app} versus α plots at different salt concentration levels (Fig. 4) are equally apparent with the second kind of graphical approach presented in Fig. 14. The separation of the curves provided by the plots of α versus $(p\{H^+\} - p\{Na^+\})$ that are presented, in this instance, is a direct

---

*Estimate of α, the degree of dissociation of the carboxylic-acid content of the CSA samples was affected by correcting for the total free-hydrogen-ion complement of its fully dissociated sulfate content;

$$= [C_b V_b + (C_{H^+})(V_0 + V_b) - f(n_{CSA})]/[(1 - f)(n_{CSA})]$$

where n_{CSA} = number of millimoles of CSA, f = fractional content of sulfate groups, 1 - f = fractional content of carboxylic-acid groups.

†Sodium form from whale cartilage, the Sigma Chemical Company, St. Louis, Missouri.

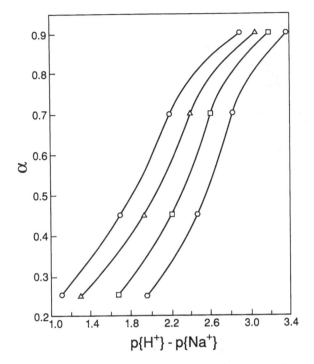

FIG. 14.   Plots of α versus $p\{H^+\} - p\{Na^+\}$ for alginic acid; NaCl concentration level: o, 0.010 M; $\triangle$, 0.030 M; $\square$, 0.10 M; O, 0.30 M.

measure of the $\Delta p\{\vec{Na}^+\}_\alpha$ value being sought for this purpose. The sizable separation of the curves in Fig. 14 reaffirm the earlier flexibility estimates for alginic acid.   The fact that the curves are separated more at the lower α values is consistent with the earlier estimate of the increased resistance to solvent uptake that is affected by higher charge density of the polyion domain as well. One must remember, however, that neglect of salt invasion leads to exaggeration of the conformational effect projected.

The plot of α versus $(p\{H^+\} - p\{Na^+\})$ that is presented in Fig. 15 for the CSA, NaCl system fully substantiates the observations drawn from analysis of the pK_a^{app} versus α plots presented in Fig. 13.   The data points obtained in 0.010 and 0.030 M NaCl almost fall on the same curve to show that the solvent content of the CSA molecule is essentially unaffected by the threefold increase in counterion

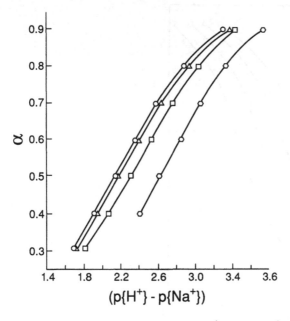

FIG. 15. Plots of α versus $p\{H^+\} - p\{Na^+\}$ for chondroitin sulfate A; NaCl concentration level: o, 0.010 M; $\triangle$, 0.030 M; $\square$, 0.10 M; $\bigcirc$, 0.50 M.

concentration. The shift in the curves when the molality of NaCl is increased from 0.030 to 0.10 and from 0.10 to 0.50 appear to correspond to a loss of approximately 40% and 60% of the water associated with the solvent sheath of the CSA in 0.010 and 0.030 M NaCl. The horizontal shift of $(p\{H^+\} - p\{Na^+\})$, on raising the salt concentration level from 0.10 to 0.50 M, corresponds to approximately a 50% reduction in the water content of the solvent sheath encompassing the CSA molecule. This estimate is probably somewhat high because of the effect of salt invasion differences at the two salt concentration levels on the observed shift.

The estimates of $\Delta(p\{\vec{Na}^+\})_\alpha$ as a function of $\Delta(p\{Na^+\})_\alpha$, affected separately with the two graphical methods, are summarized in Table 3 for the AA, NaCl and the CSA, NaCl systems. The two sets of $\Delta(p\{\vec{Na}^+\})_\alpha$ values listed should be identical since the two graphical approaches are equivalent, the extra constant $(-\log \alpha/(1-\alpha))_\alpha$ term

TABLE 3   A Comparison of $\Delta(p\{\vec{\text{Na}}\})$ Estimates Resolved
with the Two Graphical Methods

ΔM_s	$\Delta(p\{Na^+\})_\alpha$	$\Delta(p\{\vec{Na}^+\})_\alpha$	
		Method I	Method II
	System AA, NaCl		
	$\alpha = 0.25$		
0.010-0.030	0.452	0.312	0.320
0.030-0.10	0.486	0.386	0.350
0.10-0.30	0.447	0.277	0.275
0.010-0.30	1.385	0.975	0.945
	$\alpha = 0.50$		
0.010-0.030	0.452	0.272	0.210
0.030-0.10	0.486	0.286	0.275
0.10-0.30	0.447	0.257	0.265
0.010-0.30	1.385	0.815	0.750
	$\alpha = 0.70$		
0.010-0.030	0.452	0.232	0.225
0.030-0.10	0.486	0.226	0.225
0.10-0.30	0.447	0.227	0.225
0.010-0.30	1.385	0.685	0.675
	$\alpha = 0.90$		
0.010-0.030	0.452	0.222	0.175
0.030-0.10	0.486	0.156	0.150
0.10-0.30	0.447	0.197	0.175
0.010-0.30	1.385	0.575	0.500
	System CSA, NaCl		
	$\alpha = 0.40$		
0.010-0.030	0.452	0.017	0.037
0.030-0.10	0.486	0.136	0.131
0.10-0.50	0.658	0.303	0.269
0.010-0.50	1.596	0.456	0.437

TABLE 3   (continued)

ΔM_s	$\Delta(p\{Na^+\})_\alpha$	$\Delta(p\{\vec{Na}^+\})_\alpha$	
		Method I	Method II
$\alpha = 0.50$			
0.010-0.030	0.452	0.024	0.038
0.030-0.10	0.486	0.131	0.125
0.10-0.50	0.658	0.303	0.279
0.010-0.50	1.596	0.458	0.442
$\alpha = 060$			
0.010-0.030	0.452	0.032	0.038
0.030-0.10	0.486	0.126	0.128
0.10-0.50	0.658	0.298	0.288
0.010-0.50	1.596	0.456	0.454
$\alpha = 0.70$			
0.010-0.030	0.452	0.042	0.038
0.030-0.10	0.486	0.116	0.138
0.10-0.50	0.658	0.298	0.288
0.010-0.50	1.596	0.456	0.464
$\alpha = 080$			
0.010-0.030	0.452	0.052	0.038
0.030-0.10	0.486	0.106	0.119
0.10-0.50	0.658	0.298	0.256
0.010-0.50	1.596	0.456	0.413

included in the $\Delta(pK_a^{app})_\alpha$ values resolved in the first graphical scheme canceling. As was pointed out earlier, one is left with $\Delta(p\{H^+\})_\alpha$ to be subtracted from $\Delta(p\{Na^+\})_\alpha$ just as it is in the second graphical approach. The small differences observed between the two $(\Delta p\{\vec{Na}^+\})_\alpha$ estimates is a consequence of discrepancy introduced in the separate interpolations of the $(pK_a^{app})_\alpha$ and the $p\{H+\}_\alpha$ data points.

We have shown how one can obtain insight with respect to the
conformational tendencies of linear, weak-acid polyelectrolytes through
analysis of the effect that changes in counterion concentration levels
of their solution have on their measurable dissociation properties.
Such information is accessible, as well, from direct examination of
acid-dissociation data obtained for the linear, weak-acid polyelectrolyte at a single ionic strength as long as the counterion-concentration
level of their solutions is significantly lower than it is in the solvent
sheath presumed to encompass the polyelectrolyte molecule. With this
criterion met a sizable Donnan potential is operative to facilitate use
of Eq. (16) for this purpose.

The uppermost curve of Fig. 4, which is based on data compiled
during the neutralization of alginic-acid solution at an 0.010 M counterion concentration level is used here to demonstrate this capability.
In the first step of its use for this purpose the curve is interpolated
to intercept the ordinate axis at $\alpha = 0$ for the assignment of the intrinsic pK of the carboxylic-acid ensemble, $p\vec{K}_a^{int}$, repeated in the alginic
acid molecule. With such assignment of $p\vec{K}_a^{int}$ and with pK_a^{app} accessible over the complete range by interpolation the availability of
$p\{Na^+\}$ experimentally or by computation permits the assessment of
$p\{\vec{Na}^+\}_\alpha$ over the complete neutralization range with Eq. (16) as
shown.

$$p\vec{K}_a^{int} - (pK_a^{app})_\alpha + p\{Na^+\} = p\{\vec{Na}^+\}_\alpha \tag{16}$$

These values of $p\{Na^+\}_\alpha$, reduced to $\{Na^+\}_\alpha$ values, can now be
used to monitor the pattern of solvent uptake by the counterion-concentrating domain of the polyelectrolyte as a function of α. Since
$\{\vec{Na}^+\}_\alpha = (\vec{m}_{Na^+})_\alpha (\vec{\gamma}_{Na^+})_\alpha$ and since $(\vec{m}_{Na^+})_\alpha$ is, neglecting incursion
by salt, equatable to $\alpha(m_p)(1 - \phi_p)$, the millimoles of counterion en-
tering the counterion-concentrating domain of the alginic acid divided
by W_p, its water content expressed in grams, the value of $(\vec{W}_p)_\alpha / \vec{\gamma}_{Na}$
becomes accessible with Eq. (30).

$$\left[\frac{(\vec{W}_p)}{\vec{\gamma}_{Na}}\right]_\alpha = \frac{\alpha(m_p)(1 - \phi_p)}{\{\vec{Na}^+\}} \tag{30}$$

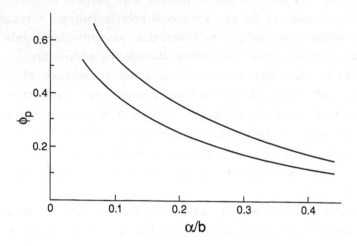

FIG. 16. The osmotic coefficient, ϕ_p, of the counterions in salt-free polyelectrolyte solutions as a function of the reciprocal intercharge distance $(b/\alpha)^{-1}$; upper curve based on data obtained for the cellulose derivatives, sodium carboxymethylcellulose, sodium cellulose sulfate and sodium alginate; lower curve based on results obtained for sodium polyacrylate, sodium polymethacrylate, sodium polyvinyl sulfate, and sodium polyphosphate [41].

Even though $(\vec{W}_p)/\vec{\gamma}_{Na}$ can be as much as 30% larger than $(\vec{W}_p)$ the variability of $\vec{\gamma}_{Na}$ with α is relatively small so that one obtains a fairly accurate estimate of the pattern of solvent uptake by the counterion-concentrating domain of the alginic-acid molecule as α is varied.

The ϕ_p values employed for these computations are interpolated from ϕ_p versus αb^{-1} plots due to Katchalsky and co-workers who compiled such data to show the dependence of osmotic coefficient properties of linear, weak-acid polyelectrolytes on both their charge density and their rigidity [41]. In these plots b is the structurally based distance between neighboring functional sites. The two separate curves that were resolved are reproduced in Fig. 16. The upper curve, associated with the more flexible polysaccharides that were studied, includes data points obtained with alginic acid and is used in this examination of water uptake by the solvent sheath of the alginic-acid molecule. The alginic-acid molecule has been assigned a b value of 4.7 Å for this operation.

In order to analyze the conformational properties of the chondroitin sulfate similarly the lower ϕ_p versus ab^{-1} curve is used. The structural distance between neighboring functional sites is estimated to be 4.7 Å in this instance as well. To obtain α_0 the overall value of α applicable for estimate of ϕ_p with this lower curve the presence of charged OSO_3^- sites had to be accounted for as shown.

$$\alpha_0 = \frac{\alpha n_{CSA}(1 - f) + n_{CSA}(f)}{n_{CSA}} \tag{31}$$

In this equation $1 - f$ is the fractional content of the carboxylic-acid group in the CSA molecule, f is the fractional content of the CSA by the sulfate group and α is the degree of dissociation of the carboxylic-acid sites defined as before.

$$\alpha = \frac{C_b V_b + (C_{H^+})(V_0 + V_b) - n_{CSA}(f)}{n_{CSA}(1 - f)} \tag{32}$$

The extrapolation technique used as shown in Fig. 4 to resolve an intrinsic $p\vec{K}_a$ value of 2.95 for alginic acid is not applicable for the CSA because a Donnan potential term attributable to the fully ionized OSO_3^- sites that occur in an alternating pattern with the carboxylic-acid sites persists at the extrapolated point of zero carboxylate ion presence. To overcome this impediment to the resolution of an intrinsic $p\vec{K}_a$ value for the carboxylic-acid moiety of the CSA a pK_a value was resolved for the repeating carboxylic-acid group at each ionic strength by extrapolation of the data points in Fig. 13 to intercept the ordinate axis at $\alpha = 0$ as shown. The curve obtained in Fig. 17 by plotting these pK_a values versus the salt concentration levels employed in their resolution was extended until a plateau was reached at a $p\vec{K}_a^{int}$ value of 2.50. The availability of enough points to define the initial shape of the curve was believed to make this a reliable projection of the point at which $p\{Na^+\}$ reaches $p\{\vec{Na}^+\}$ in value for such an assignment.

Results of such analysis of the conformational properties of both alginic acid and chondroitin sulfate A, as a function of α, at the different salt concentration levels studied are presented in Table 4.

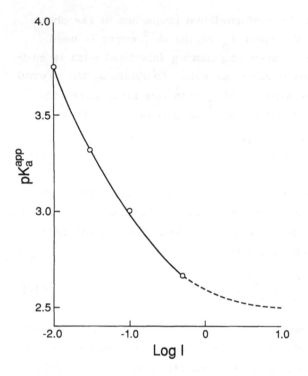

FIG. 17. A graphical estimate of the intrinsic pK of the carboxylic acid component of chondroitin sulfate A.

In this instance invasion of the solvent sheath of the polyion by salt is less of a complicating factor than before because its variation with α at a fixed ionic strength is smaller than the variation affected by changing salt concentration levels at a fixed α value.

These solvent change estimates are compared in Table 5 with those monitored by use of the two graphical schemes described earlier. The good agreement between $[\{\vec{Na}^+\}_{m_{s_2}} / \{\vec{Na}^+\}_{m_{s_1}}]_\alpha$ and $[(W_p/\vec{\gamma}_{Na^+})_{m_{s_1}} / (W_p'/\gamma_{Na^+})_{m_{s_2}}]_\alpha$ that is obtained provides strong support for the model-based description of the counterion-concentrating region of the polyelectrolyte that is employed in Eq. (30).

TABLE 4 Estimates of Water Uptake by the Solvent Sheaths of Alginic Acid and Chondroitin Sulfate A

α	ϕ_p	pK_a^{app}	$p\{Na^+\}$	$p\{\vec{Na}^+\}$	$W_p/\vec{\gamma}_{Na^+}$ $(g\ H_2O)/(\vec{\gamma}_{Na})$ AA sample	$W_p'/\vec{\gamma}_{Na^+}$ $(g\ H_2O/\vec{\gamma}_{Na})$ mmole AA
\multicolumn{7}{c}{0.010 M NaCl; 5.306×10^{-3} M AA}
0.25	0.71	3.49	2.007	1.467	0.0113	2.12
0.50	0.54	3.685	1.989	1.254	0.0219	4.13
0.70	0.44	3.81	1.981	1.121	0.0275	5.18
0.90	0.37	3.91	1.974	1.014	0.0311	5.86
\multicolumn{7}{c}{0.030 M NaCl; 5.306×10^{-3} M AA}
0.25	0.71	3.35	1.580	1.180	0.0058	1.10
0.50	0.54	3.50	1.573	1.023	0.0129	2.42
0.70	0.44	3.59	1.570	0.93	0.0177	3.34
0.90	0.37	3.665	1.568	0.853	0.0214	4.04
\multicolumn{7}{c}{0.10 M NaCl; 5.306×10^{-3} M AA}
0.25	0.71	3.25	1.100	0.800	0.0024	0.46
0.50	0.54	3.305	1.098	0.743	0.0068	1.27
0.70	0.44	3.33	1.098	0.718	0.0109	2.04
0.90	0.37	3.335	1.097	0.712	0.0155	2.92
\multicolumn{7}{c}{0.30 M NaCl; 5.306×10^{-3}}
0.25	0.71	3.09	0.657	0.517	0.00127	0.24
0.50	0.54	3.12	0.656	0.486	0.00374	0.70
0.70	0.44	3.107	0.655	0.498	0.00655	1.23
0.90	0.37	3.065	0.655	0.541	0.0105	1.97

(table continues)

TABLE 4   (continued)

α	α_0	ϕ_p	pK_a^{app}	$p\{Na^+\}$	$p\{\vec{N}a^+\}$	$\dfrac{W_p/\vec{\bar\gamma}_{Na^+}}{\text{CSA sample}}$	$\dfrac{W_p'/\vec{\bar\gamma}_{Na^+}}{\text{mmol CSA}}$
\multicolumn{8}{c}{0.010 M NaCl; 0.001083 M CSA}
0.30	0.635	0.34	3.88	2.035	0.655	0.00205	1.89
0.50	0.739	0.30	3.98	2.034	0.554	0.00201	1.85
0.70	0.844	0.275	4.065	2.034	0.469	0.00195	1.80
0.90	0.948	0.255	4.155	2.034	0.379	0.00183	1.69
\multicolumn{8}{c}{0.030 M NaCl; 0.001083 M CSA}
0.30	0.635	0.34	3.455	1.585	0.63	0.00194	1.79
0.50	0.739	0.30	3.55	1.585	0.535	0.00192	1.77
0.70	0.844	0.275	3.64	1.585	0.445	0.00185	1.70
0.90	0.948	0.255	3.72	1.585	0.365	0.00177	1.64
\multicolumn{8}{c}{0.10 M NaCl; 0.001083 M CSA}
0.30	0.635	0.34	3.06	1.104	0.544	0.00159	1.46
0.50	0.739	0.30	3.20	1.104	0.404	0.00142	1.31
0.70	0.844	0.275	3.30	1.103	0.303	0.00133	1.23
0.90	0.948	0.255	3.40	1.103	0.203	0.00122	1.13
\multicolumn{8}{c}{0.50 M NaCl; 0.001083 M CSA}
0.30	0.635	0.34	2.72	0.447	0.227	0.000764	0.71
0.50	0.739	0.30	2.845	0.447	0.102	0.000709	0.65
0.70	0.844	0.275	2.925	0.447	0.022	0.000697	0.64
0.90	0.948	0.255	3.01	0.447	-0.063	0.000662	0.61

*2.   A More Precise Computation of Solvent Uptake by the Counterion-
Concentrating Region Next to the Charged Polyion Surface*

The oversimplified approach, just described for the estimate of $W_p'/(\vec{\bar\gamma}_{Na})$
the quantity essential for eventual resolution of W_p', has, in addition to
making no attempt to evaluate $\vec{\bar\gamma}_{Na}$, neglected the extra presence of coun
terion effected by the transfer of salt to reach an equilibrium distribution between the two regions of the system.   In the following approach
to the more accurate resolution of W_p' by itself this oversight is avoided.

TABLE 5 A Comparison of Solvent Uptake Estimates with the Two Graphical Approaches and the Computational Scheme [Eq. (30)]

Δm_a	Graphical Method I		Graphical Method II		Eq. (30)
	$\Delta(p\{NA^+\})_\alpha$	$\left[\dfrac{\{\vec{Na}^+\}}{\{\vec{Na}^+\}}\right]_\alpha$	$\Delta(p\{Na^+\})_\alpha$	$\left[\dfrac{\{\vec{Na}^+\}}{\{\vec{Na}^+\}}\right]_\alpha$	$\left[\dfrac{W'_p/\vec{\gamma}_{Na^+}}{W'_p/\vec{\gamma}_{Na^+}}\right]_\alpha$
System AA, NaCl					
$\alpha = 0.25$					
0.010–0.030	0.312	2.05	0.320	2.09	1.93
0.030–0.10	0.386	2.43	0.350	2.24	2.41
0.10–0.30	0.277	1.89	0.275	1.88	1.92
0.010–0.30	0.975	9.44	0.945	8.81	8.91
$\alpha = 0.50$					
0.010–0.030	0.272	1.87	0.21	1.62	1.70
0.030–0.10	0.286	1.93	0.275	1.88	1.91
0.10–0.30	0.257	1.81	0.265	1.84	1.80
0.010–0.30	0.815	6.53	0.750	5.62	5.87
$\alpha = 0.79$					
0.010–0.030	0.232	1.70	0.225	1.68	1.55
0.030–0.10	0.226	1.68	0.225	1.68	1.64
0.10–0.30	0.227	1.69	0.225	1.68	1.65
0.010–0.30	0.685	4.84	0.675	4.73	4.20
$\alpha = 0.90$					
0.010–0.030	0.23	1.70	0.175	1.50	1.45
0.030–0.10	0.156	1.43	0.150	1.41	1.38
0.10–0.30	0.197	1.57	0.175	1.50	1.48
0.010–0.30	0.575	3.76	0.500	3.16	2.97
System CSA, NaCl					
$\alpha = 0.5$					
0.010–0.030	0.024	1.06	0.038	1.09	1.05
0.030–0.10	0.131	1.35	0.125	1.33	1.35
0.10–0.50	0.303	2.01	0.279	1.90	2.00
0.010–0.50	0.458	2.87	0.442	2.77	2.83
$\alpha = 0.7$					
0.010–0.30	0.042	1.10	0.038	1.09	1.06
0.030–0.10	0.116	1.31	0.138	1.37	1.39
0.10–0.50	0.298	1.99	0.288	1.94	1.91
0.010–0.50	0.456	2.86	0.464	2.91	2.80

First $\{\vec{N}a^+\}$, resolved through use of Eq. (16), is redefined in the following way:

$$\{\vec{N}a^+\} = (\vec{m}_{NA^+})(\vec{\gamma}_{Na^+}) = \left[\frac{\alpha(1 - \phi_p)m_p}{W_p} + \vec{m}_{NaCl}\right](\vec{\gamma}_{Na^+}) \qquad (33)$$

A path to estimate the contribution of $\vec{m}_{NaCl}$ to the counterion-concentration level of the solvated polyion sheath is then provided by first equating the activity of salt in both regimes of the polyelectrolyte, salt system as shown:

$$\{\vec{N}a^+\}\{\vec{C}l^-\} = (m_{NaCl} + m_p\alpha\phi_p)(m_{NaCl})(\gamma_{\pm NaCl})^2 \qquad (34)$$

This quantity divided by $\{\vec{N}a^+\}$ resolves $\{\vec{C}l^-\}$

$$\frac{\{\vec{N}a^+\}\{\vec{C}l^-\}}{\{\vec{N}a^+\}} = \{\vec{C}l^-\} = (\vec{m}_{NaCl})(\vec{\gamma}_{Cl}) \qquad (35)$$

The iterative procedure for resolution of W_p is then initiated by selecting values for $\vec{\gamma}_{Cl^-}$ and $\vec{\gamma}_{Na^+}$. The value of $\vec{\gamma}_{Cl^-}$ is used first in Eq. (35) to obtain a value for $\vec{m}_{NaCl}$ which is then used in Eq. (33) together with the assigned $\vec{\gamma}_{Na^+}$ value to provide a first value for W_p. New $\vec{\gamma}_{Na^+}$ and $\vec{\gamma}_{Cl^-}$ values are then computable with Eqs. (21) and (21a), as shown, at the ionic strength arrived at with the W_p and $\vec{m}_{NaCl}$ values provided by the first $\vec{\gamma}_{Na^+}$ and $\vec{\gamma}_{Cl^-}$ values selected.

$$\vec{\gamma}_{Na^+} = \frac{(\gamma_{\pm NaCl})^2}{(\gamma_{\pm KCl})} \quad \text{and} \quad \overleftarrow{\gamma}_{Cl^-} = \gamma_{\pm KCl}$$

$$\text{at} \quad \vec{I} = \frac{1}{2}\left[\frac{2 - \phi_p}{W_p}\alpha m_p + 2\vec{m}_{NaCl}\right] \qquad (21,21a)$$

Selection of a second set of $\vec{\gamma}_{Na^+}$ and $\vec{\gamma}_{Cl^-}$ values, influenced by the discrepancy between the selected and computed values, is then made to repeat the cycle. Convergence of the selected and computed $\vec{\gamma}_{Na^+}$ and $\vec{\gamma}_{Cl^-}$ values to within 1-2% of each other is the criterion employed to signal a successful resolution of W_p, $\vec{m}_{Na}$, and $\vec{m}_{NaCl}$. Such convergence is usually reached by the second repetition of the outlined program.

TABLE 6 Gibbs-Donnan-Based Estimates of the Water Content of the Solvent Sheaths of Alginic Acid and Chondroitin Sulfate as a Function of Counterion Concentration Levels and the Degree of Dissociation of Their Weak Acid Component

α	α_0	pK_a^{app}	$\{Na^+\}$	$\{Cl^-\}$	$\vec{\gamma}_{Na^+}$	$\vec{\gamma}_{Cl^-}$	$W_P\left(\dfrac{g\ H_2O}{sample}\right)$	$W'_p\left(\dfrac{g\ H_2O}{mmole\ AA}\right)$
				System 5.036×10^{-3} M AA				
				0.010 M NaCl				
0.25		3.49	0.0341	0.00260	0.792	0.778	9.7×10^{-3}	1.83
0.50		3.685	0.0557	0.00166	0.782	0.763	1.77×10^{-2}	3.33
0.70		3.81	0.0757	0.00124	0.773	0.750	2.15×10^{-2}	4.06
0.90		3.91	0.0968	0.000987	0.764	0.736	2.39×10^{-2}	4.50
				0.030 M NaCl				
0.25		3.35	0.0641	0.0105	0.762	0.734	5.49×10^{-3}	1.03
0.50		3.50	0.0948	0.00719	0.754	0.721	1.05×10^{-2}	2.09
0.70		3.59	0.117	0.00586	0.749	0.713	1.41×10^{-2}	2.65
0.90		3.665	0.140	0.00492	0.745	0.706	1.71×10^{-2}	3.22
				0.10 M NaCl				
0.25		3.25	0.155	0.0394	0.720	0.666	2.47×10^{-3}	0.465
0.50		3.305	0.176	0.0349	0.726	0.675	6.40×10^{-3}	1.21
0.70		3.33	0.190	0.0324	0.727	0.678	9.74×10^{-3}	1.84
0.90		3.335	0.189	0.0326	0.730	0.683	1.48×10^{-2}	2.69
				0.30 M NaCl				
0.25		3.09	0.304	0.150	0.711	0.632	2.09×10^{-3}	0.394
0.50		3.12	0.327	0.140	0.711	0.638	5.20×10^{-3}	0.979
0.70		3.107	0.318	0.144	0.713	0.644	8.83×10^{-3}	1.66
0.90		0.065	0.316	0.144	0.714	0.648	1.37×10^{-2}	2.58
				System 1.083×10^{-3} M CSA				
				0.010 M NaCl				
0.30	0.635	3.88	0.221	0.000375	0.723	0.669	1.48×10^{-3}	1.37
0.50	0.739	3.98	0.279	0.000298	0.716	0.652	1.44×10^{-3}	1.33
0.70	0.844	4.065	0.340	0.000245	0.713	0.640	1.40×10^{-3}	1.29
0.90	0.948	4.155	0.418	0.000199	0.711	0.624	1.30×10^{-3}	1.20
				0.030 M NaCl				
0.30	0.635	3.45	0.232	0.00286	0.721	0.666	1.43×10^{-3}	1.32
0.50	0.739	3.55	0.292	0.00227	0.714	0.647	1.35×10^{-3}	1.25
0.70	0.844	3.64	0.359	0.00185	0.711	0.637	1.33×10^{-3}	1.22
0.90	0.948	3.72	0.432	0.00154	0.711	0.627	1.26×10^{-3}	1.17
				0.10 M NaCl				
0.30	0.635	3.06	0.284	0.0209	0.706	0.657	1.23×10^{-3}	1.13
0.50	0.739	3.20	0.392	0.0152	0.712	0.629	1.06×10^{-3}	0.979

TABLE 6   (continued)

α	α_0	pK_a^{app}	$\{Na^+\}$	$\{Cl^-\}$	$\vec{\gamma}_{Na^+}$	$\vec{\gamma}_{Cl^-}$	$W_p\left(\dfrac{g\ H_2O}{sample}\right)$	$W_p'\left(\dfrac{g\ H_2O}{mmole\ AA}\right)$
				0.10 M NaCl				
0.70	0.844	3.30	0.493	0.0120	0.709	0.616	9.86×10^{-4}	0.910
0.90	0.948	3.40	0.621	0.00957	0.715	0.604	8.96×10^{-4}	0.827
				0.50 M NaCl				
0.30	0.635	2.72	0.594	0.195	0.714	0.606	8.93×10^{-4}	0.825
0.50	0.739	2.845	0.792	0.147	0.727	0.590	6.66×10^{-4}	0.615
0.70	0.844	2.925	0.952	0.116	0.742	0.582	6.21×10^{-4}	0.573
0.90	0.948	3.01	1.16	0.100	0.757	0.577	5.64×10^{-4}	0.521

The results of this more quantitative examination of the solvent uptake characteristics of the counterion-concentrating sheath enveloping the alginic acid and the chondroitin sulfate polyions are summarized in Table 6. A comparison of the solvent uptake parameters compiled for the two polysaccharides in Table 6 clearly reveals how much more spongelike the counterion concentration domain of the alginic acid is. The water uptake by the chondroitin sulfate A domain is much less sensitive to changes in its degree of dissociation and the water activity of the system. This approach for estimate of solvent uptake by the counterion-concentrating domain of linear, weak-acid polyelectrolytes, while leading to accurate assessments of their conformational properties, provides a meaningful answer to the numerous investigators who have sought explanations for the shape of curves generated when pK_a is plotted versus α [1].

In order to compare the more precise estimates of solvent uptake by the counterion-concentrating domain of linear, weak-acid polyelectrolytes that are affected by use of Eq. (33) with those obtained with Eq. (30) by neglecting salt incursion and the nonideality of the counterion the two sets of W_p' ratios, $(W_p')_1/(W_p')_2$ and

$$\frac{(W_p')_1 (\vec{\gamma}_{Na^+})_2}{(\vec{\gamma}_{Na^+})_1 (W_p')_2} \approx \frac{(W_p')_1}{(W_p')_2} \tag{36}$$

are listed next to each other in Tables 7 and 8. Careful scrutiny of
the solvent uptake estimates when α is varied and ionic strength is
kept constant (Table 7) shows that discrepancy between the solvent
uptake estimates, while never severe, is largest in the low α range,
becoming most noticeable at the highest salt concentration level. This
result shows that at high counterion concentration levels salt imbibe-
ment effects on the magnitude of $\vec{m}_{Na^+}$ and $\vec{\gamma}_{Na^+}$ are least likely to
cancel.

A parallel examination of the effect on solvent uptake estimates
that neglect of counterion invasion and nonideality introduces when
counterion concentration levels of the solution are changed while
maintaining the degree of dissociation of the polyelectrolyte constant
is provided in Table 8. Acceptable correlation of the results is restricted to the lowest salt concentration levels (0.010 to 0.030 M) in
the higher α range (α = 0.7 and 0.9) with the flexible alginic acid.
For the more rigid chondroitin sulfate sizable discrepancy between
the solvent uptake estimates occurs only when the salt concentration
level is raised to 0.50 from 0.10 M. Resistance to salt invasion and
changes in solvent uptake by the more highly counterion-concentrated
solvent sheath of the chondroitin sulfate results in the correlation of
results by the two approaches over a broader range of experimental
conditions.

3. *The Solvent Sheath-Controlled Response of the Dissociation
    Tendencies of Linear, Weak-Acid Polyelectrolytes to Changes
    of Their Concentration in Solutions of Relatively Low (0.005
    N) and High (0.10 N) Salt Content*

It has been shown by Nagasawa and co-workers [1] that whereas the
potentiometric properties of poly(acrylic acid) are noticeably affected
when its concentration level is raised from 0.00829 to 0.0419 N in a
0.0050 N NaCl solution they are barely altered in a 0.10 N NaCl solution over the same change in polymer concentration levels (Fig. 18).
We believe that these observations should be as interpretable with
the Gibbs-Donnan-based model as the sensitive response of their dissociation to changes in the salt concentration level of their solutions
has been. To examine this prognosis Eq. (16) is used first to

TABLE 7   A Comparison of Solvent Uptake Estimates as a Function of α at Constant Ionic Strength with and without Consideration of Salt Incursion of the Polyion Domain and the Nonideality of the Counterion in the Solvent Sheath of the Polyelectrolyte that Defines this Domain

$\Delta\alpha$	$[(W'_p \vec{\gamma}^{-1}_{Na^+})_{\alpha_2} / (W'_p \vec{\gamma}^{-1}_{Na^+})_{\alpha_1}]_{m_s}$	$[(W'_p)_{\alpha_2} / (W'_p)_{\alpha_1}]_{m_s}$
	5.306×10^{-3} M AA; 0.010 M NaCl	
0.25-0.50	1.95	1.82
0.50-0.70	1.25	1.22
0.70-0.90	1.13	1.11
0.25-0.90	2.76	2.46
	5.306×10^{-3} M AA; 0.030 M NaCl	
0.25-0.50	2.20	2.03
0.50-0.70	1.38	1.27
0.70-0.90	1.21	1.22
0.25-0.90	3.67	3.13
	$5,306 \times 10^{-3}$ M AA; 0.10 M NaCl	
0.25-0.50	2.78	2.60
0.50-0.70	1.61	1.52
0.70-0.90	1.43	1.46
0.25-0.90	6.39	5.78
	5.306×10^{-3} M AA; 0.30 M NaCl	
0.25-0.50	2.96	2.48
0.50-0.70	1.75	1.70
0.70-0.90	1.60	1.55
0.25-0.90	8.28	6.55
	1.083×10^{-3} M CSA; 0.010 M NaCl	
0.30-0.50	0.979	0.971
0.50-0.70	0.973	0.970
0.70-0.90	0.939	0.930
0.30-0.90	0.894	0.876
	1.083×10^{-3} M CSA; 0.030 M NaCl	
0.30-0.50	0.989	0.947
0.50-0.70	0.963	0.976
0.70-0.90	0.960	0.959
0.30-0.90	0.915	0.886

TABLE 7   (continued)

	1.083×10^{-3} M CSA; 0.10 M NaCl	
0.30-0.50	0.897	0.859
0.50-0.70	0.939	0.930
0.70-0.90	0.919	0.909
0.30-0.90	0.774	0.725
	1.083×10^{-3} M CSA; 0.50 M NaCl	
0.30-0.50	0.926	0.745
0.50-0.70	0.985	0.932
0.70-0.90	0.949	0.909
0.30-0.90	0.865	0.632

TABLE 8   A Comparison of Solvent Uptake Estimates as a Function of Ionic Strength at Constant α with and without Consideration of Salt Incursion of the Polyion Domain and the Nonideality of the Counterion in the Solvent Sheath of the Polyelectrolyte that Defines this Domain

Δm_s	$[(W'_p \vec{\gamma}_{Na^+}^{-1})m_{s_1} / (W'_p \vec{\gamma}_{Na^+}^{-1})m_{s_2}]_\alpha$	$[(W'_p)m_{s_1} / (W'_p)m_{s_2}]_\alpha$
	System AA; NaCl	
	$\alpha = 0.25$	
0.010-0.030	1.93	1.78
0.030-0.10	2.41	2.22
0.10-0.30	1.92	1.18
0.010-0.30	8.91	4.64
	$\alpha = 0.50$	
0.010-0.030	1.70	1.59
0.030-0.10	1.91	1.73
0.10-0.30	1.80	1.24
0.010-0.30	5.87	3.40
	$\alpha = 0.70$	
0.010-0.030	1.55	1.53
0.030-0.10	1.64	1.44
0.10-0.30	1.65	1.11
0.010-0.30	4.20	2.45
	$\alpha = 0.90$	
0.010-0.030	1.45	1.40
0.030-0.10	1.38	1.20

TABLE 8   (continued)

Δm_s	$\left[(W'_p\,{}^{-1}_{Na^+})m_{s_1}\big/(W'_p\,{}^{-2}_{Na^+})m_{s_2}\right]_\alpha$	$\left[(W'_p)m_{s_1}\big/(W'_p)m_{s_2}\right]_\alpha$
	$\alpha = 0.90$	
0.10–0.30	1.48	1.04
0.010–0.30	2.97	1.74
	System CSA; NaCl	
	$\alpha = 0.3$	
0.010–0.030	1.056	1.030
0.030–0.10	1.226	1.158
0.10–0.50	2.003	1.382
0.010–0.50	2.677	1.661
	$\alpha = 0.5$	
0.010–0.030	1.045	1.064
0.030–0.10	1.35	1.277
0.10–0.50	2.00	1.592
0.010–0.50	2.83	2.163
	$\alpha = 0.7$	
0.010–0.030	1.06	1.057
0.030–0.10	1.39	1.329
0.10–0.50	1.91	1.588
0.010–0.50	2.80	2.251
	$\alpha = 0.9$	
0.010–0.030	1.037	1.026
0.030–0.10	1.449	1.415
0.10–0.50	1.849	1.587
0.010–0.50	2.766	2.303

characterize the counterion-concentrating property of the solvent
sheath domain of the polyion as a function of α and the free-counterion concentration level of the solution external to it.   Potentiometric data compiled by Nagasawa et al. [1] for a 0.00829 N poly-
(acrylic acid) solution at various salt concentration levels (Fig. 1)
are the source of the $p\overset{*}{K}{}^{int}_a$ value of 4.294 and the various $(pK^{app}_a)_\alpha$
values combined with $p\{Na^+\}$ estimates in Eq. (16) to resolve the
$p\{\vec{Na}^+\}$ values sought for this purpose.   The procedure employed
for estimate of $p\{Na^+\}$ follows our earlier pattern of operations by

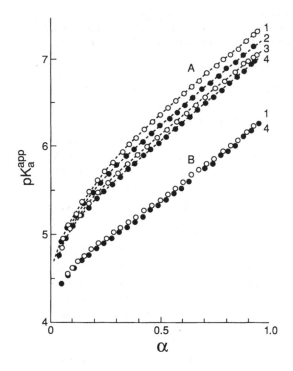

FIG. 18.   The variation of pK_a versus α plots by varying poly-(acrylic acid) concentration levels in NaCl at a concentration level of 0.0050 and 0.10 M.   A = 0.0050 M NaCl; B = 0.10 M NaCl; 1, 0.00829 M PAA; 2, 0.0193 M PAA; 3, 0.0335 M PAA; 4, 0.0419 M PAA.

resolving m_{Na^+}, γ_{Na^+}, and $p\{Na^+\}$ in the sequence of steps shown:*

$$N_{Na^+} = \phi_p N_p \alpha + N_{NaCl} = m_{Na^+} \qquad (37)$$

$$\gamma_{Na^+} = \frac{(\gamma_{\pm NaCl})^2}{(\gamma_{\pm KCl})} \quad \text{at} \quad I = N_{Na^+} = m_{Na^+} \qquad (21)$$

$$p\{Na^+\} = -\log(m_{Na^+})(\gamma_{Na^+}) \qquad (38)$$

---

*Since $I < 0.1$ values of γ_{Na^+} may be assigned by reference to estimates of this parameter provided by Kielland [42].

TABLE 9   The Counterion-Concentrating Property of the Solvent
Sheath Domain of Poly(acrylic acid) as a Function of Degree of
Neutralization and the Counterion Concentration Level of the
Solution External to It

α	ϕ_p	pK_a^{app}	m_{Na^+}	γ_{Na^+}	$p\{Na^+\}$	$p\{\vec{Na}^+\}$
		0.00829 N PAA; 0.0050 N NaCl				
0.9	0.175	7.238	0.0063	0.925	2.237	-0.707
0.7	0.20	6.797	0.00616	0.9254	2.244	-0.259
0.5	0.269	6.380	0.0061	0.9257	2.248	0.162
0.3	0.355	5.865	0.00588	0.9264	2.264	0.693
0.1	0.55	5.147	0.00546	0.9288	2.295	1.410
		0.00829 PAA; 0.010 N NaCl				
0.9	0.175	7.025	0.0113	0.9034	1.991	-0.738
0.7	0.20	6.591	0.01116	0.9042	1.998	-0.299
0.5	0.265	6.190	0.0111	0.9042	1.998	0.102
0.3	0.355	5.702	0.01088	0.9047	2.007	0.599
0.1	0.55	5.032	0.01046	0.9058	2.024	1.286
		0.00829 PAA; 0.050 N NaCl				
0.9	0.175	6.404	0.0513	0.830	1.371	-0.739
0.7	0.20	6.020	0.05116	0.830	1.372	-0.354
0.5	0.265	5.620	0.0511	0.830	1.373	0.047
0.3	0.355	5.176	0.05088	0.8304	1.374	0.492
0.1	0.55	4.688	0.05046	0.8307	1.378	0.984
		0.00829 PAA; 0.10 N NaCl				
0.9	0.175	6.139	0.1013	0.791	1.096	-0.749
0.7	0.20	5.741	0.10116	0.791	1.097	-0.350
0.5	0.265	5.353	0.1011	0.791	1.097	0.038
0.3	0.355	5.00	0.100883	0.791	1.098	0.392
0.1	0.55	4.588	0.100456	0.791	1.100	0.806

In these equations concentrations of polymer, N_p, and simple salt,
N_{NaCl}, are expressed in normality units and ϕ_p corresponds to the
practical osmotic coefficient of the salt-free polyelectrolyte that is
resolved by interpolation to the appropriate αb^{-1} value (b = 2.5 Å)
using the bottom curve of Fig. 16.   The resultant estimates of counterion concentration levels, N_{Na^+}, are low enough in the systems

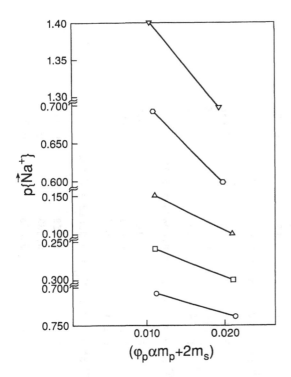

FIG. 19. A plot of the model-based estimates of $p\{\vec{Na}\}$ variation with the model-projected free-counterion concentration levels of the solution external to the polyion solvent sheath; $m_{PAA} = 0.00829$, $m_{NaCl} = 0.0050$ and 0.010.   o, $\alpha = 0.9$; □, $\alpha = 0.7$; △, $\alpha = 0.5$; O, $\alpha = 0.3$; ∇, $\alpha = 0.1$.

under consideration to permit normality and molality to be used interchangeably without introduction of serious error.

The results of these computations are summarized in Table 9. The $p\{\vec{Na}^+\}$ values compiled for the 0.005 and 0.010 N NaCl solutions at the five α values selected for study are plotted versus the total free-ion (co- and counterion) concentration of these solutions ($N_{Na^+} + 2N_{NaCl}$) in Fig. 19 to permit interpolation of $p\{\vec{Na}^+\}$ values at the free-ion concentration levels effected in the 0.0193, 0.0335, and 0.0491 N PAA, 0.0050 N NaCl systems. These $p\{\vec{Na}^+\}$ values, combined in Eq. (16) with the corresponding $p\{Na^+\}$ values resolved with Eqs. (37), (21), and (38) and the $p\vec{K}_a^{int}$ value of 4.294, resolved

TABLE 10 A Comparison of Model- and Experimentally-Based pK_a^{app} Values

N_p	$p\{Na^+\}$	$p\{\bar{Na}^+\}$	$(pK_a^{app})_{inter.}$	$(pK_a^{app})_{calc.}$	$\Delta(pK_a^{app})_{inter.}$	$\Delta(pK_a^{app})_{calc.}$
\multicolumn{7}{l}{α = 0.90; 0.0050 N NaCl; pK_a^{app} at α = 0.9, N_p = 0.00829 = 7.238}						
0.0193	2.133	-0.712	7.095	7.139	0.143	0.099
0.0335	2.037	-0.719	6.986	7.05	0.109	0.089
0.0419	1.980	-0.723	6.905	6.997	0.081	0.050
\multicolumn{7}{l}{α = 0.70; 0.0050 N NaCl; pK_a^{app} at α = 0.7, N_p = 0.00829 = 6.798}						
0.0193	2.151	-0.265	6.674	6.710	0.124	0.088
0.0335	2.058	-0.273	6.563	6.625	0.111	0.085
0.0419	2.007	-0.278	6.514	6.579	0.049	0.046
\multicolumn{7}{l}{α = 0.50; 0.0050 N NaCl; pK_a^{app} at α = 0.5, N_p = 0.00829 = 6.38}						
0.0193	2.158	0.145	6.240	6.307	0.140	0.073
0.0335	2.066	0.136	6.149	6.224	0.091	0.083
0.0419	2.020	0.130	6.072	6.184	0.077	0.040
\multicolumn{7}{l}{α = 0.30; 0.0050 N NaCl; pK_a^{app} at α = 0.3, N_p = 0.00829 = 5.865}						
0.0193	2.187	0.680	5.736	5.801	0.129	0.064
0.0335	2.106	0.664	5.673	5.736	0.063	0.065
0.0419	2.065	0.655	5.625	5.704	0.048	0.032

0.0193	2.251	1.399	5.144	5.146	0.034	0.032
0.0335	2.200	1.384	5.106	5.110	0.038	0.035
0.0419	2.173	1.378	5.082	5.089	0.024	0.021

$\alpha = 0.10;\ 0.0050\ N\ NaCl;\ pK_a^{app}$ at $\alpha = 0.1,\ N_p = 0.00829 = 5.180$

0.0419	1.077	−0.750	6.121	6.130	0.009

$\alpha = 0.90;\ 0.10\ N\ NaCl;\ pK_a^{app}$ at $\alpha = 0.9,\ N_p = 0.00829 = 6.130$

0.0419	1.0785	−0.350	5.723	5.741	0.018

$\alpha = 0.70;\ 0.10\ N\ NaCl;\ pK_a^{app}$ at $\alpha = 0.7,\ N_p = 0.00829 = 5.741$

0.0419	1.080	+0.038	5.336	5.353	0.017

$\alpha = 0.50;\ 0.10\ N\ NaCl;\ pK_a^{app}$ at $\alpha = 0.5,\ N_p = 0.00829 = 5.353$

0.0419	1.084	+0.392	4.986	5.00	0.014

$\alpha = 0.30;\ 0.10\ N\ NaCl;\ pK_a^{app}$ at $\alpha = 0.3,\ N_p = 0.00829 = 5.00$

0.0419	1.0834	+0.806	4.571	4.588	0.017

$\alpha = 0.10;\ 0.10\ N\ NaCl;\ pK_a^{app}$ at $\alpha = 0.1,\ N_p = 0.00829 + 4.588$

earlier by extrapolation of the curves in Fig. 1, lead to the estimates of pK_a^{app} that are presented in Table 10 together with the experimentally based values obtained by interpolation of the curves presented in Fig. 18. The experimentally based change affected in pK_a^{app} by the change in polyelectrolyte concentration at a particular α value is compared with the computationally based change in the last two columns of the table. A parallel prediction of the much smaller change affected in pK_a^{app} when the polyelectrolyte concentration is raised from 0.00829 to 0.0419 N in 0.10 N NaCl solution has also been included in the table.

Inspection of Table 10 shows that there is some discrepancy between the measured and the predicted pK_a^{app} values at α values, ≥ 0.3, when its concentration in dilute salt solution (0.0050 N NaCl) is increased. Most of this discrepancy is contained in the pK_a^{app} value predicted for the initial increment of change (0.00829 to 0.0193 N) by the Gibbs-Donnan-based Eq. (16). The agreement between experimentally and computationally based changes in pK_a^{app} improves with each additional concentration increment (0.0193 to 0.0335 N and 0.0335 to 0.0419 N). The barely perceptible response of pK_a^{app} to change in polyelectrolyte concentration levels (0.00829 to 0.0419 N) in the presence of a more concentrated salt solution (0.10 N NaCl) is predicted with the model as well to provide additional corroboration of the physical representation of these systems.

E.   Further Documentation of the Gibbs-Donnan Model

Another example that can be used to document the validity of the Gibbs-Donnan approach takes advantage of the special properties of the Linde-A zeolite. Because of their rigidity and high negative site density counterion concentrations in the solvated zeolite phase are sizable and constant. As a result the selectivity coefficient at any external electrolyte concentration of two mobile counterions at micro- and macroconcentration levels, respectively, becomes predictable on the basis of one selectivity measurement through application of the Gibbs-Donnan-based equations [22]. Constancy of zeolite parameters, assured by its rigidity and resistance to electrolyte invasion, ensures

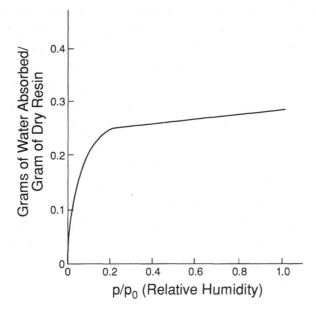

FIG. 20.   Water sorption isotherm of sodium A-zeolite.

reproducibility of the nonideality ratio of the pairs of ions in the
zeolite resolved in a single selectivity measurement.   Constancy of
the solvent content of the zeolite phase also leads to the capability
for assessment of the osmotic pressure, π, as the water activity of
the solution phase is modified by its salt content.   Justification of
these projections is based on the fact that crystallographic assessments made on both dry and moist A-zeolite samples resolve inter-
ionic distances that are the same, within the 0.1 $\overset{\circ}{A}$ sensitivity limit
of the x-ray-based measurements program.   The water adsorption
isotherm (25°C) that is presented in Fig. 20 for the sodium A-zeolite
provides corroborative support as well [43].   There is very little
change of water taken up per gram of zeolite as the value of p/p_o
($\approx a_w$) varies from unity to a value of about 0.16; the extra sorption
is attributed to an external surface effect.   The prediction of high
resistance to electrolyte invasion is consistent with the large negative charge emanating from the ring of oxygen atoms in the faces of
the corners of the cubic unit cell and the small opening available to
the exchanging ions.

To take advantage of these unique properties a research program in which the exchanging ion, N^+, was kept at radioactive tracer level concentrations in solution of MX (~0.10 M and greater) was pursued [22]. By this approach the ion fraction of M^+, at equilibrium, was essentially unity in both the zeolite and external solution phases. The constancy of water uptake by the gel phase and its resistance to electrolyte invasion over the concentration (water activity) range of study (dilute to almost saturated) ensured the constancy of $\bar{m}_{M^+}$ and $\bar{\gamma}_{M^+}$. By using microconcentration levels of N^+ the constancy of $\bar{\gamma}_{N^+}$ was preserved throughout the course of experiments as well. The $-2 \log(\gamma_{\pm MX}/\gamma_{\pm NX})$ term of Eq. (12) was immediately calculable. Since NX was present in trace quantities the activity coefficients for MX were identical with the values reported for it in the literature. The activity coefficients for trace NX, in the presence of MX, were calculated using the Harned-Cooke equation in the form

$$\log \gamma_{o(NX)} = \log \gamma_{NX(o)} + \alpha m + \beta m^2 \tag{39}$$

where $\gamma_{o(NX)}$ is the activity coefficient of a trace of NX in the presence of MX at molality m, $\gamma_{NX(o)}$ is the activity coefficient of pure NX at molality m, and α and β are experimentally determined parameters.

Our interpretation of the sorption isotherm presented in Fig. 20, strongly influenced by the evidence for zeolite rigidity, permitted calculation of the $\pi(\bar{V}_M - \bar{V}_N)/2.3RT$ term as well. The sharp decrease in water content of the zeolite that occurs below an a_w value of 0.16 was presumed to indicate that both the activity of the water component in the two phases and the pressure exerted on both phases in this low-a_w region had become identical. An a_w value of 0.16, the point of intersection of the extension of the two lines that were drawn to represent the different water sorption properties of the zeolite in Fig. 20, was identified with the composition portion of the chemical potential of the water component of the zeolite phase while the water activity of solution in equilibrium with it ranged from 0.16 to unity. The insensitivity of the zeolite water content to water activity over this range is the basis for this assessment of the situation. At a_w values larger than 0.16 the pressure differences, π, becomes calculable with Eq. (7c) by using a value of 18 mL for the partial molar volume of water. The partial molar volumes of the ions at infinite dilution were then used with this π value to arrive at the value of the $\pi(\bar{V}_M - \bar{V}_N)/2.3RT$ term. The additional approximation, constancy of partial molar volumes of ions at infinite dilution, that had to be incorporated in this step of the operation is not a serious deficiency in the approach because of the incompressibility of the ions.

With the two terms of Eq. (12) made accessible the third term, $\log \bar{\gamma}_{M^+}/\bar{\gamma}_{N^+}$, endowed with invariant properties by the design of the experimental program, became calculable using a single selectivity measurement. With this parameter, made accessible in this way, Eq. (12) could be used to anticipate the other selectivity coefficients for comparison with the experimental values.

The experimental program in which selectivity coefficients were measured, K_e, and calculated, K_c, used three systems with the same cation pair (macroquantities of Na^+ and microquantities of Cs^+), but different anions ($CX = Cl^-$, Br^-, and Ac^-) to test fully the various aspects of the model. These values of K_e and K_c are presented in Table 11 and permit assessment of the validity of the approximations made in order to facilitate use of Eq. (12) as contemplated. For example, the value of 3.390, resolved with Eq. (12) for the $\ln \bar{\gamma}_{Na^+}/\bar{\gamma}_{Cs^+}$ parameter using the K_e value measured in 2.255 M NaCl, was employed to resolve the K_c values listed in Table 11 for this assessment. Successful use of this parameter is signaled by K_c values comparable in magnitude with the experimental K_e values. Certainly, proof of the validity of our premise that the rigidity of zeolite A and its salt-exclusion properties produces constancy in the $\ln \bar{\gamma}_{Na}/\bar{\gamma}_{Cs}$ as long as a_w is larger than 0.16 has been documented by the good agreement between K_e and K_c.

For the NaCl-CsCl phase estimates of the magnitude of the $\gamma_{\pm}$ values to be used for the trace component, CsCl, were affected by using the α and β parameters compiled by Robinson and Stokes in Eq. (39). Interaction coefficients were unavailable for the trace component, CsBr, in the NaA, NaBr, CsBr system, however. Substitution of the α- and β-terms used for the NaCl-CsCl mixture [44]

TABLE 11   The Prediction of Selectivity Coefficients; the NaA-NaX-CsX (Trace) Systems

External NaCl molality	K_e	K_c	External NaBr molality	K_e	K_c	External NaAc molality	K_e	K_c
0.053	2.77	2.81	0.080	2.86	2.84	0.186	3.06	3.09
0.106	2.78	2.83	0.401	2.48	2.54	0.470	3.20	3.15
0.537	2.56	2.55	0.810	2.26	2.25	0.959	3.35	3.31
1.085	2.22	2.31	1.660	1.84	1.69	2.001	3.73	3.70
2.255	1.85	1.85	3.490	1.14	1.08	2.558	3.89	3.89
3.383	1.61	1.52	5.503	0.72	0.60	3.554	4.31	4.36
4.501	1.37	1.32	7.903	0.47	0.32			
6.068	1.18	1.09						

in Eq. (39) seemed most appropriate and were reemployed for the evaluation of $\gamma_{\pm(o)CsBr}$ as well.   The activity coefficients of the pure components at a molality as high as 3.5 were from Robinson and Stokes [26].   Beyond that concentration level activity coefficient values were available only from Landolt-Bornstein [45].   There is disagreement between the two sources below 3.5 m and the data of Robinson and Stokes [26] are believed to be more reliable because of the better correlation between K_c and K_e in the lower concentration range.   The values reported at 5.503 and 7.903 m are included in Table 11 only to demonstrate that the selectivity trend continues to be predictable by the Gibbs-Donnan-based approach.

Even though there were no interaction coefficients published for the NAAc, CsAc mixtures study of the NaA-NaAc-CsAc system was included because the trend in activity coefficients of the acetates are opposite those of the chlorides and bromides.   Anticipation, with the model, of the effect of this aspect on the selectivity coefficient trend was expected to provide additional corroboration of the concepts developed.   In order to approximate the effect of NaAc molality on the activity coefficient of the CsAc present in micro amounts as it was in this study the absence of interaction coefficients had to be remedied.   It was by presuming that, in the system of interest, the Harned rule, stated as follows:

$$\log \gamma_B = \log \gamma_{B(o)} - \alpha_B m_C \tag{40a}$$

$$\log \gamma_C = \log \gamma_{C(o)} - \alpha_C m_B \tag{40b}$$

is obeyed [46]. Equation (41), developed on this basis [26], was used to resolve the interaction coefficient needed to calculate $\gamma_{\pm o(CsAc)}$ values.

$$\frac{0.4343(\phi_B - \phi_C)}{m} = \alpha_B \tag{41}$$

In Eq. (41) ϕ_B is the osmotic coefficient of NaAc at molality m and α_B is the interaction coefficient. Reference to Table 11 shows that the computations of K_c, so facilitated, led to numbers in excellent agreement with the experimental values to justify our use of Eq. (41).

Equation (41) was arrived at [26] by applying the Gibbs-Duhem equation to the two 1:1 electrolytes of the aqueous phase under consideration. By maintaining such a system at constant pressure and temperature the following expression is obtained.

$$-55.51 \ d \log a_w = 2m_B \ d \log m_B \gamma_B + 2m_C \ d \log m_C \gamma_C \tag{42}$$

The assumption that Harned's rule is obeyed eliminates the β parameter ($\beta_B = \beta_C = 0$) when substituting for γ_B and γ_C in Eq. (42). By putting m_B equal to xm $m_C = (1 - x)m$, where m is the total salt molality and x represents the mole fraction of B, the following equation results as long as m remains constant.

$$-55.51 \ d \log a_w = m^2\{2(\alpha_B + \alpha_C)x - 2(\alpha_C)\} \ dx \tag{43}$$

Integration of Eq. (43) from x = 0 when $\log a_w = \log a_{w(C)}$, the value in a solution containing only electrolyte C at a molality of m, yields Eq. (44):

$$\frac{-55.51}{xm^2} \log \frac{a_{w(x)}}{a_{w(C)}} = x(\alpha_B + \alpha_C) - 2\alpha_C = \frac{0.8686}{xm}(\phi_B - \phi_C) \tag{44}$$

where $a_{w(x)}$ is the water activity of a solution xm molal in B and (1 - x)m molal in C. When x = 1, $\log a_{w(x)} = \log a_{w(B)}$, $\phi = \phi_B$, and

$$\frac{-55.51}{m^2} \log \frac{a_{w(B)}}{a_{w(C)}} = \alpha_B - \alpha_C = 0.8686 \frac{\phi_B - \phi_C}{m} \qquad (45)$$

To obtain Eq. (41) we added one more constraint by equating α_B to $-\alpha_C$.

The K_e values that were compiled for the three systems are plotted in Fig. 21 versus the corresponding solution molality for each sequence of experiments to show how the three separate curves converge as zero molality (infinite dilution) is approached. The fact that they merge at this point is also consistent with model-based expectation. The variation of the selectivity coefficient with external molality, according to the model, is a reflection of the change affected in the activity-coefficient ratio of the exchanging counterions and the osmotic pressure since all the other parameters remain invariant. At infinite dilution the activity-coefficient ratio of the counterions in solution reaches a value of unity and this term in Eq. (12) ($-2 \ln \gamma_{\pm NaX}/\gamma_{\pm CsX}$)

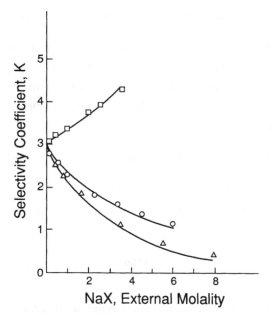

FIG. 21.   Selectivity data for systems NaZ–NaX–CsX (trace)–H_2O: □, X = Ac^-; △, X = Br^-; o, X = Cl^-.

is reduced to zero. Concurrently, the osmotic pressure increases
to its maximum value of 2491 atm; this estimate of π is reached using
Eq. (7c) as shown

$$\frac{RT}{\bar{V}_w} \ln \frac{a_w}{\bar{a}_w} = \pi = \frac{0.0821 \text{ L atm}}{0.018 \text{ L mole}^{-1} \text{ deg mole}} (298 \text{ deg}) \ln \frac{1.0}{0.16} \quad (7c)$$

The value of K at infinite dilution, accessible with Eq. (12) as shown
in the footnote,* is equal to 2.97. Its value, so determined, is in excellent agreement with the value of 3.0 that is extrapolated using the
three separate sets of selectivity coefficient measurements.

Additional documentation of the applicability of the Gibbs-Donnan
model for the interpretation of the equilibrium properties of charged
polymer gel-simple salt mixtures has been provided in similar studies
of two additional systems, KA-KCl-CsCl (trace) and KA-KCl-NaCl
(trace). The selectivity coefficients, experimental and calculated, that
were compiled are listed in Table 12. The values of $\ln \bar{\gamma}_{K^+}/\bar{\gamma}_{Cs^+}$
(2.340) and $\ln \bar{\gamma}_{K^+}/\bar{\gamma}_{Na^+}$ (0.1793), based on K_e values measured at
2.090 and 1.831 m KCl, respectively, were used in Eq. (12) to esti-
mate the K_c values listed. The $\gamma_\pm$ values, as well as the interaction
parameters, were taken from the data of Robinson [44,47].

An interesting aside with respect to the KA-KCl-NaCl (trace) system is the concern that was felt in our laboratory initially when our
K_c values were only in fair agreement with the K_e values using the
interaction parameters published at that time. In earlier research of
this system the Harned rule had been presumed to be applicable to
this system and only α parameters were reported. At the height of

---

*
$$\ln K = \ln \frac{\bar{\gamma}_{Na^+}}{\bar{\gamma}_{Cs^+}} + \frac{\pi(\bar{V}_{Na^+} - \bar{V}_{Cs^+})}{RT}$$

$$= 3.390 + \frac{2491 \text{ atm } (-0.0057 \text{ L mole}^{-1} - 0.0169 \text{ L mole}^{-1})}{0.0821 \text{ L atm mole}^{-1} \text{ degree}^{-1} (298 \text{ deg})}$$

$$= 3.390 - 2.301 = 1.089$$

and $K = 2.97$

TABLE 12   The Prediction of Selectivity Coefficients;
the KA-KCl-CsCl (Trace) and the KA-KCl-NaCl
(Trace) Systems

External KCl molality	K_e	K_c	External KCl molality	K_e	K_c
0.049	2.92	2.80	0.1089	3.42	3.41
0.098	2.77	2.80	0.4394	3.52	3.48
0.499	2.67	2.72	0.8905	3.49	3.51
1.000	2.58	2.66	1.8308	3.59	3.59
2.090	2.48	2.48	2.8291	3.75	3.69
3.220	2.45	2.39	3.8851	3.80	3.77
4.414	2.45	2.36			

our concern Robinson reevaluated his earlier study and published β
interaction coefficients as well.   Use of these numbers in the Harned-
Cooke equation [Eq. (39)] to recalculate the $\gamma_\pm$ ratio of the ions in
solution led to the agreement observable between K_e and K_c in Table
12.

We believe that the results of this analysis of the exchange of
pairs of univalent ions, one present in macro amounts and the other
in micro quantities, between the Linde A-zeolite and simple salt solutions has shown unequivocally that the Gibbs-Donnan model provides
a quantitative description of these systems.   It has, as a consequence,
become the basis of our description of all charged polymers in solution.   We showed, right at the start of this presentation, that the
response of all charged polymers to counterion concentration levels
is very similar, no matter whether the polymer is present as an easily
discernible separate phase or appears to be completely and uniform-
ally dissolved, to justify taking this path.

F.   The Effect of Linear Polyelectrolyte Presence on the Distribution
     of Pairs of Ions between its   Crosslinked Ion-Exchange Gel
     Analog and Simple Salt Solutions

Differences in the equilibrium distribution of trace divalent metal ions
between a cation-exchange resin and $NaClO_4$ in the presence and

absence of the linear polyelectrolyte analog of the crosslinked resin, sodium polystyrene sulfonate, are consistent with the picture proposed earlier. The linear polyelectrolyte in solution does indeed simulate the counterion concentrating properties of the crosslinked gel analog. By presuming that the distribution of trace divalent cations in the polyelectrolyte-salt mixtures is indeed controlled by a Donnan potential and by assigning a volume to the counterion-concentrating regions defined by the polyelectrolyte the apparent enhancement of divalent metal-ion binding to the polyion by increasing its presence in a solution at a fixed salt concentration level has been fully explained [23].

In this application of the model the electrochemical potential of each of the diffusible components, MX_2 and MX, at equilibrium is identical at every point in the system. By dividing the system into regions large enough to buffer disturbance by ion fluctuations but small enough to be representative of differential volume elements of the system, one can equate the electrochemical potential of the diffusible components in the volume element at the surface of the polymer which is representative of the polymer domain, with any other differential volume element throughout the solution. These, in turn, can be equated with the average macroscopic electrochemical potential measured for these components [48]. Electrical potentials cancel as before and Eq. (11) is once again applicable. In this case, however, the osmotic pressure term is small enough to neglect and the activity ratio of the ions in the solution and the polyelectrolyte domain can be equated.

$$\frac{\overset{\leftarrow}{a}_{M^+}^2}{(\overset{\leftarrow}{a}_{M^+})^2} = \frac{a_{M^+}^2}{(a_{M^+})^2} = \frac{m_{MX_2}(\gamma_{\pm MX_2})^3}{(m_{MX})^2(\gamma_{\pm MX})^4} \tag{11a}$$

An arrow is used to identify the counterion-concentrating region of the polyelectrolyte as before.

With Eq. (11a) rearranged as shown

$$\frac{\overset{\leftarrow}{a}_{M+2}}{(\overset{\leftarrow}{a}_{M+})^2} \frac{(a_{M+})^2}{(a_{M+2})} = \frac{(\overset{\leftarrow}{a}_{M+2})}{(\overset{\leftarrow}{a}_{M+})^2} \frac{(m_{MX})^2(\gamma_{\pm MX})^4}{(m_{MX_2})(\gamma_{\pm MX_2})^3} = 1 \qquad (11b)$$

To test the above rationalization of the physical-chemical properties of linear polyelectrolyte, salt mixtures, mean molal activity coefficients were assigned to MX_2 and MX. At an ionic strength of 0.3
they were taken directly from the literature [26] to estimate the
activity ratio of the ions in the external solution. In those experiments where the ionic strength equaled 0.10 the single-ion activity
coefficients due to Kielland [42] were used. No correction was made
for the effect of the macroion presence on the activity coefficient of
the ion at trace level concentrations.

To obtain a comparable evaluation of $\overset{\leftarrow}{a}_{M+2}/(\overset{\leftarrow}{a}_{M+})^2$ ion-exchange
selectivity data available for the various pairs of ions [49] were used
to assign values of $\overset{\leftarrow}{\gamma}_{M2+}/(\overset{\leftarrow}{\gamma}_{M+})^2$ in the PSS systems studied. These
ratios were obtained directly by simply extrapolating such ion-selec-
tivity data, plotted as a function of resin molality, to the molality
value estimated for the polymer domain [19,49-51].

The distribution data obtained for the trace-metal-ion-NaPSS-
$NaClO_4$ system in earlier studies by us were reduced to molality terms
that could be combined with the activity coefficients assigned by the
procedures just described in the following way. The distribution data,
D, when combined with the D_o value measured for the polyelectrolyte-
free but otherwise equivalent system, relate the quantity of trace divalent metal ion confined to the polyelectrolyte domain in one gram of
solvent to the quantity of the trace divalent ion in the same 1 g of
solvent as shown: $D_o = \bar{m}_{M+2}$ and $D = \bar{m}_{M+2}/(m_{M+2} + \Sigma \vec{M}^{+2})$; by
dividing the numerator and denominator of the expression for D by
m_{M+2}:

$$D = \frac{\bar{m}_{M^{+2}}/m_{M^{+2}}}{m_{M^{+2}}/m_{M^{+2} + \Sigma \overrightarrow{M}^{+2}/m_{M^{+2}}}} = \frac{D_o}{1 + \Sigma \overrightarrow{M}^{+2}/m_{M^{+2}}}$$

Rearrangement of the resultant expression for D leads to Eq. (29) [37].

$$\frac{D_o - D}{D} = \frac{\Sigma \overleftarrow{M}^{+2}}{m_{M^{+2}}} \tag{29}$$

To obtain the corresponding inverse ratio for the ionic strength controlling counterion the molality of the salt and the polyelectrolyte were combined as follows:

$$\frac{m_{M^+}}{\Sigma \overrightarrow{M}^+} = \frac{m_{NaClO_4} + m_{NaPSS}(\phi_p)}{m_{NaPSS}(1 - \phi_p)} \tag{29a}$$

In Eq. (29a) ϕ_p is the measured [49,52,53], or computed [54], practical osmotic coefficient of the M^+ ion form of the polyelectrolyte. It corrects for the fraction of counterion that escapes from the polymer domain. Its use in this manner is justified by the fact that the osmotic properties of polyelectrolyte and salt mixtures are additive. Each of the components of such a system behaves as it would by itself [55-57].

The only term missing at this point is the solvent content of the polyelectrolyte domain. It is inaccessible to measurement and for our estimate of this parameter in these studies we have used the theoretically based equation due to Manning. This equation, obtained by minimizing the free energy of the polyelectrolyte system that is defined by Manning's ion-condensation model [10], is presented below.

$$V_p = 4.11(\xi - 1)(b)^3 \tag{46}$$

In this equation ξ is the nondimensional charge density parameter ($\xi = e^2/\epsilon kTb$), b is the average structurally based distance between polyion charges, ε is the dielectric constant of pure solvent, and the other symbols have their conventional meaning. By expressing b in angstrom units V_p has the units of cm^3/mole. For our purpose, the difference between mL/mole and g H_2O/mole, the appropriate dimension of V_p in our treatment, was considered to be within the limits of the uncertainty in this parameter.

The distribution data and the activity coefficient and V_p estimates were combined in Eq. (11b) as shown below:

$$\frac{D_o - D}{D} \frac{(m_{MX} + \phi_p(m_{MPSS}))^2(V_p)}{(m_{MPSS}(1 - \phi_p))^2} \frac{(\vec{\gamma}_{M+}^{~2})}{(\vec{\gamma}_{M+})^2} \frac{(\gamma_{\pm MX})^4}{(\gamma_{\pm MX_2})^3} = 1 \quad (11a)$$

The D_o values employed in these computations were used in interpolation of log D_o versus log $a_{\pm MX}$ plots, obtained by measuring the distribution of the trace divalent metal ion of interest between solution and resin phases as a function of the mean molal activity of the simple salt. To facilitate the resolution of suitable D_o values by interpolation the mean molal activity coefficient of the simple salt in the trace MX_2-MX-MPSS mixture, the source of the corresponding D values, was estimated with Eq. (38a)

$$a_{\pm MX} = \{m_{MX} + (m_{MPSS})(\phi_p)\}\{\gamma_{\pm MX} \text{ at } I = (m_{MX} + (m_{HPSS})(\phi_p))\}$$
$$(38a)$$

A representative presentation of the equilibrium distribution behavior of a trace metal ion, Co^{+2}, between resin and polyelectrolyte-free $NaClO_4$ solution as the electrolyte concentration is varied is given in Fig. 22. In this figure the logarithm of D^o (counts/s/g of dry resin/counts/s/g of soln) is plotted versus the logarithm of the mean molal activity of $NaClO_4$. Such examination of the distribution data provides a stringent assessment of the salt concentration range in which the ion-exchange distribution method can be employed to measure quantitatively the extent of sequestering of trace metal ion by the polyelectrolyte under investigation in our systems. As long as (1) curvature is absent in the resultant line and (2) the straight line has a slope of -2.00 for the trace divalent ions the fundamental

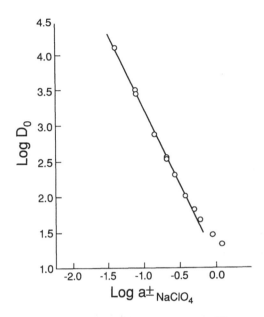

FIG. 22.   Distribution, D_o, of ^{60}Co between NaPSS resin and sodium perchlorate solutions.

assumption that is the basis of the method can be assumed to be obeyed.   As soon as deviation from the straight line occurs this basic assessment of the situation, i.e., the constancy of the macro-counterion composition of the resin phase, is no longer valid due to significant imbibement of the resin by $NaClO_4$.   Such deviation from ideal behavior is noticeable in Fig. 22, curvature from the straight line observed (slope = -2.10) occurring when the concentration level of the $NaClO_4$ reaches 0.60 m.   In the other trace activity (^{65}Zn, ^{45}Ca) $NaClO_4$ systems examined straight lines were also obtained [a slope of -2.08 for ^{45}Ca(II) and -2.04 for ^{65}Zn(II)] much beyond the limited (≤0.3 M) $NaClO_4$ concentration range of this study.   In every case examined D_o is well-behaved and useful for our purpose.

The modus operandi for evaluation of the $(\vec{\gamma}_{Na^+})^2/(\vec{\gamma}_{M^{+2}})$ parameter used in Eq. (11c) has been briefly described but needs further clarification since the data used for its estimate were from a comparable system in which H^+ ion, not Na^+ ion, was the

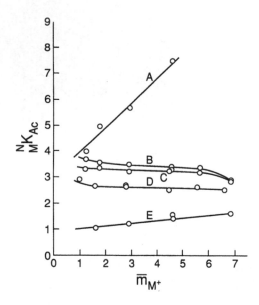

FIG. 23. Plots of $_N^M K_{Ac}$ versus resin-phase molality [49,51].

A: $_{H^+}^{45Ca^{+2}} K_{Ac} = (\bar{\gamma}_{H^+})^2 / (\bar{\gamma}_{Ca^{+2}})$, 0.168 M HClO$_4$;

B: $_{H^+}^{60Co^{+2}} K = (\bar{\gamma}_{H^+})^2 / \bar{\gamma}_{Co^{+2}}$; 0.168 M HClO$_4$; C: $_{H^+}^{65Zn^{+2}} K_{Ac} =$

$(\bar{\gamma}_{H^+})^2 / (\bar{\gamma}_{Zn^{+2}})$; 0.168 M HClO$_4$; D: $_{H^+}^{65Zn^{+2}} K_{Ac} = (\bar{\gamma}_H)^2 / \bar{\gamma}_{Co^{+2}}$;

0.0168 M HClO$_4$; E: $_{H^+}^{Na^+} K_{Ac} = \bar{\gamma}_{H^+} / \bar{\gamma}_{Na^+}$; 0.10 M NaCl.

concentration-determining counterion. Such clarification, as well as
justification for such substitution is therefore provided at this point.
If one expresses the exchange of a trace-level concentration of divalent metal ion, M^{+2}, present in a solution of univalent acid, HX, in
equilibrium with ion-exchange resins whose counterion molality, $\bar{m}_{H^+}$,
is controlled by varying its degree of crosslinking with the following modification of Eq. (11a) the K_{Ac} parameter resolved can be

equated to $(\bar{\gamma}_{H^+})^2/\bar{\gamma}_{M^{+2}}$. Use of such data in plots of K_{Ac} versus $\bar{m}_H$ can then

$$K_{Ac} = \frac{(\bar{m}_{M^{+2}})}{(m_{H^{+2}})} \frac{(m_{H^+})^2}{(\bar{m}_{H^+})^2} \frac{(\gamma_{HX})^4}{(\gamma_{\pm MX_2})^3} = \frac{(\bar{\gamma}_H)^2}{(\bar{\gamma}_{M^{+2}})} \tag{11d}$$

be used to interpolate the value of $(\bar{\gamma}_{H^+})^2/\bar{\gamma}_{M^{+2}})$, the reciprocal of the activity-coefficient ratio being sought for the linear polyelectro-lyte analog of the resin used in the resolution of the K_{Ac} values. The molality of H^+ in the counterion-concentrating domain of the polyelectrolyte is found to equal 0.9 by employing the V_p parameter evaluated with Manning's equation as shown:

$$\vec{m}_{H^+} = \frac{1 - \phi_p}{V_p} = \frac{0.8}{0.879} = 0.91 \tag{33a}$$

The fact that a K_{Ac} value of unity is resolved at this molality from interpolation of the K_{Ac} versus $\bar{m}_H$ plot that was also available from an earlier study of the NaX-HX-HPSS resin system showed that at this molality $\bar{\gamma}_{H^+} = \bar{\gamma}_{Na^+}$ and our use of the data from the trace-level MX_2-HX-HPSS resin system to obtain the value of $(\vec{\gamma}_{Na^+})^2/(\vec{\gamma}_{M^{+2}})$ is fully justified on this basis. The various plots of K_{Ac} versus m_H that were employed for the estimates of $(\vec{\gamma}_{Na})^2/(\vec{\gamma}_{M^{+2}})$ as well as its justification are presented in Fig. 23.

There are in Fig. 23 two plots of K_{Ac} for the ^{65}Zn-HClO$_4$-resin (HPSS) system, one set of numbers having been derived from selec-tivity studies in 0.016 m HClO$_4$ while the other set was obtained from measurements in 0.168 m HClO$_4$. Because some error has to be introduced into the assessment of K_{Ac} by assigning the mean molal activity coefficient of the pure divalent metal perchlorate at the ionic strength of the system to the trace metal ion, the deduced value of $^M K_{Ac}$ must be less affected by neglect of ion-ion interaction at the lower ionic strength. We have as a consequence used the ^{65}Zn selectivity data obtained in 0.0160 m HClO$_4$ to assign a value of $(2.8)^{-1}$ to both $\vec{\gamma}_{Co}/(\vec{\gamma}_{Na})^2$ and $\vec{\gamma}_{Zn}/(\vec{\gamma}_{Na})^2$ for use in Eq. (11c).

TABLE 13   Test of Donnan Model Using Ion-Exchange Distribution Data

NaPSS(m_p)	$\dfrac{D_o - D}{D}$	$\dfrac{\Sigma \, \vec{Na}^+/g \; H_2O}{(m_p(1 - \phi_p))}$	$\dfrac{[(\vec{a}_{M^{2+}})(a_{Na^+})^2]}{[(\vec{a}_{Na^+})^2(a_{M^{+2}})]}$

System $^{60}Co^{2+}$, NaPSS, 0.30 m $NaClO_4$; V_p = 0.879 mL/0.0010 mole; $\vec{\gamma}_{Co^{2+}}/(\vec{\gamma}_{Na^+})^2 = (2.8)^{-1}$; $(\gamma_{\pm o}(NaClO_4))^4/(\gamma_{\pm o}(Co(ClO_4)_2))^3 = 1.318$

0.00094	0.0194	0.000752	1.202
0.00130	0.0291	0.00104	1.302
0.00250	0.0438	0.00200	1.019
0.0052	0.0961	0.00416	1.075
0.0080	0.137	0.00640	0.996
0.0105	0.179	0.00840	0.991

System $^{60}Co^{2+}$, NaPSS, 0.10 m $NaClO_4$; V_p = 0.879 mL/0.0010 mole; $\vec{\gamma}_{Co^{2+}}/(\vec{\gamma}_{Na})^2 = (2.8)^{-1}$; $(\gamma_{Na^+})^2/(\gamma_{Co^{2+}}) = 1.521$

0.00084	0.113	0.000672	0.991
0.0017	0.172	0.00136	0.743
0.0026	0.198	0.00208	0.561
0.0054	0.553	0.00432	0.754
0.0092	1.186	0.00736	0.950
0.0123	1.698	0.00984	1.017

System $^{65}Zn^{2+}$, NaPSS, 0.30 m $NaClO_4$; V_p = 0.879 mL/0.0010 mole; $\vec{\gamma}_{Zn^{2+}}/(\vec{\gamma}_{Na^+})^2 = (2.8)^{-1}$; $(\gamma_{\pm o}(NaClO_4))^4/(\gamma_{\pm o}(Zn(ClO_4)_2))^3 = 1.318$

0.00059	0.00899	0.000472	0.886
0.00093	0.0154	0.000744	0.963
0.0013	0.0236	0.00104	1.056
0.0025	0.0404	0.0020	0.941
0.0050	0.0869	0.0040	1.011
0.0079	0.139	0.00632	1.024
0.010	0.183	0.0080	1.064

TABLE 13   (continued)

System $^{45}Ca^{2+}$, NaPSS, 0.30 m $NaClO_4$, V_p = 0.879 mL/0.0010 mole

$\vec{\gamma}_{Ca^{+2}}/(\vec{\gamma}_{Na^+})^2 = (3.8)^{-1}$; $(\gamma_{\pm o}(NaClO_4))^4/(\gamma_{\pm o}(Ca(ClO_4)_2))^3 = 1.397$

0.0052	0.117	0.00416	1.023
0.0070	0.141	0.00560	0.916
0.0099	0.241	0.0079	1.107
0.0213	0.642	0.0234	0.996
0.0513	1.153	0.0410	1.022

Overall average   = 0.984
Standard deviation   = 0.151
Probable error   = 0.102

Selectivity data at this lower ionic strength was unavailable for the
$^{45}Ca-NaClO_4$-resin (HPSS) system and we have used the value of
$(3.8)^{-1}$ extrapolated from the $^{Ca}_{H}K_{Ac}$ versus m_{HPSS} plot presented
for the higher-ionic-strength system to test the Donnan-model applicability in this instance.

The results of the above treatment of the ion-distribution data
compiled to test the Donnan model are summarized in Table 13. It
can be observed that even when polyelectrolyte concentration is
varied more than tenfold and salt concentration a factor of 3 there
is no real deviation from the Donnan-predicted behavior. The overall average of 0.984 is very close to the value of unity expected.
Because error in the estimate of $\Sigma \vec{M}^{+2}/m_{M+2}$ can be sizable in those
experiments where D_o - D is a small number, the standard deviation
of 15% is understandable.

C.   Use of the Gibbs-Donnan Model for Interpretation of the
     Equilibrium Distribution Properties of Pairs of Ions between
     a Crosslinked (8% Divinyl Benzene by Weight) and Its Linear
     Polyelectrolyte Analog

The equilibrium distributions of trace-level concentrations of Co^{+2}
ion and macro quantities of H^+-ion between salt-free polystyrene
sulfonic-acid solutions and its crosslinked analog (8% DVB) that

have been measured in our laboratory [49] are used next in the
Gibbs-Donnan model to provide a meaningful interpretation of the
observed results. To facilitate this exercise the equilibrium distributions of ^{60}Co and H$^+$ ion between HClO$_4$ solution and the same
crosslinked resin have also been measured.

These data when expressed for both systems as the product of
the ^{60}Co^{+2} and the H$^+$ ion distribution ratios in Eq. (12)

$$_{H^+}^{^{60}Co^{+2}}K_{Ex} = \frac{(\bar{m}_{Co^{+2}})}{(m_{Co^{+2}})}\frac{(m_{H^+})^2}{(\bar{m}_{H^+})^2} \tag{12}$$

where

$$\frac{\bar{m}_{Co^{+2}}}{m_{Co^{+2}}} = K_{D_o}$$

$$= \frac{D_o - D_e}{D_e}\frac{(mL\ of\ solution)}{\left(\dfrac{g\ of}{stock\ HR}\right)_{\%DVB}\left(\dfrac{g\ dry\ HR}{g\ stock\ HR}\right)_{\%DVB}\left(\dfrac{g\ H_2O}{g\ dry\ HR}\right)_{\%DVB}}$$

and

$$\frac{(m_{H^+})}{(\bar{m}_{H^+})} = \frac{\left(\dfrac{meq\ H^+}{g\ H_2O}\right)}{\dfrac{\left(\dfrac{meq\ H^+}{g\ dry\ HR}\right)_{\%DVB}}{\left(\dfrac{g\ H_2O}{g\ dry\ HR}\right)_{\%DVB}}}$$

with D representing the counts per minute per gram of solution
before and after equilibrium (subscripts o and e), lead to
$_{H^+}^{^{60}Co^{+2}}K_{Ex}$ values that are sizably different. The selectivity coefficient in perchloric acid is practically independent of solution-phase
concentration. In salt-free polystyrenesulfonic acid, however,
$_{H^+}^{Co^{+2}}K_{Ex}$ is markedly lower than that measured in perchloric acid
solutions and is a remarkably strong function of polyelectrolyte

TABLE 14 A Comparison of Selectivity Coefficients Measured for Trace Cobalt-Hydrogen Ion Exchange with an 8% DVB Crosslinked Polystyrene Sulfonic-Acid Ion-Exchange Resin in Solutions of Perchloric Acid and Polystyrene Sulfonic Acid

m_{HClO_4}	m_{HR}	$\bar{m}_{Co^{+2}}/m_{Co^{+2}}(K_{D_o})$	$^{60}Co^{+2}_{H^+}K_{Ex}$
		$HClO_4$	
1.68×10^{-1}	4.62	1342	1.78
1.086×10^{-1}	4.58	3016	1.67
1.114×10^{-1}	4.58	2859	1.68
6.30×10^{-2}	4.54	9271	1.77
6.72×10^{-2}	4.54	7548	1.64
4.13×10^{-2}	4.52	22800	1.89
4.06×10^{-2}	4.52	23700	1.90

m_{HPSS}	m_{HR}	$\bar{m}_{Co^{+2}}/m_{Co^{+2}}(K_{D_o})$	$^{60}Co^{+2}_{H^+}K_{Ex}$
		$HPSS$	
0.973×10^{-3}	4.42	27060	1.312×10^{-3}
0.978×10^{-3}	4.42	26567	1.298×10^{-3}
2.41×10^{-2}	4.425	3277	9.74×10^{-2}
2.42×10^{-2}	4.425	2991	8.97×10^{-2}
4.61×10^{-2}	4.44	1494	1.61×10^{-1}
5.82×10^{-2}	4.45	1098	1.88×10^{-1}
1.06×10^{-1}	4.47	642	3.61×10^{-1}
0.910×10^{-1}	4.46	649	2.86×10^{-1}
1.628×10^{-1}	4.49	317	4.14×10^{-1}
1.666×10^{-1}	4.49	299	4.12×10^{-1}

concentration, decreasing rapidly with decreasing polyelectrolyte concentration (Table 14).

The rapid decrease in $^{60}Co^{+2}_{H^+}K_{Ex}$ with decrease in the molality, monomer basis, of polystyrene sulfonic acid is indeed

consistent with expectations based on the earlier rationalization of
the dissociation properties of linear, weak-acid polyelectrolytes.
Whereas the concentration of counterions in the gel phase is accurately known the concentration of the trace metal ion in the solution
phase is, according to our model, masked by the fact that almost all
of the Co^{+2} ion found in the solution is really associated with the
counterion-concentrating domain of the polyelectrolyte. When varying the molality of the polyelectrolyte the distribution pattern is
misinterpreted to produce the variability observed in the selectivity
coefficient value.

With both sets of data available it has been a simple matter to
employ the model for assessment of the concentration and activity
ratio of this pair of ions in the resin phase and that portion of the
solution that is presumed to be separate from the counterion-concentrating solvent sheath of the linear polyelectrolyte. By dividing one
ratio into the other the acceptability of such an apportionment of the
polyelectrolyte solution into two separate regions can be judged from
the constancy of the result and from the closeness of its value to
unity. The same criteria are applied to the other portion of the
solution phase that is assignable to the counterion-concentrating solvent sheath presumed to define the polyion domain. In this instance
the reasonableness of the solvent-content estimates that need to be
made in order for an activity ratio of the pair of ions in the polyion
domain to reach their ratio in the resin phase is supportive of the
model as well. Both exercises provide a satisfying explanation of
the observed difference in ion-exchange behavior of polyelectrolyte
and simple electrolyte within the context of the physical picture that
has evolved from the Gibbs-Donnan treatment of linear polyelectrolytes and their crosslinked gel analogs.

The parameters used to facilitate the characterization of the 8%
divinyl benzene crosslinked polystyrene sulfonic-acid ion-exchange
resin in the first step of the Gibbs-Donnan-based approach are presented in Table 15. The perchloric-acid molality in a particular experiment and the corresponding H^{+}-ion molality of the resin at equilibrium are listed in the first two columns of this table. The

TABLE 15   Solution Parameters Used to Facilitate Characterization of the 8% Divinyl Benzene Crosslinked Polystyrene Sulfonic-Acid Ion-Exchange Resin

m_{HClO_4}	$\bar{m}_{H^+}$	γ_{H^+}	$\gamma_{Co^{+2}}$	$\log(a_{H^+})^2/\gamma_{Co^{+2}}$	$\log K_{D_o}$
0.0406	4.52	0.866	0.511	−2.616	4.375
0.0413	4.52	0.865	0.507	−2.599	4.358
0.0672	4.54	0.855	0.442	−2.127	3.878
0.0630	4.54	0.850	0.451	−2.197	3.967
0.1114	4.58	0.825	0.389	−1.663	3.456
0.1086	4.58	0.826	0.393	−1.689	3.479
0.168	4.62	0.820	0.338	−1.251	3.128

single-ion activity coefficients assigned to the H^+ and Co^{+2} ions at these $HClO_4$ concentration values are given in columns 3 and 4 and are based on graphical projections of the single-ion activity coefficients proposed for these ions by Kielland [42] at ionic strength values of 0.0010, 0.0050, 0.010, 0.050, and 0.10.   The $\log(a_{H^+})^2/\gamma_{Co^{+2}}$ values, made accessible in this way, and the corresponding $\log K_{D_o}$ values, presented in the last two columns, are then plotted against each other in Fig. 24.   The resultant plot yields a straight line with a negative slope of unity as the following analysis of Eq. (12), rearranged as shown, tells us it should.

$$\log K_{D_o} = \log \frac{\bar{m}_{Co^{+2}}}{m_{Co^{+2}}} = \log \frac{(\bar{m}_{H^+})^2 (\bar{\gamma}_{H^+})^2 (\gamma_{Co^{+2}})}{(m_{H^+})^2 (\gamma_{H^+})^2 (\bar{\gamma}_{Co^{+2}})} + \frac{\pi(2V_H - V_M)}{2.3\ RT} \quad (12)$$

The resin parameters $\bar{m}_{H^+}$, $\bar{\gamma}_{H^+}$, $\bar{\gamma}_{M^{+2}}$, and $\pi(2V_H - V_M)/2.3\ RT$ remain essentially unchanged in the series of resin characterization experiments so that

$$\log K_{D_o} = \log(\text{constant}) + \log \frac{\gamma_{Co^{+2}}}{(m_{H^+})^2(\gamma_{H^+})^2} = \log(\text{constant})$$
$$+ \log \frac{\gamma_{Co^{+2}}}{(a_{H^+})^2}$$

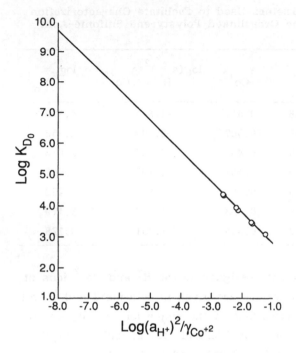

FIG. 24.   A plot of log K_{D_o} versus log $(a_{H^+})^2/\gamma_{Co^{+2}}$ using resin characterization data summarized in Table 14.

In order to interpolate the $\dfrac{m_{^{60}Co^{+2}}}{m_{^{60}Co^{+2}}}$ ($K_{D_o}^*$) values expected for

the colligatively active region of the $^{60}Co^{+2}$-ion containing HPSS solutions studied with Fig. 24, $(a_{H^+})^2/\gamma_{^{60}Co^{+2}}$ had to be evaluated. For this purpose m_{H^+} was obtained with Eq. (37)

$$m_{H^+} = (\phi_{HPSS})(m_{HPSS}) \tag{37}$$

by using osmotic coefficient data available from vapor-pressure osmometry studies [52].   The log $K_{D_o}^*$ values, so obtained, are presented in the last column of Table 16.   The $\log(a_{H^+})^2/\gamma_{Co^{+2}}$

estimates are placed next to them in column 8. Column 1 lists the molality, on a monomer basis, of the linear HPSS used in each equilibration. The H^+ ion molality of the gel phase in each experiment is given in column 2; K_{D_o}, the measured distribution of trace $^{60}Co^{+2}$ ion between resin and HPSS, is listed in column 3. The osmotic coefficients, presented in the next column of the table, are used to provide column 5 which lists the molality, (ϕ_{HPSS}) (m_{HPSS}), of the H^+ ion that is accessible to these measurements. The single-ion activity coefficients that are listed in columns 6 and 7 were obtained through use of Kielland's single-ion activity-coefficient tables [42].

To examine the internal consistency of the first segment of this exercise the resin characterization data compiled for the $^{60}Co^{+2}$, $HClO_4$, 8% by weight DVB crosslinked resin system (Tables 14, 15) were used in Eq. (12); as shown, to evaluate $\log(\bar{\gamma}_{H^+})^2/\bar{\gamma}_{60_{Co}+2}) +$ $\pi(2V_H - V_M)/2.3 \ RT$:

$$\log K_{Ac} = \log \frac{^{60}Co^{+2}}{H^+} K_{Ex} + \log \frac{(\gamma_{\pm HClO_4})^4}{(\gamma_{\pm Co(ClO_4)_2})^3}$$

$$= \log \frac{(\bar{\gamma}_{H^+})^2}{(\bar{\gamma}_{60_{Co}+2})} + \frac{\pi(2V_H - V_M)}{2.3 \ RT} \tag{12}$$

The value of 0.461 ± 0.05 resolved for this parameter remains essentially invariant over the $HClO_4$ concentration range examined and is expected to be duplicated in the $^{60}Co^{+2}$, HPSS, 8% DVB-crosslinked HPSS resin system. It has been used in Eq. (11) to determine whether $\log K_{th} + \pi(V_{Co} - 2V_H)/2.3 \ RT$ is equal to zero as it should be. By definition

$$\log K_{th} + \frac{\pi(V_{Co} - 2V_M)}{2.3 \ RT} = 0 \tag{11a}$$

TABLE 16 Estimates of Trace-Metal-Ion Ratios (log $K_{D_o}^*$) for the Characterized Dowex-50 Resin and Different Polyelectrolyte Solutions (HPSS) at Equilibrium

m_{HPSS}	$\bar{m}_{H^+}$	K_{D_o}	ϕ_p	m_{H^+}	γ_H	$\gamma_{Co^{+2}}$	$\log \dfrac{(a_{H^+})^2}{\gamma_{Co^{+2}}}$	$\log K_{D_o}^*$
0.973×10^{-3}	4.42	27060	0.22	2.14×10^{-4}	0.98	0.977	-7.344	9.06
0.978×10^{-3}	4.42	26567	0.22	2.15×10^{-4}	0.98	0.972	-7.340	9.06
2.41×10^{-2}	4.425	3277	0.22	5.30×10^{-3}	0.939	0.74	-4.475	6.23
2.42×10^{-2}	4.425	2991	0.22	5.32×10^{-3}	0.932	0.74	-4.475	6.23
4.61×10^{-2}	4.44	1494	0.225	1.02×10^{-2}	0.915	0.674	-3.889	5.64
5.82×10^{-2}	4.45	1098	0.23	1.34×10^{-2}	0.905	0.65	-3.647	5.41
1.06×10^{-1}	4.47	642	0.245	2.60×10^{-2}	0.879	0.575	-3.042	4.81
9.10×10^{-2}	4.46	649	0.24	2.18×10^{-2}	0.882	0.597	-3.208	4.98
1.628×10^{-1}	4.49	317	0.28	4.56×10^{-2}	0.861	0.494	-2.506	4.29
1.666×10^{-1}	4.49	299	0.28	4.66×10^{-2}	0.860	0.490	-2.484	4.26

$$= \log \frac{(\bar{m}_{Co^{+2}})}{(a_{Co^{+2}})} \frac{(a_{H^+})^2}{(\bar{m}_{H^+})^2} + \log \frac{(\bar{\gamma}_{Co^{+2}})}{(\bar{\gamma}_{H^+})^2} + \frac{\pi(V_{Co} - 2V_H)}{2.3\ RT} \qquad \begin{array}{l}(11a)\\ \text{cont.}\end{array}$$

Substitution of

$$\log \frac{K_{D_o}}{Co^{+2}} \qquad \text{for} \qquad \log \frac{(\bar{m}_{Co^{+2}})}{a_{Co^{+2}}}$$

$$\log \frac{(m_{HPSS})(\phi_{HPSS})(\gamma_{H^+})^2}{(\bar{m}_{H^+})^2} \qquad \text{for} \qquad \log \frac{(a_{H^+})^2}{(\bar{m}_{H^+})^2}$$

and

$$0.461 \qquad \text{for} \qquad \log \frac{(\bar{\gamma}_{H^+})^2}{(\bar{\gamma}_{Co^{+2}})} + \frac{\pi(2V_H - V_{Co})}{2.3\ RT}$$

leads to the estimates of $\log K_{th} + \pi(V_{Co} - 2V_H)/2.3\ RT$ that are
presented in the last column of Table 17. The standard deviation
of the model-based estimates of $\log K_{th} + \pi(V_{Co} - 2V_H)/2.3\ RT$
from 0.0 is 0.018. This is quite satisfying.

The relationship between the two separate regions proposed for
the linear polyelectrolyte in solution as well as between the resin
phase and the counterion-concentrating region of the polyelectrolyte
has been considered next to examine the self-consistency of the model
further. In both instances the Gibbs-Donnan-based equations have
been extended to provide estimates of solvent uptake,* $W'_p m_{HPSS}$, by
the counterion-concentrating domain of the linear polystyrene sulfonic
acid. In the first approach the equilibrium distribution of $^{60}Co^{+2}$ and
and H^+-ion between the resin phase and the counterion-concentrating

---

*$W'_p m_{HPSS}$ = g H_2O uptake by solvent sheath of HPSS solute associ-
ated with 1 g H_2O of solution phase; W'_p = g H_2O uptake per millimole
HPSS.

TABLE 17 The Projection of $[\log K_{th} + \pi(V_{Co} - 2 V_H)/2.3\ RT]$ for the Equilibrium Distribution of $^{60}Co^{+2}$ and H^+ Ions between the External Portion of the HPSS Solution and the 8% Crosslinked HPSS Resin

m_{HPSS}	$\bar{m}_{H^+}$	γ_{H^+}	$\gamma_{Co^{+2}}$	$\log \dfrac{(m_{HPSS})^2(\phi_{HPSS})^2(\gamma_{H^+})^2}{(\bar{m}_H)^2}$	$\log \dfrac{K_{D_o}}{\gamma_{Co^{+2}}}$	$\log K_{th} + \dfrac{\pi(V_{Co^{+2}} - 2V_{H^+})}{2.3RT}$
9.73×10^{-4}	4.42	0.980	0.972	-8.679	9.072	-0.068
9.78×10^{-4}	4.42	0.980	0.972	-8.643	9.072	-0.032
2.41×10^{-2}	4.425	0.939	0.740	-5.898	6.361	+0.002
2.42×10^{-2}	4.425	0.932	0.740	-5.901	6.361	-0.001
4.61×10^{-2}	4.44	0.915	0.674	-5.340	5.811	+0.010
5.82×10^{-2}	4.45	0.905	0.650	-5.130	5.597	+0.006
1.06×10^{-1}	4.47	0.879	0.575	-4.584	5.050	+0.005
9.10×10^{-2}	4.46	0.882	0.597	-4.729	5.204	+0.014
1.628×10^{-1}	4.49	0.861	0.494	-4.091	4.596	+0.044
1.666×10^{-1}	4.49	0.860	0.490	-4.098	4.570	+0.011
						$\sigma = 0.028$

domain of the polyelectrolyte in contact with it is analyzed. In the second their equilibrium distribution between the two regions assigned to the polyelectrolyte is examined for this purpose.

With the first approach[†]

$$\frac{(\vec{a}_{Co})}{(\vec{a}_{H^+})^2} \frac{(a_{H^+})^2}{(a_{Co^{+2}})} = \frac{K^*_{D_o} - K_{D_o}}{K_{D_o} W'_p m_{HPSS}} \left[\frac{\phi_{HPSS}}{1 - \phi_{HPSS}}\right]^2 (W'_p)^2 (m_{HPSS})^2$$

$$\times \frac{(\vec{\gamma}_{Co^{+2}})}{(\vec{\gamma}_{H^+})^2} \frac{(\gamma_{H^+})^2}{(\gamma_{Co^{+2}})} = K_{th} = 1 \qquad (11b)$$

and

$$\frac{K^*_{D_o} - K_{D_o}}{K_{D_o}} \frac{(\phi_{HPSS})^2}{(1 - \phi_{HPSS})^2} (m_{HPSS}) \frac{(\gamma_{H^+})^2}{(\gamma_{Co^{+2}})} = \frac{1}{W'_p} \frac{(\vec{\gamma}_{H^+})^2}{(\vec{\gamma}_{Co^{+2}})}$$

By taking the inverse of this equation the following expression is obtained for $W'_p (\vec{\gamma}_{Co^{+2}} / (\vec{\gamma}_{H^+})^2)$:

$$\frac{(K_{D_o})}{- \,_{\mathrm{?SS}} \, K^*_{D_o} - K_{D_o}} \frac{(1 - \phi_{HPSS})^2}{(\phi_{HPSS})^2} \frac{\gamma_{Co^{+2}}}{(\gamma_{H^+})^2} = W'_p \frac{(\vec{\gamma}_{Co^{+2}})}{(\vec{\gamma}_{H^+})^2}$$

[†] $\dfrac{K^*_{D_o} - K_{D_o}}{K_{D_o}} = \dfrac{\Sigma \, \vec{Co}^{+2}}{(1 \text{ g } H_2O)(m_{Co^{+2}})}$

$\dfrac{(\phi_{HPSS})^2}{(1 - \phi_{HPSS})^2} \dfrac{(m_{HPSS})^2}{(m_{HPSS})^2} = \dfrac{(\phi_{HPSS})^2}{(1 - \phi_{HPSS})^2} \dfrac{(m_{H^+})^2}{(\Sigma \, \vec{H}^+ / g \, H_2O)^2}$

$\dfrac{\Sigma \, \vec{Co}^{+2}}{1 \text{ g } H_2O W'_p (m_{HPSS})} = \vec{m}_{Co^{+2}}$

$\dfrac{(\Sigma \, \vec{H}^+)^2}{(1 \text{ g } H_2O)^2 (W'_p m_{HPSS})^2} = (\vec{m}_{H^+})^2$

In the second approach[†]

$$\frac{\bar{a}_{Co^{+2}} \, (\vec{a}_{H^+})^2}{(\bar{a}_{H^+})^2 \, (\vec{a}_{Co^{+2}})} = K_{th} = \frac{W'_p \, m_{HPSS} K^*_{D_o} K_{D_o} (1 - \phi_{HPSS})^2}{(K^*_{D_o} - K_{D_o})(\bar{m}_{H^+})^2}$$

$$\times \, \frac{(m_{HPSS})^2}{W'^2 (m_{HPSS})^2} \frac{1}{K_{Ac}} \frac{(\vec{\gamma}_{H^+})^2}{(\vec{\gamma}_{Co^{+2}})} \qquad (11c)$$

This reduces to

$$K_{Ac}^{-1} \frac{K^*_{D_o} K_{D_o}}{K^*_{D_o} - K_{D_o}} \frac{(1 - \phi_{HPSS})^2}{(\bar{m}_{H^+})^2} \, m_{HPSS} = W'_p \frac{\vec{\gamma}_{Co^{+2}}}{(\vec{\gamma}_{H^+})^2}$$

for assessment of the same parameter. In these equations the arrow
placed above a symbol is used as before to designate its association
with the counterion-concentrating domain of the polyelectrolyte.

The $W'_p \, \vec{\gamma}_{60 \, Co^{+2}}/(\vec{\gamma}_{H^+})^2$ value computed with these two approaches
are compared in Table 18. Agreement between the two sets of values
is excellent to demonstrate the internal consistency of the approach
once again.

[†] $$\frac{K^*_{D_o} K_{D_o}}{K^*_{D_o} - K_{D_o}} = \frac{\bar{m}_{Co}/\Sigma \, \vec{Co}^{+2}}{g \, H_2O \text{ in solution}}$$

$$(1 - \phi_{HPSS})^2 (m_{HPSS})^2 = \frac{(\Sigma \, \vec{H}^+)^2}{(g \, H_2O \text{ in solution})^2}$$

$$W'_p m_{HPSS} = \frac{g \, \vec{H}_2O \text{ sheath}}{(g \, H_2O \text{ in soln})}$$

$$K_{Ac} = \frac{(\bar{\gamma}_{H^+})^2}{(\bar{\gamma}_{Co^{+2}})} + \text{antilog} \frac{\pi (2V_H - V_{Co})}{2.3 \, RT} = 2.89$$

TABLE 18 A Comparison of $W_p'(\overrightarrow{\gamma}_{Co^{+2}})/(\overrightarrow{\gamma}_{H^+})^2$ Estimates

m_p	K_{D_o}	$K_{D_o}^*$	ϕ_p	$\bar{m}_{H^+}$	$\dfrac{\gamma_{Co^{+2}}}{(\gamma_{H^+})^2}$	K_{Ac}^{-1}	$W_p'\dfrac{(\overrightarrow{\gamma}_{Co^{+2}})}{(\overrightarrow{\gamma}_{H^+})^2}$	
							Path 1	Path 2
0.000973	27060	1.24×10^9	0.22	4.42	1.01	0.346	0.288	0.284
0.000978	26570	1.23×10^9	0.22	4.42	1.01	0.346	0.283	0.280
0.0241	3277	1.69×10^6	0.22	4.425	0.839	0.346	0.858	0.814
0.0242	2991	1.70×10^6	0.22	4.425	0.852	0.346	0.787	0.780
0.0461	1494	4.41×10^5	0.225	4.44	0.805	0.346	0.753	0.730
0.0582	1098	2.52×10^5	0.23	4.45	0.794	0.346	0.674	0.668
0.106	642	6.35×10^4	0.245	4.47	0.744	0.346	0.726	0.679
0.0910	649	9.33×10^4	0.24	4.46	0.767	0.346	0.598	0.597
0.1628	317	1.866×10^4	0.28	4.49	0.666	0.346	0.472	0.482
0.1666	299	1.778×10^4	0.28	4.49	0.663	0.346	0.450	0.451

Values of W_p', the essential feature of the Gibbs-Donnan-based description of solutions of linear, fully ionized polyelectrolytes now become accessible with either of the equations developed for the evaluation of $W_p \bar{\vec{\gamma}}_{60_{Co}+2}/(\vec{\gamma}_{H^+})^2$. The iterative procedure for eventual evaluation of W_p' has already been described in our earlier examination of linear, weak-acid polyelectrolytes. In this instance one needs only to select a first value of W_p' to provide a first estimate of the ionic strength of the counterion-concentrating domain of the polyelectrolyte. The values of $\vec{\gamma}_{Co}$ and $\vec{\gamma}_{H^+}$ are then evaluated with Eq. (21) using the mean-molal-activity-coefficient values available in the literature for HCl, KCl, and $CoCl_2$ at that ionic strength. The value of W_p' is then reassessed with either of the two equations. The procedure is repeated until the selected value of W_p' agrees with the value computed using the W_p'-based estimates of $\vec{\gamma}_{H^+}$ and $\vec{\gamma}_{Co^{+2}}$. A sample calculation of W_p' is presented next.

1. Sample Calculation of W_p'

System-Dowex-50 (8% DVB) in H^+-ion form, polystyrene sulfonic acid, $^{60}Co^{+2}$. System parameters: $\bar{m}_{H^+} = 4.425$; $m_{HPSS} = 0.0241$; $\phi_{HPSS} = 0.22$; $K_{D_o} = 3.277 \times 10^3$; $K_{D_o}^*$ (interpolated, Fig. 24) $= 1.73 \times 10^{-6}$;

$$\log \frac{\bar{\gamma}_{Co^{+2}}}{(\bar{\gamma}_{H^+})^2} + \frac{\pi(V_{Co^{+2}} - 2V_{H^+})}{2.3\ RT} = -0.461 = \log(K_{Ac}^{-1})$$

Step 1. Evaluation of $W_p' \dfrac{\vec{\gamma}_{Co^{+2}}}{(\vec{\gamma}_{H^+})^2}$ with Eq. (11c)

$$\log \frac{K_{D_o}^* K_{D_o}}{K_{D_o}^* - K_{D_o}}\ \frac{(1 - \phi_{HPSS})^2 m_{HPSS}}{(\bar{m}_{H^+})^2} + \log \frac{\bar{\gamma}_{Co^{+2}}}{(\bar{\gamma}_{H^+})^2}$$

$$+ \frac{\pi(V_{Co^{+2}} - 2V_{H^+})}{2,3\ RT} = \log W_p' \frac{\vec{\gamma}_{Co^{+2}}}{(\vec{\gamma}_{H^+})^2}$$

$$-0.070 = \log W'_p \frac{(\vec{\gamma}_{Co^{+2}})}{(\vec{\gamma}_{H^+})^2} \; ; \qquad 0.849 = W'_p \frac{(\vec{\gamma}_{Co^{+2}})}{(\vec{\gamma}_{H^+})^2}$$

Step 2. Evaluation of W'_p

Let $W'_p = 2.0$ g H_2O/mmole HPSS; then

$$\vec{I} = \frac{(1 - \phi_{HPSS})}{W'_p} = \frac{0.78}{2.0} = 0.39$$

and

$$\vec{\gamma}_{H^+} = \frac{(\gamma_{\pm HCl})^2}{(\gamma_{\pm KCl})^2} \quad \text{at} \quad \vec{I} = 0.39 = \frac{(0.755)^2}{0.6682} = 0.853$$

$$\vec{\gamma}_{Co^{+2}} = \frac{(\gamma_{\pm CoCl_2})^3}{(\gamma_{\pm KCl})^2} \quad \text{at} \quad \vec{I} = 0.39 = \frac{(0.5098)^3}{(0.6682)^2} = 0.297$$

and

$$W'_p = 0.849 \frac{(0.853)^2}{0.297} = \frac{2.08 \text{ g } H_2O}{\text{mmole HPSS}}$$

Now let $W'_p = 2.062$ g H_2O/mmole HPSS; then

$$\vec{I} = \frac{0.78}{2.062} = 0.3783$$

and

$$\vec{\gamma}_H = \frac{(\gamma_{\pm HCl})^2}{(\gamma_{\pm KCl})} \quad \text{at} \quad \vec{I} = 0.3783 = \frac{(0.755)^2}{0.6726} = 0.8475$$

$$\vec{\gamma}_{Co^{+2}} = \frac{(\gamma_{\pm CoCl_2})^3}{(\gamma_{\pm KCl})^2} \quad \text{at} \quad \vec{I} = 0.3783 = \frac{(0.5115)^3}{(0.6726)^2} = 0.296$$

and

$$W_p = 0.849 \frac{(0.8475)^2}{0.296} = \frac{2.060 \text{ g } H_2O}{\text{mmole HPSS}}$$

TABLE 19 The Water Uptake, in Grams, by the Counterion-Concentrating Region of One Millimole (Monomer Basis) of Polystyrene Sulfonic Acid

m_{HPSS}	$K^*_{D_o}$	K_{D_o}	ϕ_{HPSS}	γ_{H^+}	$\gamma_{Co^{+2}}$	$\vec{\gamma}_{H^+}$	$\vec{\gamma}_{Co^{+2}}$	W'_p
9.73×10^{-4}	1.24×10^9	2.706×10^4	0.22	0.98	0.972	0.980	0.270	1.023
9.78×10^{-4}	1.23×10^9	2.654×10^4	0.22	0.98	0.972	0.982	0.2695	1.005
2.41×10^{-2}	1.73×10^6	3.277×10^3	0.22	0.939	0.74	0.848	0.296	2.062
2.42×10^{-2}	1.70×10^6	2.991×10^3	0.22	0.932	0.74	0.856	0.296	1.930
4.61×10^{-2}	4.41×10^5	1.494×10^3	0.225	0.915	0.674	0.865	0.293	1.800
5.82×10^{-2}	2.52×10^5	1.098×10^3	0.23	0.905	0.650	0.869	0.291	1.718
1.06×10^{-1}	6.35×10^4	6.42×10^2	0.245	0.879	0.575	0.863	0.293	1.745
9.10×10^{-2}	9.33×10^4	6.49×10^2	0.24	0.882	0.597	0.877	0.288	1.582
1.628×10^{-1}	1.866×10^4	3.17×10^2	0.28	0.861	0.494	0.893	0.281	1.339
1.666×10^{-1}	1.778×10^4	2.99×10^2	0.28	0.860	0.490	0.899	0.278	1.308

$$W'_p \text{ selected} = \frac{2.062 \text{ g } H_2O}{\text{mmole HPSS}} \qquad W'_p \text{ returned} = \frac{2.060 \text{g } H_2O}{\text{mmole HPSS}}$$

Estimates of the W'_p parameter that were made with this approach are listed in the last column of Table 19. The values are reported as grams of water associated with one millimole of HPSS, monomer basis.

2. A Rationalization of the Model-Based W'_p Estimates

Because the reduced viscosity of polyelectrolytes is expected to be proportional to their shear volume in solution we have compared the reduced-viscosity measurements, η_{sp}/c, reported in the literature for sodium polystyrene sulfonate [58] with our estimates of W'_p for polystyrene sulfonic acid over the same molality (monomer basis) range in Fig. 25. Except for the most dilute region examined the

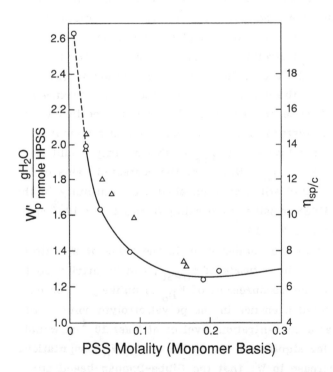

FIG. 25. The correlation of polyelectrolyte reduced viscosity with solvent content of the polyelectrolyte solvent sheath: o, η_{sp}/c; Δ, W'_p (g H_2O/mmole polyelec.).

trend of both reduced viscosity and W_p' with molality is very close to the same. Both the reduced viscosity of polyelectrolyte solutions and the bulk of the polyelectrolyte sphere of influence increase with decreasing molality; the increase with decreasing molality of solvent uptake that is predicted by our model for the solvent sheath of the charged polyelectrolyte molecule is, in fact, in agreement with expectations based on the reduced-viscosity pattern.

The sudden drop in the value of W_p' at the lowest molality examined with the Gibbs-Donnan-based approach merits scrutiny. It should be noted that the ϕ_{HPSS} estimate at this lowest concentration level had to be based on linear extrapolation of the data. Any error in its extrapolated value would be reflected in the extrapolated value of $K_{D_o}^*$ which would in turn affect the value of W_p' that is eventually resolved.

A review of the literature does indeed indicate that the smaller than expected value of W_p may, in part, be due to overestimate of ϕ_{HPSS}. The assessment is based on a plot of counterion activity coefficient versus polyelectrolyte molality (monomer basis) that has been reported for a salt-free sodium polystyrene sulfonate [59]. Linear extrapolation of these data to parallel the earlier treatment of the ϕ_{HPSS} data [52] leads to a single-ion activity coefficient of 0.21 at 9.7×10^{-4} m instead of the 0.11 value actually measured. Such an error in our estimate of ϕ_{HPSS} at this molality would lead to a 40% underestimate of W_p'. However, the corrected value of 1.4 g H_2O/millimole HPSS still leaves an abrupt discontinuity in the plot of W_p versus HPSS molality on passing from a 2.4×10^{-2} m solution to one that is 9.7×10^{-4} m.

The possibility that the sudden drop in the value of W_p', after correction for error in the estimate of ϕ_{HPSS}, can be attributed to experimental bias in the measurement of K_{D_o} is unlikely. For example, the unsuspected presence in the polyelectrolyte solution of simple electrolyte at a concentration level of at least 10^{-4} m would have been needed for significant disruption of the W_p' computation.

The abrupt decrease in W_p' that the Gibbs-Donnan-based approach projects for very dilute polystyrene sulfonic-acid solutions

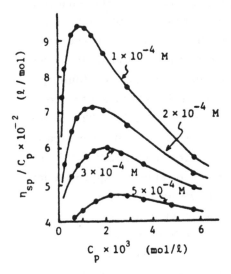

FIG. 26. Experimentally obtained values of η_{sp}/C_p of Na-polyvinyl sulphate solutions plotted versus C_p at various salt concentrations (C_s).

is apparently real. Support for this conclusion is provided by the recent study of Imai and Gekko [60]. Their plot of the specific viscosity of sodium polyvinyl sulfate solutions as a function of concentration at different salt concentration levels is presented in Fig. 26 to justify this prognosis. The specific viscosity decreases abruptly just like the W_p' parameter once the polymer concentration is reduced sufficiently.

IV. CONCLUSIONS

The Gibbs-Donnan-based description of charged polymeric gels, e.g., ion-exchange resins, and the solutions in contact with them, have been shown to be extremely useful for the accurate interpretation of their equilibrium properties. We believe that the Gibbs-Donnan-based description of charged polymers, seemingly dissolved in simple salt solution, but present as solvent-sheathed entities colloidally dispersed in such solutions has been fully documented in the above presentation as well. Its use for a better understanding of all solution phenomena found to be sensitive to salt concentration levels is, as a consequence, strongly recommended. For example, one can expect the puzzling redox properties of simple salt solutions of natural organic substances such as fulvic and humic acids to be amenable to analysis by such an approach. Certainly, a better understanding of the susceptibility of their metal-ion-binding properties to salt concentration levels of their solutions has been facilitated by the employment of such a model [61,62]. The somewhat successful rationalization of the influence of salt concentration levels on the response of polymer-coated redox electrodes [63] is supportive, as well, for the incorporation of such an approach for the investigation of redox phenomena in natural systems.

The strong dependence of metal-ion binding to clays and glasses on the concentration level of the salt in contact with them [64] provides other obvious examples of Gibbs-Donnan-based control of such equilibria. To demonstrate this consider a glass with aluminum and boron substituted for the silicon of its framework. The excess negative charge that is a consequence of such substitution is compensated for by the accumulation of mobile counterion, M^+, in these solvated regions of the glass. Accompanying the high concentration of counterion is the equilibrium concentration of H^+ ion arising from dissociation of hydroxylated silica. A pK_a value of approximately 9.4 characterizes the dissociation of H_4SiO_4 [65] the activity ratio of M^+ to H^+ in the solvated glass regions must approach a value of 10^7 to 10^8. Equilibration of this glass with dilute $(10^{-3}$ N), CO_2-free salt (MX) solution is predicted by Eq. (13) to lead to a rise in p(H) from the initial value of 7.0 to a value of 10 to 11. The resultant disturbance of electroneutrality in the aqueous phase has to result in the release of M^+ ion from the glass until electroneutrality is reached once again. Such loss of counterion from the solvated glass phase has to be compensated. Protonation of the hydroxylated silica surface of the glass, affected as shown, is the most likely path to this end. One can

$$Si(OH)_4 + H_3O^+ \rightleftarrows HSi(OH)_4^+ + H_2O \qquad (47)$$

envision initial dissolution of the glass in the basic solution to be
followed by recrystallization of the glass charge-compensating reaction zones.

With this operational picture it can be concluded that after repeated batch equilibrations of a single glass sample with the above
solution (10^{-3} N MX at pH = 7), the resultant gain in positive
charge must facilitate the diffusion of electroneutral species through
the glass barrier. As a consequence one can expect a glass used
to encapsulate radioactive waste for underground burial to become
increasingly susceptible to leakage with time.

On the basis of this analysis the contemplated use of glass to
encapsulate high-level radioactive waste for its eventual underground disposal merits careful reconsideration. Since any contact
with the water table must eventually result in leakage of the
radioactivity into the environment its use for radioactive waste con-
tainment offers only temporary advantage.

The protonation phenomenon, introduced in the above Gibbs-
Donnan-based prognosis of the use of glass encapsulation of radioactive waste for the prevention of its release to the environment
upon underground burial is a characteristic of hydrous metal oxides
as well. Their p(H) response to change in the salt concentration
level of their suspensions is also associated with the dissociation of
hydroxylated species. An excellent example of this is contained in
the scholarly study by Paterson and Rahman of a protonated β-
$FeO(OH_2)^+$ gel prepared in microcrystalline state by acid hydrolysis
of ferric chloride [66]. In the course of its neutralization with
standard base they showed that the pH of the system at a particular degree of neutralization, equatable with the fractional transfer
of Cl ion out of the gel phase, was inversely proportional to the
p(Cl) of the salt solution. They also showed that a singular,
unique line was resolved by plotting $p\{H^+\} + p\{Cl^-\}$ versus the
corresponding Cl^- ion content of the gel. Use of chloride-ion
content in place of α is permissible because chloride-ion content is
equatable to α by a constant parameter, the initial chloride-ion
content of the fully protonated gel prior to its neutralization.

These results were rationalizable through recourse to Gibbs-
Donnan-inspired Eq. (13).

$$p\{OH\}_\alpha = p\{\bar{O}H\}_\alpha - p\{\bar{Cl}^-\}_\alpha + p\{Cl\}_\alpha \qquad (13)$$

Because of the rigidity of the microcrystalline gel $p\{\bar{O}H\}_\alpha - p\{\bar{Cl}\}_\alpha$
is expected to be invariant at each α value. On this basis $p\{OH\}_\alpha$
should be directly proportional to $p\{Cl\}$, and $p\{H\}_\alpha$ should increase when $p\{Cl\}$ decreases with the extent of the increase being
directly proportional to the extent of the decrease of the concentration level of the salt (NaCl) solution as observed. In addition,
since $p\{OH\}_\alpha - p\{Cl\}_\alpha = p\{\bar{O}H\}_\alpha - p\{\bar{Cl}\}_\alpha$ and since $p\{\bar{O}H\}_\alpha -$
$p\{\bar{Cl}\}_\alpha$ is uniquely defined at each α value because of the rigidity
of the microcrystalline phase one should expect a plot of $p\{OH\} -$
$p\{Cl\}$ or $p\{H\} + p\{Cl\}$ versus α to resolve a single, unique line as
it did.

It is interesting to note that when the β-$FeO(OH_2)^+$ gel, in its
chloride ion form, was suspended in sodium perchlorate solutions at
several different concentration levels, its neutralization with standard base led to the resolution of $p\{H\}$ values that were a unique
function of the chloride content of the gel. The absence of any
noticeable response of $p\{H\}_\alpha$ to change in the sodium perchlorate
concentration level is attributable to the fact that the perchlorate
ion is too large to enter the 5-A channel of the microcrystalline gel
phase. The absence of ClO_4^- ion exchange for Cl^- ion results in
the Cl^- ion content of the sodium perchlorate solution being repro-
ducibly defined by its exclusive transfer to the solution through
removal of charge from the gel surface in the course of its controlled neutralization with standard base. Any variability of $p\{H\}_\alpha$
as a function of the sodium perchlorate concentration level in a
particular experiment thus can only arise from differences in the
activity coefficient of the chloride ion transferred to the solutions
of different ionic strength.

With bromide or nitrate salts there is a $\{pH\}_\alpha$ response to change
in their concentration levels during neutralization of β-$FeO(H_2O)^+$
with standard base but it is less than the proportionate response

observed with the chloride salt. This is attributed to the incom-
plete transfer of these sterically impeded bulky anions to the fully
protonated surface of the oxide gel prior to their controlled removal
to the solution phase during neutralization of the surface charge by
the addition of standard base.

With complete removal of the oxide gel's positive charge by the
standard base added the sensitivity of $p\{H\}$ response to change in
salt concentration levels of the salt solution in contact with the gel
is completely lost. Further addition of base does not lead to a re-
vival of $p\{H\}$ sensitivity to salt concentration levels to indicate the
absence of negative charge buildup on the gel surface and/or the
inaccessibility of the charged gel phase to counterions.

The acid/base titration data compiled for titanium oxide (rutile)
in 0.10, 0.010, and 0.0010 M KNO_3 by Yates [5] while providing
another example of the above phenomenon yield a different pattern
of $p\{H\}$ response to changes in salt concentration level. When base
is added to the TiO_2 samples in contact with each of the three KNO_3
solutions the H^+ ion concentration, initially the same in the three
solutions prior to this addition, diverge, with the pH being smallest
at the highest ionic strength (Fig. 27). This result, while consistent with Gibbs-Donnan-based logic developed for the negatively
charged surface provided by dissociation of the $Ti(OH)_4$ surface

$$Ti(OH)_4 + OH^- \rightleftarrows H_3TiO_3^- + H_2O \qquad\qquad (48)$$

was not obtainable with the β-$FeO(H_2O)OH$ gel as we have noted
above. The maximum separation reached by one curve from the other
next to it approaches the corresponding change in $p(KNO_3)$ to signal
full accessibility of the Donnan potential determining counterion to
the increasingly negatively charged TiO_2 gel phase in this system.

When, on the other hand, the uncharged TiO_2 is protonated by
the addition of nitric acid

$$Ti(OH)_4 + H_3O^+ \rightleftarrows HTi(OH_4)^+ + H_2O \qquad\qquad (47)$$

the $p\{H\}$ response to changes in salt concentration level is reversed,
increasing with increasing salt concentration level as it did during

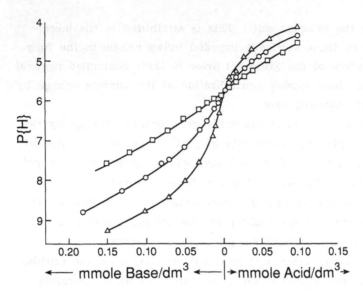

FIG. 27. Acid/base titration of 12.07 g/dm^3 rutile suspended in
KNO$_3$ solutions of different concentration: KNO$_3$ concentration:
□, 0.1 M; o, 0.01 M; Δ, 0.001 M (Experimental data, Yates, 1975;
in reference [5].)

protonation of the β-FeO(H$_2$O)$^+$ gel. The response, while consistent
with Gibbs-Donnan-based projection, is much smaller than the corresponding change in p(KNO$_3$) which was approached in the protonated
iron oxide gel system. Apparently the channels of the TiO$_2$ gel are
narrowed by the protonation reaction and the bulky NO$_3^-$ ion becomes
increasingly unaccessible to the charged gel phase.

    An interesting aspect of the potentiometric profile obtained for
titanium oxide as a function of ionic strength over the large H$^+$ ion
concentration range examined is the close resemblance that it bears
to similar profiles obtained in parallel examinations of the potentiometric properties of proteins as a function of salt concentration levels
[67]. One can immediately conclude from this that analysis of protein
chemistry with the Gibbs-Donnan-based approach should prove fruitful
as well.

    The physical chemical properties of proteins in aqueous media seem
ideally suited to clarification through use of Gibbs-Donnan-based concepts. The charged surfaces of their ion-concentrating domains are

initially highly positive. During neutralization with base the charge reduces to zero. Beyond this point their surfaces become increasingly negatively charged. Since such data can, through the employment of Gibbs-Donnan-based logic, be used to deduce the extent of the solvent sheath of a protein molecule and the concentration of counterion associated with this sheath the various functions of a protein such as catalysis, oxidation and reduction, complexation of metal ions, and adsorption that are facilitated through the agency of this sheath should be susceptible to analysis.

There are numerous salt concentration sensitive phenomena that remain to be reassessed by using Gibbs-Donnan-based logic.

REFERENCES

1. M. Nagasawa, T. Murose, and K. Kondo, *J. Phys. Chem.*, 69:4005 (1965).

2. K. Gekko and H. Naguchi, *Bipolymers*, 14:2555 (1975).

3. S. Alegret, J. A. Marinsky, and M. T. Escalas, *Talanta*, 31:683 (1984).

4. F. G. Lin, in *Studies of Hydrogen and Metal Ion Equilibria in Polysaccharide Systems—Alginic Acid and Chondroitin Sulfate*, Ph.D. thesis, Chemistry Department, State University of New York at Buffalo, Buffalo, New York, 1981.

5. J. Westall and H. Hohl, *Adv. Colloid Interface Sci.*, 12:265 (1980).

6. D. E. Yates, S. Levine, and T. W. Healy, *Trans. Faraday Soc.*, 70:1807 (1974).

7. J. A. Davis, R. O. James, and J. O. Leckie, *J. Colloid Interface Sci.*, 63:480 (1978).

8. A. Katchalsky and S. Lifson, *J. Polymer Sci.*, 11:809 (1953).

9. R. A. Marcus, *J. Chem. Phys.*, 23:1057 (1955).

10. G. S. Manning, *J. Phys. Chem.*, 85:870 (1981).

11. R. A. G. Friedman and G. S. Manning, *Bipolymers*, 23:2671 (1984).

12. Y. Merle and J. A. Marinsky, *Talanta*, 31:199 (1984).

13. P. Slota and J. A. Marinsky, in *Ions in Polymers* (A. Eisenberg, ed.), Advances in Chemistry Series 187, Amer. Chem. Soc., Washington, DC, 1980, p. 311.

14. J. A. Marinsky, *J. Environ. Sci. Technol.*, 20:349 (1986).

15. J. A. Marinsky, T. Miyajima, E. Högfeldt, and M. Muhammed, *Reactive Polymers*, 11:279 (1989).

16. J. A. Marinsky, T. Miyajima, E. Högfeldt, and M. Muhammed, *Reactive Polymers, 11*: 291 (1989).

17. H. P. Gregor, *J. Am. Chem. Soc., 73*:642 (1951).

18. E. Glueckauf, *Proc. Roy. Soc. (London), A214*:207 (1952).

19. F. Helfferich, *Ion Exchange*, McGraw Hill, New York, 1962.

20. J. A. Marinsky and W. M. Anspach, *J. Phys. Chem., 79*:433 (1975).

21. C. Travers and J. A. Marinsky, *J. Polym. Sci., Symp. no. 47*:285 (1974).

22. S. Bukata and J. A. Marinsky, *J. Phys. Chem., 68*:994 (1964).

23. J. A. Marinsky, R. Baldwin, and M. M. Reddy, *J. Phys. Chem., 89*:5303 (1985).

24. J. A. Marinsky and M. M. Reddy, presented for publication, *J. Phys. Chem., 95*:10208 (1991).

25. P. Mukerjee, *J. Phys. Chem., 65*:740 (1961).

26. R. A. Robinson and R. H. Stokes, in *Electrolyte Solutions*, 2nd ed., Butterworths, London, 1959.

27. J. W. Moore, in *Physical Chemistry*, Prentice Hall, Englewood Cliffs, NJ, 1955, p. 484.

28. E. P. Serjeant and B Dempsey, Ionization Constants of Organic Acids in Aqueous Solution, Pemagnon Press, Oxford, (a) p. 45, (b) p. 78, 1979.

29. A. Chatterjee and J. A. Marinsky, *J. Phys. Chem., 67*:41 (1963).

30. L. S. Goldring, in *Ion Exchange and Solvent Extraction* (J. A. Marinsky and Y. Marcus, eds.), Marcel Dekker, New York, 1966, p. 205.

31. R. Schlögl and H. Schurig, *Z. Elektrochem., 65*:863 (1961).

32. T. Miyajima, K. Ishida, Y. Kanegae, H. Tohfuku, and J. A. Marinsky, *Reactive Polymers, 15*:55 (1991).

33. L. G. Sillén and A. Martell, *Stability Constants*, Supplement No. 1, The Chemical Society, London, 1971; (a) Ni^{+2}, p. 253; (b) Zn^{+2}, p. 254; (c) Cu^{+2}, p. 251; (d) Co^{+2}, p. 114.

34. M. Lloyd, V. Wycherley, and C. B. Monk, *J. Chem. Soc.*, Part III, 1786 (1951).

35. J. A. Marinsky and W. M. Anspach, *J. Phys. Chem., 79*:439 (1975).

36. J. A. Marinsky, *J. Phys. Chem., 86*:3318 (1982).

37. J. Schubert, E. R. Russell, and L. S. Meyers, *J. Biol. Chem., 185*:387 (1950).

38. L. K. Jang, N. Haupt, T. Uyen, D. Grasmuck, and G. G. Geesey, *J. Polym. Sci. B: Polym. Phys., 27*:1301 (1989).

39. J. A. Marinsky, F. G. Lin, and K. S. Chung, *J. Phys. Chem.*, 87:3139 (1983).

40. G. G. Geesey and L. K. Jang, in *Bacterial Interactions with Metallic Ions* (T. Beveridge and R. Doyle, eds.), Wiley, New York, 1989.

41. A. Katchalsky, Z. Alexandrowicz, and O. Kedem, in *Chemical Physics of Ionic Solutions* (B. E. Conway and R. G. Barradas, eds.), Wiley, New York, 1966, p. 295.

42. J. Kielland, *J. Am. Chem. Soc.*, 59:1675 (1937).

43. G. W. Breck, *J. Am. Chem. Soc.*, 78:5963, 5972 (1956).

44. R. A. Robinson, *J. Am. Chem. Soc.*, 74:6035 (1952).

45. Landolt-Bornstein, in *Physikalisch-Chemisohen Tabellen Auflage*, Vol. 2, Part 2, Verlag von Julius Springer, 1931, p. 125.

46. H. S. Harned and A. B. Gancy, *J. Phys. Chem.*, 62:627 (1958).

47. R. A. Robinson, *Trans. Faraday Soc.*, 49:1147 (1953).

48. I. Prigogine, P. Mazur, and R. Defay, *J. Chem. Phys.*, 50:146 (1953).

49. M. M. Reddy and J. A. Marinsky, *J. Macromol. Sci. Phys.*, B5(1):135 (1971).

50. M. S. Patterson and R. Greene, *Anal. Chem.*, 37:854 (1965).

51. G. E. Meyers and G. E. Boyd, *J. Phys. Chem.*, 60:521 (1956).

52. M. M. Reddy and J. A. Marinsky, *J. Phys. Chem.*, 74:3884 (1970).

53. M. M. Reddy, J. A. Marinsky, and A. Sarkar, *J. Phys. Chem.*, 74:3891 (1970).

54. G. S. Manning, *J. Chem. Phys.*, 51:924 (1969).

55. Z. Alexandrowicz, *J. Polym. Sci.*, 43:325 (1960).

56. Z. Alexandrowicz, *J. Polym. Sci.*, 43:327 (1960).

57. Z. Alexandrowicz, *J. Polym. Sci.*, 56:115 (1962).

58. R. F. Prini and A. E. Lagos, *J. Polym. Sci.*, A-2:2917 (1964).

59. F. Oosawa, *Polyelectrolytes*, Marcel Dekker, New York, 1971, p. 89 (bottom curve of Fig. 29b).

60. N. Imai and K. Gekko, *Biophys. Chem.*, 41:31 (1991).

61. J. A. Marinsky, J. H. Ephraim, A. Mathuthu, S. Alegret, M. Bicking, and R. Malcolm, *J. Environ. Sci. Tech.*, 20:354 (1986).

62. J. H. Ephraim and J. A. Marinsky, *J. Environ. Sci. Tech.*, 20:367 (1986).

63. R. Noxgell, J. Ledepenning, and F. C. Anson, *J. Phys. Chem.*, 90:6227 (1986).

64. J. A. Marinsky, in *Aquatic Surface Chemistry: Chemical Processes at the Particle-Water Interface* (Werner Stumm, ed.), Wiley, New York, 1987, p. 49.

65. R. H. Busey and R. E. Mesmer, *Inorg. Chem.*, *16*:2444 (1977).

66. R. Paterson and H. Rahman, *J. Colloid Interface Sci.*, *97*:423 (1984).

67. S. Iida and N. Imai, *J. Phys. Chem.*, *73*:75 (1969).

6

Influence of Humic Substances on the Uptake of Metal Ions by Naturally Occurring Materials

JAMES H. EPHRAIM and BERT ALLARD   Linköping University, Linköping, Sweden

I. INTRODUCTION

Natural materials in terrestrial and aquatic environments are closely related. These relationships are pictorially presented in Fig. 1. In this figure natural matter in the aquatic environment is classified as either animate or inanimate. The first category is made up of species like algae, fungi, bacteria, while the second category is further divided into dissolved and particulate components. The particulate component is comprised of inorganic substances like alumina, manganese dioxide, hydroxy oxides of iron and silicon. The dissolved component may be either organic or inorganic. Anions (NO_3^-, Cl^-, SO_4^{2-}, etc.) and cations (Cu^{2+}, Na^+, Hg^{2+}, etc.) comprise the inorganic component. The organic component may be characterized as either hydrophilic or hydrophobic. Such an assessment is based on whether or not they adsorb onto XAD-8 Amberlite resins in a well-documented method for the extraction of humic substances [1]. Such substances like carbohydrates, amino acids, simple monomeric acids, e.g., benzoic acid and citric acid are not adsorbed and are thus placed in the

335

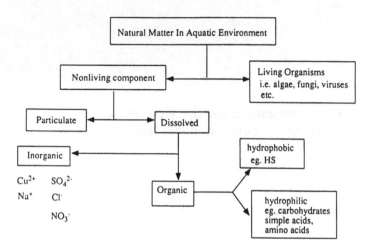

FIG. 1. The distribution of natural matter in the aquatic environment.

hydrophilic category. The adsorbed acidic components, removed by
alkali, are placed in the hydrophobic category and are defined as
humic substances. It is immediately apparent that the simultaneous
presence of the components of natural substances in terrestrial and
aquatic environments that are separated and classified for identification in Fig. 1 must result in sizable interactions that complicate interpretation of the puzzling properties of these systems [2].

II. ACID-BASE PROPERTIES OF HUMIC SUBSTANCES

Extensive research of the neutralization properties of fulvic and
humic acids has been complicated by the physical complexities of
these substances [2-11]. The stability of millivolt or pH readings
has often been influenced by their irreversible sorption tendencies.
The absence of a clear inflection point in their acid-base potentiometric titration curves has prompted early researchers to arbitrarily
select a particular pH as the equivalence point [6]. Recently, a
modified form of the Gran approach has been employed [3] and in
situations where assumptions in the Gran analysis are violated,
linear programming has been recommended [4] to reach this goal.

Hysteresis in fulvic-acid potentiometric titrations [8,9] has been attributed to auto-oxidation [10]. Combination of the above factors makes interpretation of the acid-base properties of humic substances difficult.

Even without the introduction of such uncertainty insights sought with respect to the acid-base properties of fulvic and humic acids, the resolution of acid-dissociation constants, and assessments of the effects of changing bulk electrolyte and humic substance concentrations remain difficult to achieve. Despite tremendous advances, the nature of the perturbations encountered in acid-base potentiometric titrations of humic substances still need further amplification.

Numerous models have been proposed [12-14] to account for the effect of the functional group heterogeneity of the humic-acid substances on the shape of their titration curves. These models, proposed with the understanding that these substances consist of a mixture of chemically nonidentical acid sites, have sought either to fit experimental data empirically [15] or to offer explanations on a molecular level [16]. The paths available to researchers fall into two broad categories identified as the discrete ligand approach [15] or the continuous distribution approach [16].

It has been suggested that the discrete ligand approach provides a greater potential for practical application whereas the continuum approach is believed to provide insights on a molecular level. Both approaches have been reviewed in great detail [17-27]. A brief description of the two methods follows.

A. The Discrete Ligand Approach

In this approach, it is presumed that the observed potentiometric properties are a consequence of the different acid strengths of a limited number of functional sites that constitute the humic-acid molecule [23-26]. Different numbers of such sites have been employed by different researchers using humic substances from different sources [23-26]. Sposito and Holtzclaw [23], in studies of dissociation of fulvic acid extracted from sewage sludge, concluded that

four separate classes of dissociable functional groups, each consti-
tuted by a continuum of acid groups defining a particular class,
control the shape of the titration curve over a pH range from 1 to 11.
Paxeus [24], describing the acid-base properties of aquatic fulvic acid
concentrated from surface water, postulated six monoprotic acids to be
dissociable over a pH range from 2.5 to 10.4. Gregor and Powell [25]
described the protonation properties of fulvic acids by assuming a mixture of n monoprotic acids. Their nonlinear least-squares method of
program assessment showed that the best fit was obtained when n, the
number of monoprotic acids, was assigned a value of 4. Recently,
Tipping et al. [26] and Falck [27] have separately used a three-site
model to describe the protonation behavior of a number of fulvic acid
sources. Consideration of intersite interaction in the various projections has normally been excluded because of the inaccessibility of
suitable methodology for the quantification of such interactions.

B. The Continuum Approach

The underlying assumption in this approach projects the existence of
a continuum of sites with overlapping properties which arise from
statistical and electrostatic effects [28-30]. Perdue and Lytle [28],
employing a Gaussian distribution, assumed that the relative concentration of each of the mixtures of ligands follows the normal pK_a
distribution of organic acids. Though this method approaches the
known complexity of humic substances, its application can only serve
as a first approximation in describing the most probable distribution
of acidic functionalities in humic substances [28]. There is no way
that such statistics can account for site occurrence distortions by
environmental factors. One continuum approach, the so-called affinity
spectrum continuous distribution model [29], has been found to facilitate the identification of discrete sites from experimental data [30].

The sensitivity of the acid-dissociation properties of fulvic and
humic acids to bulk-electrolyte concentration levels, as well as to
their heterogeneity, has only recently been recognized as a separate
aspect of the problem in the development of these approaches. It is
a consequence of their developing counterion-concentrating regions

next to their charged surface. In the last few years procedures have been developed to account for both effects separately [31].

The fact that the sensitivity of the dissociation equilibria of fulvic and humic acids to counterion concentration levels of their solution can be attributed to a counterion-concentrating solvent sheath next to the charged surface of these naturally occurring molecules defines the direction taken in the development of this chapter. The steps in the application of this "separate phase" approach to the interpretation of proton and metal-ion binding by humic substances are emphasized and discussion of the role played by these natural organic acids on the adsorption of environmentally important metal ions like Hg^{2+}, Cd^{2+}, and Zn^{2+} is presented.

III. CONCEPTUALIZATION OF HUMIC SUBSTANCES

Examples of how these substances are conceptualized in the literature are presented below:

1. Strongly associated aggregates of acids, each of comparatively low molecular weight [32].

2. An assembly of identical *mean fulvic-acid units*; a fulvic-acid unit is a hypothetical macromolecule that contains one or more distinct classes of acidic functional groups and is able to associate with other mean fulvic acid units or with part of itself (self-association) through the mechanism of hydrogen bonding, van der Waals interactions and π-bonding [33].

3. An assemblage of relatively small hydrophobic moieties which are slightly different but composed of four to five predominant separate acidic sites with each site characterized by a distribution of acid strengths which may be averaged [31].

In the two-phase approach, questions to which answers are sought are the following:

a. What is the dependence of the acid-dissociation titration curves $[pK_a = pH - \log(\alpha/(1 - \alpha))$ versus $\alpha]$ on changes in the bulk electrolyte concentration?

b. What is the minimum number of acidic sites necessary to describe the observed potentiometric behavior in all situations?

c. What are the identities of the functional groups associated with the acidic sites envisaged?

d. How do these sites interact individually with metals?

e. Is it possible to describe the interactions of the sites with metal ions as a reflection of the functionality of the site?

Of course, in those instances where dependence on electrolyte concentration is negligible or nonexistent a "two-phase" model is not applicable and only questions b to e apply.

IV. METAL-HUMATE INTERACTIONS—A REVIEW OF EXPERIMENTAL TECHNIQUES

The early emphasis of metal-humate interactions research centered on the determination of the so-called binding capacity of the natural organic acid. This parameter is defined as the maximum binding per unit mass of the humic substance [34]. Such research has since been extended to provide a more detailed assessment of metal-humate interaction. The achievement of such an objective depends on the availability of reliable methods for determining separately either the free-metal-ion concentration or the concentration of metal bound to the humic substance in the metal-humate mixture. The fact that analytical methods for such metal-ion speciation studies are highly dependent on the complexity of the aqueous chemistry and the redox properties of the metal ion has influenced the development of various techniques. The methods that are most often employed include anodic stripping voltammetry, fluorescence spectroscopy, ion-selective techniques, equilibrium dialysis, ultrafiltration, spectroscopic (FTIR, UV) and ion-exchange distribution methods.

A. Anodic Stripping Voltammetry

This method has been employed for metal-humate studies primarily involving the following metal ions: Cd^{2+}, Cu^{2+}, and Zn^{2+} [35-37]. The advantage of the method is the ability to determine labile, free-metal-ion concentrations in the presence of the ligand. A disadvantage is the potential disruption of the metal-humate complex or the ligand (humic substance) itself.

B. Fluorescence Spectroscopy

The fluorescence of humic substances has been used to study their complexation with a number of metal ions [38-40]. Unfortunately only 1% of most HS has been determined to have these fluorescent properties so the method has limited potential. The fluorescent properties of selected metal ions, e.g., Eu^{3+}, have been followed in studies of Eu^{3+}-fulvate interactions [41].

C. Potentiometry Using Ion-Selective Electrodes

Metal ions for which ion-selective electrodes have been developed include Cu^{2+}, Pb^{2+}, Ca^{2+}, Cd^{2+} [42-45]. The advantage of this method is that it permits the in situ measurement of the metal ions without destruction of the sample. Factors like pH, temperature, and bulk-electrolyte concentrations need to be carefully controlled to ensure meaningful results. In addition, possible "poisoning" of the electrodes by the humic substance has to be avoided.

D. Chromatographic Methods

Gel permeating/filtration has been employed in studies of metal-humate interactions [46,47].

E. Ion Exchange Distribution

This approach has been employed for a number of metal ions, such as Cu^{2+}, Ni^{2+}, Co^{2+}, Pb^{2+}, Ca^{2+}, Zn^{2+}, Mn^{2+}, Mg^{2+}, and Al^{3+} [48-50] to determine metal-bound to metal-free ratios. For this approach to provide reliable answers the FA molecule and the FA-metal complex must not interact in any way with the ion exchanger being employed. Errors will arise if the above requirement is violated by the sorption of hydrophobic humic acid fragments or hydroxy species of the metal ion by the ion exchanger.

F. Ultrafiltration Method

The relatively simple method of separating species on the basis of their size has been employed in metal-humate binding studies where

the FA molecule and the FA-metal complex are separated from the
free metal ion [51-53]. This method has been used for studies involving the following metal ions: Zn^{2+}, Cd^{2+}, Eu^{3+}, Cu^{2+}, Sr^{2+}
[54-56]. Disadvantages of this approach include the possible disruption of the metal-humate complex in the course of the filtration and
the difficulty of excluding the filtration of low-molecular-weight fragments of the fulvic acid through the membrane. However, corrective
measures have been suggested for useful application of the method
[56].

G. Equilibrium Dialysis/Dialysis Titration

This approach also involves the separation of the free metal ion from
the ligand and the complex. For material balance considerations, it
is essential to minimize (eliminate) the sorption of species onto the
surface of the dialysis tube [57,58].

V. STEPS OF THE APPROACH

A. Potentiometric Titrations in Aqueous Medium as a Function of Ionic Strength

Traditionally, plots of pK_a obtained with the Henderson-Hasselbalch
equation ($pK_a = pH - log(A^-/HA)$) versus α, where α is the degree
of neutralization, have been used to measure nonideality in polymeric
systems [59,60]. Analysis of the acid-dissociation behavior of a number of linear, weak-acid polyelectrolytes as a function of counterion
concentration using the Gibbs-Donnan model [61,62] has indicated
that these analogs of crosslinked, weak-acid polyelectrolyte gels
appear to develop a counterion-concentrating region next to their
surface that simulates the separate phase properties of the gels [62].
In both systems, the distribution of diffusible electrolyte components,
e.g., NaCl, HCl, at equilibrium is controlled by their respective electrochemical potentials which are identical at every point. Apparently,
the only difference between polyelectrolyte gels and their linear analogs is in their charge distribution. The separate well-defined
phases of the gel-salt systems are electroneutral but in their linear
analogs, the counterion-enriched domain and the solution external to

it are not. The negative charge due to the fractional release of counterions from the polyelectrolyte domain is balanced by the equal positive charge they produce in the solution phase [63].

In the application of such plots to humic substances it has been stated that their "usefulness is limited to obtaining insight into ionic strength effects on their potentiometric titrations in an attempt to describe their polymeric nature in solution" [64]. That this is indeed the case may be seen from the fact that one immediately observes that the sensitivity to ionic strength of the measurable acid-dissociation properties of humic acid substances, is considerably smaller than for the high-molecular-weight, weak-acid polyelectrolytes. This result is rationalizable by presuming that only a small fraction of a typical fulvic-acid source contains molecules large enough to exhibit separate phase properties. Insight is thus provided with respect to the molecular size heterogeneity of the particular humic substance source. While the measurable acid dissociation properties of the low-molecular-weight fulvic-acid molecules are expected to be affected only by nonideality factors (long-range Debyean interactions) normally encountered with simple organic-acid, simple salt systems when ionic strength is changed the acid-dissociation properties of the high-molecular-weight fulvic-acid molecules are expected to be subjected to the additional perturbation of a Donnan potential term. Its source is the counterion-concentrating solvent sheath the higher-molecular-weight molecules presumably develop. Because the hydrogen-ion activity at the site of reaction in the polyion domain is not directly measurable the absolute insensitivity to change in ionic strength of the polyion domain hydrogen-ion activity at a particular degree of dissociation is not observable. Instead the change in hydrogen activity, with salt concentration level, that one can expect is by a factor that roughly corresponds to the ratio of counterion activity levels at the two ionic strength values $(\{a_{H^+}\}_2 = \{a_{H^+}\}_1 \{a_{c^+}\}_2 /\{a_{c^+}\}_1)$.

In summary then, the presence of such regions in a fraction of the fulvic-acid molecules in a source is responsible for the variability of hydrogen-ion and multivalent metal ion, M^{+2}, response to ionic strength. The fact that fulvic-acid sources, whose number average

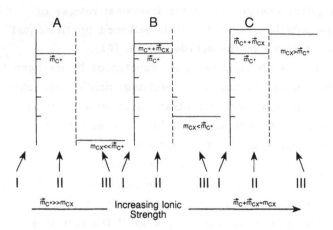

FIG. 2. The operational model; a representation of counterion concentration in the polyion domain as bulk electrolyte increases (I = surface of molecule, II = polyion domain, III = bulk electrolyte).

molecular weight may be as low as 600 to 700 still exhibit salt-sensitive properties relatable to "two-phase" behavior is probably a consequence of their hydrophobic nature [1,11]. One can attribute the development of a counterion-concentrating domain in aggregates that are smaller than those that normally give rise to this phenomenon to the hydrophobic regions normally present in humic substances. With this rationalization the uptake of solvent in the hydrophilic portions of the fulvic-acid molecule can be presumed to be a fairly unique function of the degree of dissociation, with the hydrophobic regions controlling the boundary limits of the solvent-sheathed regions.

Figure 2 pictures the counterion-concentrating, solvent-sheathed region of the fulvic acid conceptualized above. In the three experimental cases selected to illustrate the source of salt sensitivity in these systems the solvent content is pictured to be unaffected by drawing the top and bottom boundary of the counterion-concentrating regions at the same levels. This conceptualization is consistent with the suggestion that the boundary limits of these polyion domains are determined by their hydrophobic boundaries. In this figure the three experimental situations considered to illustrate model concepts are identified by letters A, B, and C. In the first case, designated by A, the external electrolyte concentration levels, m_{CX}, in zone III is much smaller than $\vec{m}_{C^+}$ the concentration of the counterion in the solvent sheath (zone II) that separates the fulvic-acid molecule's surface (zone I) from the bulk electrolyte. The salt concentration level raised, but still kept smaller than the fixed concentration level of the polyion domain in the second case, is identified by B. In this instance the limited invasion of the polyion solvent sheath by the salt, $\vec{m}_{CX}$, is large enough to be noticeable when μ^e_{CX}, the electrochemical potential of the diffusible salt component reaches the same value throughout the system at equilibrium $(\vec{\mu}^e_{C^++CX} = \mu^e_{CX})$. In the third instance, case C, the bulk-electrolyte concentration, m_{CX}, is initially larger than the counterion concentration level, $\vec{m}_{C^+}$, of the polyion domain. This thermodynamically unstable condition is remedied by salt invasion of the polyion domain until $(\vec{m}_{C^+} + \vec{m}_{CX})$ barely exceeds m_{CX} when equilibrium is reached. At these sizable bulk-electrolyte concentration levels the sensitivity of $\{a_{H^+}\}$ or $\{a_{M^{+2}}\}$ to change in the salt concentration is no longer discernible. The response measured at a particular acid-dissociation value is the one associated with the polyion domain at the site of the reaction since the Donnan potential term, $\{\vec{a}_{C^+}\}/\{a_{C^+}\}$, has reached a value very close to unity. In experiments carried out to measure the dependence of acid-dissociation properties of a soil-derived fulvic acid this "critical" ionic strength was determined to occur at I = 1.00 M KNO_3 [64,68]. The acid-dissociation properties presented as plots of pK_a [pK_a = pH - log(A$^-$/HA) versus α], where α is the degree of neutralization were observed to be insensitive to a change in ionic strength from 1.00 to 5.00 M KNO_3 [64]. To quantify the sensitivity of its acid dissociation properties to changes in ionic strength, the observed pK_a versus α curve at the "critical" ionic strength was employed as the salt-insensitive reference and deviations from it at any given ionic strength were considered to provide an estimate of the Donnan-potential-inspired term, which may be expressed as follows:

$$\Delta pK_{a(I,\alpha)} = [pK_{a(I,\alpha)} - pK_{a(I=1.00,\alpha)}]$$

where $pK_{a(I,\alpha)}$ is the negative logarithm of the acid dissociation constant at a given ionic strength, I, and degree of neutralization, α, and $pK_{a(I=1.00,\alpha)}$ is the corresponding term at the same α but at I = 1.00.

The ΔpK_a term has been used to facilitate correction of the measurable concentration of free metal ion in the bulk electrolyte to its value in the polyion domain for the correct assessment of metal-humate interactions. For a metal ion with a valency of z, the ΔpK_a term is raised to the power z, i.e., $(\Delta pK)^z$, to ensure proper assessment of charge effects.

B.   Potentiometric Titrations in Presence of Heavy-Metal Ions

There are very weak acid groups present in humic-acid substances. They are identifiable with the phenolic-OH groups that are present in these materials. These phenolic-OH groups are often located next to the stronger, directly titratable acid groups (carboxyl and alcohol) in the fulvic-acid molecule. When this situation prevails advantage can be taken of the chelate formation tendencies of the bidentate ligand that this geometry produces. In the presence of a suitable multivalent metal ion the complete or selective release of protons from the phenol groups so situated becomes accessible to assay by titration with standard base [64]. Examples of such differences in proton-release behavior are provided in separate examinations of the proton release of a soil fulvic acid by Cu(II) and Eu(III)/La(III).

In the presence of Cu(II), the extra acid release has been attributed to reactions like the following:

In reaction I, a salicylic-like arrangement, where the very weak
–OH group is adjacent to the stronger –COOH group, reacts with Cu^{2+},
as shown, to release the –OH proton. The stability constant of Cu-
(Salicy) (4.6×10^{10}) is large enough to ensure completeness of the
envisaged reaction. In reaction II, a dihydroxy-like moiety with its
weak acid alcohol (–OH), already protonated, has the much weaker
phenolic (–OH) acid situated adjacent to it react with Cu^{2+} to release
a proton. Again the stability of the resultant complex (of Cu cate-
chol $\approx 2 \times 10^9$) lends credence to such a postulate.

When Eu(III) is added to the same fulvic acid source [64], however, proton release is approximately fifty percent of the release observed with Cu(II). This reduction of titratable protons in the soil
fulvic acid source was attributed to the small likelihood of reaction I
occurring. The low value of $\sim 2.75 \times 10^4$ that is reported for the stability constant of Eu(salic)$^+$ led to this conclusion.

Assignment of the extra titratable protons produced in the humic
substance to proton release from the dihydroxy-like moiety present in
the sample, as shown, is consistent with the abundance estimates of
the carboxylic acid and the alcohol assignable to phenolic OH neighbors.

In spite of the uncertainties with respect to real chelation possibilities and because of the likelihood of chelation by groups different
from those designated, the above approach is felt to be useful. The
procedure offers a means of gaining insight with respect to the acidic
spectra of humic substances by utilizing chelating potentials of its
most likely components.

C. Potentiometric Titrations in Nonaqueous Medium

Whereas the carboxylic-acid capacity of natural organic matter has
been affected by measuring its complexation by cobalthexamine, CoHM
[69], potentiometric titrations in a nonaqueous medium have facilitated

estimate of both the carboxylic acid and the alcohol capacity of natural organic matter simultaneously. Earlier problems of determining −COOH and −OH groups in nonaqueous media [70] have been solved by the application of a suitable internal reference compound [64,71]. The criteria for the selection of such an internal reference is listed below.

1. The internal reference compound must contain the weak-acid functionalities of interest, i.e., the COOH and OH groups.
2. The titration curve of the internal reference compound must show clearly resolved inflections that identify the carboxylic and hydroxyl groups for comparison with slope changes apparent in the titration curve of the humic material.
3. The model compound must not react with the humic material and must be soluble in the titration medium.

In these titrations even though the inflection points are a function of acid strength [72], they are not easily amenable to pK assignments. The usefulness of nonaqueous titrations is thus limited to gaining insight with respect to the acidic spectrum and in providing the means for comparing humic substances from different sources [73].

D. Interpretation of Protonation Equilibria in Aqueous Media

As noted earlier, one interpretation of the protonation equilibria of fulvic acids in aqueous medium is based on the conceptualization that humic substances are composed of four to five separate families of acid sites with different number averaged pK_a values. This site heterogeneity leads to the increase in pK_a with α discernible at a "critical" ionic strength. Such disturbance of acid dissociation properties is accounted for by assigning pK_a values and abundances to the sites. Fitting of the potentiometric titration curve at the "critical" ionic strength has been facilitated through knowledgeable interpretation of aqueous and nonaqueous titration data in the manner already described. Anticipation of the potentiometric titration data at any ionic strength may then be effected by the incorporation of the ΔpK term which accounts for the difference between the H^+ ion measurement in the external solution and its higher value at the inaccessible acid-dissociation site of the fulvate-ion domain.

The steps involved in the approach are as follows:

1. A maximum of four to five "average" sites are projected and their abundances are deduced from the nonaqueous titrations data and the titration results compiled in the presence of increasing Cu(II) and Eu(III)/La(III).

2. First estimates of their respective pK_a values are made taking into consideration the projected configuration of the selected sites.

3. At each experimental pH, the degree of neutralization of each of the "average" acid sites is computed using the following equation:

$$\alpha_i = \frac{1}{1 + 10^{(pK_i - pH)}}$$

where α_i is the degree of neutralization of the i-th site, pK_i is the dissociation constant assigned to the site, and pH is the experimentally determined value. The above expression, valid for fitting at the "critical" ionic strength, must be modified by the inclusion of the ΔpK term, as shown, for fitting data at ionic strengths other than the "critical" ionic strength:

$$\alpha_i = \frac{1}{1 + 10^{(pK_i + \Delta pK - pH)}}$$

4. The ultimate objective is to obtain the following relationship:

$$\sum_i \alpha_i Ab_i = \alpha^{computed} = \alpha^{experimental}$$

This relationship is obtained in an iterative procedure which involves changing the pK_i's until the residual

$$| \alpha^{computed} - \alpha^{experimental} |^2$$

reaches a minimum for the set of data points.

Four to five "average" sites have been employed in the fitting of fulvic-acid samples. In curve-fitting exercises of similar nature by other researchers [23-26] a range of three to six sites has been employed.

E.   Description of Metal-Ion Binding by Fulvic Acid

Once correction for metal ion response to salt concentration levels is available the description of metal-ion binding by humic substances still remains subject to considerable uncertainty because of their functional site heterogeneity.   This situation has led to multisite modeling in an effort to account for the functional group heterogeneity of the humic substances [20-30].   Various methods to achieve this objective have been reported [23-29] but so far the choice of method has been dictated by the objective of the researcher.

In the "two-phase" approach, a number of methods for the description of the metal-ion uptake by humic substances have been employed [33,69].   The choice of method has been influenced by the binding potential for the metal ion of the various sites assigned, as described earlier, to the fulvic-acid molecule.

In this approach, a multisite model was used to predict the sequencing of Cu^{2+} by a soil FA [73] as a function of degree of neutralization of the FA molecule, ionic strength, and metal ion to FA concentration ratios.   To achieve the objective, a set of mass-action equations for all possible reactions of Cu^{2+} with each "average" site was generated and complex formation constants obtained from the literature for the Cu^{2+}-functional group anticipated [74] were employed.   At high pH values and in instances where a high ratio of metal ion to FA existed, the inability of the model to account for the complete removal of Cu^{2+} from solution was attributed to ion-dipole interaction between the fulvic acid and the $Cu(OH)^+$ species which forms at high pH.   The possibility of metal-induced aggregation of FA in the presence of high metal concentration has been observed by other researchers [75].

Examination of the complexation of cobalt, zinc, and europium by fulvic acid in the same manner by employing literature-based complex formation constants in a series of mass-action equations has indicated that these metal ions are selectively complexed by the weak-acid enol group of the fulvic-acid molecule [54].

Modification of the above program to provide an algorithm capable of predicting the pattern of multivalent metal-ion removal from solution by a soil or aquatic fulvic acid as a function of pH when the concentrations of metal, humic substance, and total salt are known has been shown to be feasible as well [63].

A less informative description of metal-humate interaction involves the resolution of an overall interaction parameter, β_{ov}, from the mass-action equation for the following "average" behavior equation [75] $[M^{z+} + FA^- \rightleftharpoons (M-FA)^{(z-1)+}]$ as follows:

$$\beta_{ov} = \frac{\{(M-FA)^{(z-1)+}\gamma_{\pm,(M-FA)^{(z-1)+}}\}}{\{(M^{z+})\gamma_{\pm,M^{z+}}(10^{z\Delta pK})(Fa^-)\gamma_{\pm,FA}\}}$$

To facilitate the use of the above expression, the activity coefficients of the metal-fulvate complex, $\gamma_{\pm,(M-FA)(z-1)}$, and the FA, $\gamma_{\pm,FA}$, are equated to permit their cancellation. Additionally, it is presumed that the metal-ion experiences the same electric field as the proton to allow use of the ΔpK term resolved in the metal-free titrations by raising it to the power z, the charge on the metal ion. It is once again presumed that single ion activities for the metal ion may be employed as well.

It has to be recognized that the terms (FA^-) and $M-FA^{(z-1)+}$ are summations incorporating all the various "average" sites. The overall interaction parameter, β_{ov}, may be used to compare on a macroscopic level, results obtained from different experimental techniques [76] and to facilitate the description of the effects of humic substances on the uptake of metal ions by geologic substances [77].

Numerous investigators, instead of confining their studies to such a restrictive approach, have sought to quantify the extent of metal-ion interaction with each of the various "average" sites [23-30]. This objective, successfully reached with the Gibbs-Donnan-based approach [33,73] already described, has also been sought [77,78] by assuming one-to-one binding between the metal-ion and the "average" binding sites of the humic substance; a relation between the overall interaction function, β_{ov}, and the individual interaction parameters, β_i, may on this basis be expressed as follows:

$$\beta_{ov} = \frac{\Sigma \beta_i \alpha_i Ab_i}{\Sigma \alpha_i Ab_i}$$

where Ab_i is the abundance of the i-th site with α_i its corresponding degree of neutralization and β_i is the interaction parameter for the metal-i-th site complex. Arguments for classifying the individual interaction functions resolved as thermodynamically based have been offered [79]. The inability to quantify site-site interaction and the nonavailability of activity coefficient values for the fulvic acid molecules in solution would, however, support the classification of these parameters as non-thermodynamic in nature. They may, however, be used to facilitate both the assignment of particular functionality characteristics to the predominant sites in the fulvic- or humic-acid molecule and the design of experimental techniques aimed at quantitative characterization of the projected functional groups [80].

VI. ADSORPTION OF METAL IONS ONTO NATURAL SOLIDS

The transport of metal ions in natural aquatic systems is significantly affected by their (metal ions) adsorption on oxide surfaces. Such adsorption is a function of a number of factors including the following:

1. The pH of the system
2. Speciation of the metal ion, i.e., whether it is hydrolyzed or not
3. Surface type, i.e., charge on surface or the active surface sites

The pH of the aqueous system is the most important criterion because it not only affects the surface of the solid but, in addition, it determines the speciation of the metal ion. To understand completely metal adsorption on geologic media as a function of pH, the changes of the surface properties that accompany pH change must be related to the speciation changes of the metal ion. Descriptions of the types of reaction that occur with oxides have been presented [81,82] and are classified as (a) hydrolysis (acid-base), (b) complexation, and (c) ligand exchange.

The extent of adsorption of the metal ions has been described [83] as percent (%) adsorbed but such assessment of adsorption is very much dependent on the adsorbent-adsorbate concentration ratios. Distribution coefficients (K_D) provide better representations of metal-ion adsorption than the percent adsorbed or left in solution description because they are independent of the concentration of suspended solids in water and of the total concentration of metal ions (as long as metal concentrations are small in relation to the concentration of surface groups). However, representation of the adsorption of multivalent ions by distribution coefficients is also faulted because of the sensitivity of this parameter to counterion concentration levels of the solution in contact with the oxides.

The interaction of metal ions with hydrous oxide surfaces has been studied extensively and models have been proposed to explain the observations and to facilitate quantitative predictions of metal-ion sorption behavior in natural systems [83-92]. The surface-complexation model [93,94] which treats the binding of metal ions to oxides in a manner similar to the complexation of metal ions by ligands in solution has been employed to facilitate this objective by compiling distribution coefficients.

Many factors influence metal sorption by oxide surfaces and complicate their analysis by these methods. Benjamin and Leckie [93], investigating the adsorption of Cd(II) onto amorphous iron oxyhydroxide, found that, where the ratio of initial moles of cadmium to available surface area was very low, fractional adsorption was independent of the initial cadmium concentration. They suggested that the observed phenomenon could be attributed to a distribution of site types with different adsorption characteristics. In a recent study of Cd(II) adsorption on geothite, the adsorption edge was observed to be positively dependent on the initial adsorbate concentration and negatively dependent on temperature; i.e., the adsorption edge moved to a higher pH with an increase in adsorbate concentration and moved to a lower pH with temperature increase [94]. Studies of actinide adsorption on igneous rocks have shown that the extent of adsorption was a positive function of the degree of hydrolysis of the actinides, the oxidation state of the actinide and the nature of the igneous rock [95,96].

VII. EFFECTS OF LIGANDS ON METAL ADSORPTION

The adsorption of metal ions on a hydrous oxide surface has been
observed to be decreased or increased by soluble complexing ligands
[97-100]. The decrease in metal ion adsorption on the surface has been
attributed either to the formation, in solution, of a strong ligand-metal-
ion complex which is not surface active [97] or to a simple competition
between metal ion and ligand for the surface [99,100] while the increase
in adsorption has been attributed to the formation of ternary surface
complexes [100]. In studies of chromate adsorption on amorphous iron
oxyhydroxide, a decrease in adsorption was observed in the presence
of silicic acid [101]. This decrease was rationalized by suggesting that
interaction of the silicic acid with surface sites resulted in a charged
surface complex that reduced positive interfacial charge and thus decreased the electrostatic attraction for the CrO_4^{2-} anion. In studies
of Cu(II) and Ag(I) uptake on amorphous iron oxide [99], it was observed that some metal-ligand complexes are strongly bound by oxide
surface while others form nonadsorbing complexes in solution. For
example, the uptake of Ag(I) on iron oxide was decreased by the
presence of chloride ion, with the decrease increasing with increasing
chloride concentration, while the uptake of Cu(II) was not significantly affected by the presence of salicylic acid and protocatechuic acid.
In both instances, strong complexes are formed between the copper ion
and the ligands. Column studies, followed by liquid chromatography,
have shown that EDTA under certain conditions increased the retention of certain radionuclides, such as Sr, on the column, whereas,
in other cases, e.g., Co, a decrease in retention was observed [102].
The sorption of Sr, Cs, I, and Np by glacial sand was not significantly affected by natural organic material, EDTA and acetate, but
the distribution coefficient of Eu was reduced by 99.5% in the presence of 3×10^{-4} M EDTA [103].

VIII. HUMIC SUBSTANCES AS LIGANDS

A literature survey yields very few reports of the use of humic substances as ligands in the ternary, metal-ligand-solid system. The
adsorption of Cd^{2+} on alumina (Fig. 3), which is a function of pH,

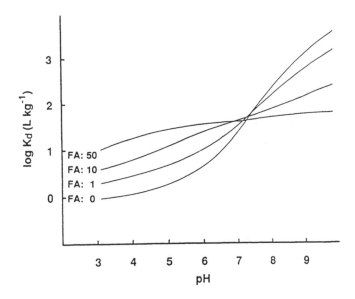

FIG. 3.   Adsorption of Cd on Al_2O_3 as a function of pH and concentration of FA (0, 1, 10, and 50 mg/L); [Cd] = 10^{-6} M.   (Adapted from Ref. 73.)

the property of the solid surface, and the concentration of fulvic acid, has been observed to be enhanced by fulvic acid at pH values below the point of zero charge, pzc (pH 7.8), and reduced at pH values above the pzc [77].   The enhancement of adsorption at lower pH values was attributed to bridging reactions with the solid surface while the reduction at higher pH values was assigned to the formation of fulvic-acid complexes in solution that were not adsorbable.

The uptake of mercury on alumina has similarly been observed to be influenced by the presence of an aquatic fulvic acid, Fig. 4 [104]. The adsorption of Hg(II) alumina was enhanced by fulvic acid over the entire pH range studies (2.5-9.5).

The adsorption of Am(III) onto alumina has been observed to be affected by humic acid [105].   In the presence of humic substances, the uptake of americium was enhanced at low pH values, whereas at higher pHs, the adsorption was lowered and depended on the concentration of the humic substance.

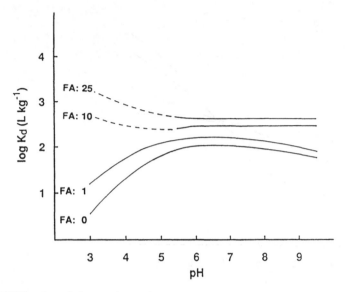

FIG. 4. Adsorption of Hg(II) as a function of pH, and FA ([Hg] =
2.8×10^{-8} M, solid/liquid = 10 g/L; I = 0.10 M). (Adapted from
Ref. 104.)

IX. MODELING OF METAL UPTAKE BY NATURAL SOLIDS

A conceptual model for metal-ligand-surface interactions during adsorption has been proposed by considering complexed species to be
either "metal-like" or "ligand-like," depending on whether adsorption
of the complex increases or decreases with increasing pH [106]. This
model uses the surface complexation approach, with the percent adsorbed or percent remaining in solution description being employed
to illustrate its tenets. According to the model, if adsorption of the
complex is similar to that of the free metal, the percent adsorbed-
pH curves in the absence and presence of ligand should be approximately parallel to each other and an increase in the ligand concentration can be expected to produce a monotonic effect on the fractional
adsorption. On the other hand, if the adsorption of the complex is
similar to that of the free ligand, then the complex is the dominant
surface species at low pH and the free-metal ion is dominant at high
pH. In addition, in this situation, there is a pH (pH_c) at which the
fractional adsorption of the complex and free metal are equal.

When the surface-complexation model, used to account for the effect of ligands on the uptake of metal ions on natural oxides [100], employs distribution coefficients instead of percent adsorbed, more analytical relationships appear to develop. With this approach, for example, the uptake of a metal ion by geologic solids, which is a function of the pH and the nature of the solid, may, in the absence of humic substances, be represented by the following reaction:

$$M^{z+} + S\text{-}H_n \rightleftharpoons M\text{-}S + nH^+$$

where S-H is a protonated geologic surface, M-S is the metal ion complexed to the surface, and M^{z+} and H^+ are the respective free metal ion with charge, z, and the proton ion. At equilibrium the reaction constant may be expressed as follows:

$$K_{eq} = \frac{\{M\text{-}S\}\{H^+\}^n}{\{M\}\{S\text{-}H\}} = \frac{K_D\{H\}^n}{\{S\text{-}H\}}$$

where K_D, the distribution coefficient of the metal ion is defined as the ratio of surface bound to free-metal-ion species, $\{M\text{-}S\}/\{M\}$, with all the various species of the metal ion, $[M^{z+}, M(OH)_n^{(z-n)+}$ for n = 1 to ∞] represented by $\{M\text{-}S\}$ and $\{M\}$.

From the above expression, the following relationship between the easily determined K_D parameter and pH is obtained:

$$\log K_D = \log(K_{eq}\{S\text{-}H\}) + npH$$

Examination of the above relationship, however, indicates that it has little predictive value. The dependence of $\log K_D$ on pH, experimentally attainable for any metal-solid system would be linear if the product of K_{eq} and $\{S\text{-}H\}$ were constant, but such a possibility is obviously nonexistent and no other simple relationship has been observed for these systems as well.

In the presence of a ligand, e.g., fulvic acid, the overall reaction may be perceived as shown:

$$M^{z+} + S\text{-}H_n + FAH_q \rightleftharpoons M\text{-}S + FA\text{-}S + M\text{-}FA\text{-}S + (n + q)H^+$$

where FAH_q is the fulvic acid in solution, and FA-S and M-FA-S are the respective fulvic-acid and fulvic-acid-metal complex species adsorbed onto the solid. Because assessment of an equilibrium constant for this reaction is too complex to contemplate such an approach has been avoided. A more convenient path to consideration of the adsorption phenomenon on this system is provided by expressing the distribution coefficient of the metal ion, K_D. Such an exercise, effected by taking into account material balance considerations, leads to the following expression:

$$K_D = \frac{[\{M\text{-}S\} + \Sigma\{M\text{-}(OH)_n\text{-}S\} + \Sigma\{M\text{-}FA\text{-}S\}]}{\{M^{z+}\} + \Sigma\{M\text{-}(OH)_n\} + \Sigma\{M\text{-}FA\}}$$

In the above equation, the metal species adsorbed by the solid have been identified as the free metal, M-S, the hydroxy complexes of the metal ion, $\Sigma M\text{-}(OH)_n\text{-}S$ and the fulvic-acid-complexed metal, $\Sigma M\text{-}FA\text{-}S$, while the metal species left in solution are their counterparts, M^{z+}, $\Sigma\{M\text{-}FA\}$ and $\Sigma\{M\text{-}(OH)_n\}$. The above expression for K_D remains valid as long as the properties of the solid surface remain unchanged and as long as an equilibrium situation prevails.

The overall reaction of the ternary system, i.e., metal ion-fulvic acid-solid, may be visualized in terms of the various "primary" reactions as follows:

a. $M^{z+} + S\text{-}H_n \rightleftharpoons M\text{-}S + nH^+$ $K_D^o = \{M\text{-}S\}/\{M\}$

b. $M^{z+} + nOH^- \rightleftharpoons \Sigma M(OH)_n^{(z-n)+}$

$\qquad \beta_{ov} = \{M(OH)_n^{(z-n)+}\}/(\{M^{z+}\}\{OH\}^n)$

c. $M^{z+} + FA \rightleftharpoons M\text{-}FA$ $\beta_{ov} = \{M\text{-}FA\}/(\{M^{z+}\}\{FA\})$

d. $FA + S\text{-}H_n \rightleftharpoons FA\text{-}S + nH^+$

$\qquad K_{eq,FA} = (\{FA\text{-}S\}\{H^+\}^n)/(\{FA\}\{S\text{-}H_n\})$

e. $M\text{-}FA + S\text{-}H_n \rightleftharpoons M\text{-}FA\text{-}S + nH^+$

$\qquad K_{eq,MFA} = (\{M\text{-}FA\text{-}S\}\{H^+\}^n)/(\{M\text{-}FA\}\{S\text{-}H_n\})$

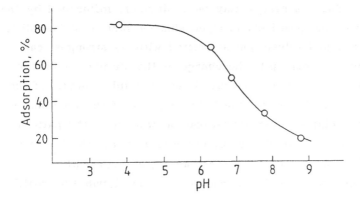

FIG. 5. Adsorption of FA onto alumina as a function of pH (I = 0.10 M $NaClO_4$; FA = 10 mg L^{-1}; solid/liquid = 10 g L^{-1}; contact time = 5d).

If the hydroxy complexes of the metal ion adsorb onto the solid, the reaction may be expressed as follows:

$$M(OH)_n^{(z-n)+} + S-H_t \rightleftharpoons \Sigma M(OH)_n-S + tH^+$$

Reactions (a) and (b) are expected to occur in the absence of ligand, FA, while all the reactions, i.e., (a)-(e), are likely to prevail in the presence of the fulvic-acid molecule. The value of K_D^o obtainable for ligand-free systems as a function of pH at a particular salt concentration level is applicable at that salt concentration level to systems with ligand when the effect of ligand on the solid is resolvable. The value of β_{OH} is obtainable from the literature [107] while β_{ov} needs to be obtained from earlier study of humic substances-metal systems in the absence of solid [76]. The parameter, $K_{D,FA}$, defined as {FA-S}/{FA}, may be obtained from FA-solid experiments in the absence of metal ions (Fig. 5).

A closer inspection of the expression for K_D in the presence of fulvic acid, reveals that the magnitude of K_D relative to the magnitude of K_D^o is dependent on the nature of the fulvic-acid-metal complex, M-FA. Fulvic acid adsorption on alumina (Fig. 5) decreases with increasing pH. This behavior may be attributable to the fact that the fulvic-acid molecule is hydrophobic at lower pH and hydrophilic at higher pH [1]. The observation may be additionally influenced by the extra attraction of the positively charged alumina surface at low pH for the negatively charged fulvate ion associated with its strongest carboxylic-acid site. At high pH, the charge of the alumina surface is reversed and must repel the now fully dissociated fulvic acid. If the FA-metal complex should assume the bulk property of the FA molecule at the pH of the solution, then its adsorption will follow the pattern of the FA-only system. In this case, one would expect the K_D to be higher than K_D^o at lower pH and lower than K_D^o at higher pH.

The difference between the expression for the distribution coefficient, K_D, in the absence and presence of FA, consists of the addition of the $\Sigma\{M\text{-}FA\text{-}S\}$ and $\Sigma\{M\text{-}FA\}$ terms in the numerator and denominator. If the metal-fulvic-acid complex is adsorbed more by the solid, then $\Sigma\{M\text{-}FA\text{-}S\}$ is much greater than $\Sigma\{M\text{-}FA\}$ with the resultant K_D being larger than K_D^o. If, on the other hand, the metal-fulvic-acid complex tends to stay in solution $\Sigma\{M\text{-}FA\}$ exceeds $\Sigma\{M\text{-}FA\text{-}S\}$ and a K_D smaller than K_D^o results.

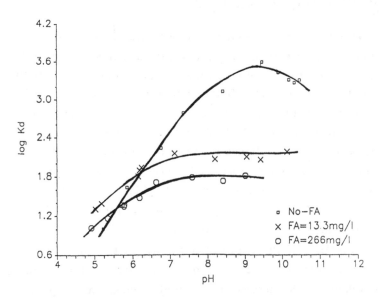

FIG. 6. Adsorption of Zinc onto alumina as a function of pH and FA concentration ($I = 0.10$ M $NaClO_4$; contact time = 2d).

X. INTERPRETATION OF ZINC AND CADMIUM ADSORPTION
    ON ALUMINA

The adsorption of zinc on alumina as a function of pH and fulvic-acid
concentration (Fig. 6) shows that the presence of fulvic acid reduces
Zn adsorption after a pH of 7 is reached. At pH's below 7, the presence of FA appears to have little effect on the adsorption pattern of
zinc, which is similar to that of cadmium. These observations may be
explained by presuming that at lower pH values, the metal fulvate adsorbs appreciably onto the solid while at higher pH values, the metal
fulvate is nonadsorbing. The complexation of Zn by FA(ZnFA) keeps
the Zn from adsorbing as it would in the absence of FA. In these
two instances, it is obvious that the role of the hydroxy complexes is
insignificant since their free energy of formation at the higher pH
values is relatively large. The adsorption of zinc fulvate and cadmium
fulvate onto alumina may be considered ligand-like; i.e., it decreases
with increasing pH [106].

XI. INTERPRETATION OF MERCURY ADSORPTION ON ALUMINA

The adsorption of Hg(II) on alumina was enhanced by fulvic acid in
the whole pH range of study (2.5-9.5). The observation, which differs from the one involving Cd and Zn adsorption behavior, implies
that the Hg(II)-fulvic-acid complex that forms is strongly adsorbed
by the alumina over the whole pH range. This result is considered
to indicate that at low pH the fulvic-acid-metal complex formed assumes
the bulk properties of the fulvic-acid molecule, adsorbing onto alumina,
while at high pH, the complex, does not. This tendency to continue
to adsorb on the alumina in the high pH range has been attributed to
the configuration of the fulvic-acid-metal complex being such that adsorption is still possible [97]. However, another explanation may be
that the metal-fulvate complex is not strong enough to resist interaction of metal-hydroxy species with the alumina.

XII. SUMMARY

Humic substances are acidic in nature with molecular weights ranging from a few hundred to a couple of thousand. Their functional group content is highly heterogeneous and their solution chemistry suggests that they are amphilic. The influence of these substances on the uptake of metal ions onto natural solids has been sought by considering the following factors:

1. The ionization of the humic substance molecule as a function of ionic strength and proton concentration
2. Their interaction with metal ions as a function of pH
3. Their interaction with natural geologic solids as a function of pH

   The extent of humic substances influence on metal uptake may be postulated to be dependent on the strength of the humic substance-metal complex, the configuration of the complex, the strength of the metal-hydroxy complex, and the nature of the interaction between the humic substance and the geologic material.

   A more detailed assessment of the nature of these systems is unlikely, however, until an effort is made to characterize the separate species encountered in these systems. Factor 1 has been neglected in the studies carried out so far. By expressing the equilibria encountered as described the effect of ionic strength is canceled. Any advantage that might have been gained from a more detailed examination of metal complex speciation has been lost as well by using combined terms in their place. A more informative analysis of these systems awaits synchronization of the more detailed Gibbs-Donnan approach, outlined earlier in describing the equilibrium properties of metal-humate systems.

ACKNOWLEDGMENTS

The authors are grateful to the Swedish Nuclear Waste Management Company and the Swedish Natural Science Research Council for their financial support.

REFERENCES

1. G. R. Aiken, in *Humic Substances in Soil, Sediment and Water.
    Geochemistry, Isolation and Characterization* (G. R. Aiken, D.
    M. McKnight, R. L. Wershaw, and P. MacCarthy, eds.), Wiley,
    New York, 1985, pp. 363-385.

2. C. Pettersson, I. Arsenie, J. Ephraim, H. Borén, and B. Allard,
    Sci. Total Environ., *81/82*:287-296 (1989).

3. D. S. Gamble, *Can. J. Chem.*, *50*(6):2680-2690 (1972).

4. P. Brassard, J. R. Karmer, and P. V. Collins, *Environ. Sci.
    Technol.*, *24*:195-201 (1990).

5. A. M. Posner, *J. Soil Sci.*, *17*:65 (1966).

6. A. E. Martin and R. Reeve, *J. Soil Sci.*, *9*:89 (1958).

7. H. R. Geering and J. F. Hodgson, *Soil Sci. Soc. Am. Proc.*,
    33:54 (1969).

8. N. Paxeus and M. Wedborg, *Anal. Chim. Acta*, *169*:87-98 (1985).

9. H. Davis and C. J. B. Mott, *J. Soil Sci.*, *32*:379-391 (1981).

10. A. M. Posner, in *Trans 8th Int. Congr. Soil Sci.*, Vol. 3, 1964,
    p. 161.

11. F. J. Stevenson, *Humus Chemistry*, Wiley, New York, 1982.

12. D. S. Gamble, *Can. J. Chem.*, *48*:2662-2669 (1970).

13. E. M. Perdue, J. H. Reuter, and R. S. Parrish, *Geochem.
    Cosmochim. Acta*, *48*:1257-1263 (1984).

14. B. A. Dempsey and C. R. O'Melia, in *Aquatic and Terrestrial
    Humic Substances* (R. F. Christman and E. Gjessing, eds.),
    Ann Arbor Science, Ann Arbor, MI, 1983, pp. 239-273.

15. I. M. Klotz and D. L. Hunston, *J. Biol. Chem.*, *259*:10,060-
    10,062 (1984).

16. E. M. Purdue, in *Humic Substances in Soil, Sediment and Water.
    Geochemistry, Isolation and Characterization* (G. R. Aiken, D.
    M. McKnight, R. I. Wershaw, and P. MacCarthy, eds.), Wiley
    New York, 1985, Chap. XI.

17. D. A. Dzombak, W. Fish, and F. F. Morel, *Environ. Sci. Tech-
    nol.*, *20*:669 (1986).

18. W. Fish, D. A. Dzombak, and F. F. Morel, *Environ. Sci. Tech-
    nol.*, *20*:676 (1986).

19. W. E. Falck, British Geological Survey Technical Report WE/88/49,
    1988.

20. S. Boggs, Jr., D. Livermore, and M. G. Seitz, Humic Substances
    in Natural Water and Their Complexation with Trace Metals and
    Radionuclides: A Review, Argonne National Laboratory Report,
    ANL-87-78, 1985.

21. J. Buffle and R. S. Altmann, in *Aquatic Surface Chemistry* (W. Stumm, ed.), Wiley, New York, 1987, Chap. 13.

22. G. Sposito, *CRC Crit. Rev. Environ. Control*, 16:193-229 (1986).

23. G. Sposito and K. M. Holtzclaw, *Soil Sci. Soc. Am. J.*, 41:330-336 (1977).

24. N. Paxeus, Studies on Aquatic Humic Substances, Ph.D. Thesis, Chalmers University of Technology and University of Gothenborg, Sweden, 1985.

25. J. E. Gregor and H. K. J. Powell, *J. Soil Sci.*, 39:243-252 (1988).

26. E. Tipping, C. A. Backes, and M. A. Hurley, *Wat. Res.*, 22(5): 597-611 (1988).

27. W. E. Falck, in *Proc. Int. Symp. Humic Substances Aquatic Terrestrial Environment*, Linköping, Sweden, 1989.

28. E. M. Perdue and C. R. Lytle, *Environ. Sci. Technol.*, 17:654-660 (1983).

29. M. S. Shuman, B. J. Collins, P. J. Fitzgerald, and D. L. Olsen, in *Aquatic and Terrestrial Humic Materials* (R. F. Christman and E. T. Gjessing, eds.), Ann Arbor Science, Ann Arbor, MI, 1983, pp. 349-370.

30. A. K. Thakur, P. J. Munson, D. L. Hunston, and D. Rodbard, *Anal. Biochem.*, 103:240-254 (1980).

31. J. H. Ephraim, M. M. Reddy, and J. A. Marinsky, in *Proc. Int. Symp. Humic Substances Aquatic Terrestrial Environment*, Linköping, Sweden, 1989.

32. R. L. Wershaw, *J. Contamin. Hydrology*, 1:29-45 (1986).

33. G. Sposito, K. M. Holtzclaw, and D. A. Keech, *Soil Sci. Soc. Am. J.*, 41:1119-1125 (1977).

34. R. E. Truitt and J. H. Weber, *Environ. Sci. Technol.*, 15:1204-1208 (1981).

35. T. A. O'Shea and K. H. Mancy, *Anal. Chem.*, 48(11):1603-7 (1976).

36. P. Figura and B. McDuffie, *Anal. Chem.*, 51(1):120-125 (1979).

37. T, Y, Aualituta and W. F. Pickering, *Water Res.*, 20(11):1397-1406 (1986).

38. N. Senesi, *Anal. Chim. Acta*, 232(1):77-106 (1990).

39. R. A. Saar and J. H. Weber, *Anal. Chem.*, 52:2095 (1980).

40. A. W. Underdown, C. H. Langford, and D. S. Gamble, *Can. J. Soil Sci.*, 61:469 (1981).

41. G. Bidoglio, I. Grenthe, P. Qi, P. Robouch, and N. Omeneto, *Talanta*, 38:999-1008 (1991).

42. W. T. Bresnahan, C. L. Grant, and J. H. Weber, *Anal. Chem.*, 50(12):1675-1679 (1978).

43. R. A. Saar and J. H. Weber, *Can. J. Chem.*, *57*(11):1263-1268 (1979).

44. N. Paxeus and M. Wedberg in *Lecture Notes in Earth Sciences* (B. Allard, H. Borén, and A. Grimvall, eds.), *33*:287-296 (1991).

45. A. Fitch, F. J. Stevenson, and Y. Chen, *Org. Geochem.*, *9*(3): 109-116 (1986).

46. R. F. C. Mantoura and J. P. Riley, *Anal. Chim. Acta*, *78*:193-200 (1975).

47. G. Petruzzelli, G. Guidi, and L. Lubrano, *Environ. Technol. Lett.*, *1*(4):201-208 (1980).

48. J. A. Marinsky and M. M. Reddy, *Org. Geochem.*, *7*:215-221 (1984).

49. M. S. Ardakani and F. J. Stevenson, *Soil Sci. Soc. Amer. Proc.*, *36*(6):884-890 (1972).

50. M. Schnitzer and S. I. M. Skinner, *Soil Sci.*, *103*:247 (1967).

51. C. Staub, J. Buffle, and W. Haerdi, *Anal. Chem.*, *56*:2843 (1984).

52. J. Buffle and C. Staub, *Anal. Chem.*, *56*:2837-2842 (1984).

53. R. L. Wershaw and G. R. Aiken, in *Humic Substances in Soil, Sediment and Water. Geochemistry, Isolation and Characterization* (G. R. Aiken, D. M. McKnight, R. L. Wershaw, and P. MacCarthy, eds.), Wiley, New York, 1985, p. 477.

54. J. H. Ephraim, S. J. Cramer, and J. A. Marinsky, *Talanta*, *36*(4):437-443 (1989).

55. M. Nordén, J. H. Ephraim, and B. Allard, in *Proc. Int. Symp. Humic Substances Aquatic Terrestrial Environment*, Linköping, Sweden, 1989.

56. J. H. Ephraim and J. A. Marinsky, *Anal. Chim. Acta*, *232*(1): 171 (1990).

57. R. E. Truitt and J. H. Weber, *Anal. Chem.*, *53*(2):337-342 (1981).

58. D. P. Rainville and J. H. Weber, *Can. J. Chem.*, *60*:1-5 (1982).

59. J. A. Marinsky, F. G. Lim, and C. J. Chung, *J. Phys. Chem.*, *87*(16):3139-3145 (1983).

60. M. Nagasawa, T. Murase, and K. Kondo, *J. Phys. Chem.*, *69*:4005-4012 (1965).

61. J. A. Marinsky, *J. Phys. Chem.*, *89*:5294-5303, 5303-5307 (1985).

62. J. A. Marinsky, in *Aquatic Surface Chemistry* (W. Stumm, ed.), Wiley, New York, 1987, pp. 49-81.

63. J. A. Marinsky, M. M. Reddy, J. Ephraim, and A. Mathuthu, Ion Binding by Humic Substances. A Computational Procedure Based on Functional Site Heterogeneity and Separate Phase Behavior, SKB Technical Report, 88-04, 1988.

64. J. Ephraim, S. Alegret, A. Mathuthu, M. Bicking, R. L. Malcolm, and J. A. Marinsky, *Environ. Sci. Technol.*, *20*:354-366 (1986).

65. J. A. Marinsky and M. M. Reddy, *Anal. Chim. Acta*, *232*:123-130 (1990).

66. E. H. Hansen and M. Schnitzer, *Anal. Chim. Acta*, *46*:247 (1967).

67. M. K. Chaudhurg and G. M. Whitesides, *Science*, *256*:1539 (1992).

68. J. Ephraim, H. Borén, I. Arsenie, C. Petterson, and B. Allard, *Environ. Sci. Technol.*, *23*:356-362 (1989).

69. A. Maes, J. Vancluysen, F. Van Elewijck, J. Tits, and A. Cremers, in *Proc. Int. Symp. Humic Substances Aquatic Terrestrial Environment*, Linköping, Sweden, 1989.

70. J. R. Wright and M. Schnitzer, in *Trans. Int. Congr. Soil Sci. 7th*, Vol. 2, 1960, pp. 112-118.

71. H. Probiser, *Anal. Chim. Acta*, *156*:59-65 (1983).

72. J. H. Ephraim, *Talanta*, *36*(3):379-382 (1989).

73. J. Ephraim and J. A. Marinsky, *Environ. Sci. Technol.*, *20*(4); 367-376 (1986).

74. L. G. Sillen and A. E. Martell, *Stability Constants of Metal-Ion Complexes*, The Chemical Society, Burlington House, London, 1964.

75. D. S. Gamble, C. H. Langford, and A. W. Underdown, *Org. Geochem.*, *8*:35-39 (1985).

76. J. H. Ephraim and H. Xu, *Sci. Total Environ.*, *81/82*:625-634 (1989).

77. H. Xu, J. Ephraim, A. Ledin, and B. Allard, *Sci. Total Environ.*, *81/82*:653-660 (1989).

78. J. H. Ephraim, *Sci. Total Environ.*, *108*:261-273 (1991).

79. D. S. Gamble, *Environ. Sci. Technol.*, *22*:1325-1336 (1988).

80. J. H. Ephraim, H. Borén, I. Arsenie, C. Pettersson, and B. Allard, *Sci. Total Environ.*, *81/82*:615-624 (1989).

81. J. C. Westall, in *Aquatic Surface Chemistry* (W. Stumm, ed.), Wiley, New York, 1987.

82. W. Stumm and J. J. Morgan, *Aquatic Chemistry*, 2nd ed., Wiley, New York, 1981.

83. L. Sigg and W. Stumm, *Colloids Surf.*, *2*:101 (1981).

84. W. Stumm, H. Hohl, and F. Dalang, *Croat. Chem. Acta*, *48*:491 (1976).

85. P. W. Schindler, B. Fuerst, R. Dick, and P. U. Wolf, *J. Colloid Interface Sci.*, *55*:469 (1976).

86. N. J. Barrow, J. W. Bowden, A. M. Posner, and J. P. Quirk, *Aust. J. Soil Res.*, *19*:309-321 (1981).

87. W. F. Bleam and M. B. McBridge, *J. Colloid Interface Sci.*, *103*:124-132 (1985).

88. R. O. James and T. W. Healy, *J. Colloid Interface Sci.*, *40*(1):42-81 (1972).

89. H. Hohl and W. Stumm, *J. Colloid Interface Sci.*, *55*:281-288 (1976).

90. P. H. Tewari, A. B. Campbell, and W. Lee, *Can. J. Chem.*, *50*:1642-1648 (1972).

91. P. H. Tewari and W. Lee, *J. Colloid Interface Sci.*, *52*:77-88 (1975).

92. H. Tamura, E. Matijefic, and L. Meites, *J. Colloid Interface Sci.*, *92*:303-314 (1983).

93. M. M. Benjamin and J. O. Leckie, *J. Colloid Interface Sci.*, *79*:209-221 (1981).

94. B. B. Johnson, *Environ. Sci. Technol.*, *24*:112-118 (1990).

95. B. Allard, G. W. Beall, and T. Krajcswski, *Nucl. Tech.*, *49*:474 (1980).

96. B. Allard, U. Olofsson, B. Torstenfelt, H. Kipatsi, and K. Andersson, in *Scientific Basis for Radioactive Waste Management. V* (W. Lutze, ed.), Elsevier, 1982.

97. J. A. Davis and J. O. Leckie, *Environ. Sci. Technol.*, *12*:1309-1315 (1978).

98. M. G. MacNaughton and J. O. Leckie, *J. Colloid Interface Sci.*, *47*:431-440 (1974).

99. M. M. Benjamin and J. O. Leckie, *Environ. Sci. Technol.*, *16*:162 (1982).

100. A. C. M. Bourg and P. W. Schindler, *Chimia*, *32*:166 (1978).

101. J. M. Zachara, D. C. Girvin, R. I. Schmidt, and C. T. Reach, *Environ. Sci. Technol.*, *21*:589-594 (1987).

102. L. Carlsen, *Europ. Appl. Res. Rept.-Nucl. Sci. Technol.*, *6*:1419 (1985).

103. D. Haigh et al., DOE Report: DOE/RW/89.067, 1989.

104. H. Xu and B. Allard, *Water, Air and Soil Pollution*, *56*:709-717 (1991)

105. V. Moulin and D. Stammose, in *Sci. Basis for Nuclear Waste Management. X*, Materials Res. Soc., Pittsburgh, 1989.

106. M. M. Benjamin and J. O. Leckie, *Environ. Sci. Technol.*, *15*(9):1050-1057 (1981).

107. E. Högfeldt, *Stability Constants of Metal-Ion Complexes, Part A: Inorganic Ligands*, Pergamon Press, New York, 1983.

Index

[Humic substances]
  ultrafiltration method, 341-342
  modeling of metal uptake by
    natural solids, 357-361
  steps of the approach, 342-352
    description of metal-ion
      binding by fulvic acid,
      350-352
    interpretation of pronota-
      tion equilibria in aqueous
      media, 348-350
    potentiometric titrations in
      aqueous medium as func-
      tion of ionic strength,
      342-346
    potentiometric titrations in
      nonaqueous medium, 348
    potentiometric titrations in
      presence of heavy-metal
      ions, 346-348

Inner-sphere surface complexes,
  212, 213-214
Inorganic ion exchangers,
  143-145
Ion-exchange equilibria, description of, 151-209
  comparison of experimental
    and predicted equilibria,
    191-202
    binary equilibria, 191-198
    multicomponent equilibria,
      198-202
  derivation of resin-specific
    equilibrium parameters,
    166-176
    generalized separation fac-
      tor in binary systems,
      166-169
    generalized separation fac-
      tors in multicomponent
      systems, 169-171
    graphical representation of
      equilibria, 173-176
    parameters in multicom-
      ponent systems, 171-172
    superposition of several
      exchange equilibria,
      172-173
  evaluation of equilibrium
    data, 176-191

[Ion-exchange equilibria]
  evaluation of binary equi-
    libria, 176-191
  evaluation of multicomponent
    equilibria, 191
  theory, 153-165
    binary exchange equilibria,
      157-163
    model assumptions, 153-156
    multicomponent exchange
      equilibria, 164-165
Ionic-solute activity, adjusting
  exchange-equilibrium-
  constant values for, 93-97

Ligands:
  effects on adsorption of metal
    ions, 354-355
  humic substances as, 355-357
Linear polyelectrolytes, 141-143
  effect on the distribution of
    pairs of ions between its
    crosslinked ion-exchange
    gel analog and simple salt
    solutions, 298-307
  evidence of resemblance to
    their crosslinked-gel ana-
    logs through use of Gibbs-
    Donnan-based concepts,
    256-261
  in solution, counterion-concen-
    trating solvent sheath of,
    261-290
Liquid ion exchangers, three-
  parameter model for data
  summary, 119-127
  free energy, 119-125
  water uptake, 125-127
Logarithmic excess Gibbs energy
  model, 58-59

Mercury adsorption on alumina,
  362
Metal-humate interactions, techniques for study of, 340-342
  anodic stripping voltammetry,
    340
  chromatographic methods, 341
  equilibrium dialysis/dialysis
    titration, 342
  fluorescence spectroscopy, 341
  ion exchange distribution, 341